Elementary Structural Design of Steelwork to BS 449

P.C.L. Croxton, L.H. Martin
and J.A. Purkiss

University of Aston in Birmingham

Edward Arnold

© P.C.L. Croxton, L.H. Martin and J.A. Purkiss, 1984

First published in Great Britain 1984 by
Edward Arnold (Publishers) Ltd, 41 Bedford Square, London WC1B 3DQ

Edward Arnold, 300 North Charles Street, Baltimore, Maryland, U.S.A.

Edward Arnold (Australia) Pty Ltd, 80 Waverley Road, Caulfield East, Victoria 3145, Australia.

ISBN 0 7131 3531 X

Printed in Great Britain by Butler & Tanner Ltd, Frome and London

Preface

During several years of teaching steelwork design it has become clear that there is a need for a single book which not only describes methods of design, up to the level required by final year students in universities and polytechnics, but which also contains enough background material for an understanding of the provisions of the current code of practice, and which deals adequately with the methods of analysis required in the design process. This book is an attempt to satisfy these criteria.

Chapter 1 is a review of general principles, and of the processes that go into the production of a complete steel structure and thus affect its design. Chapters 2 and 3 are concerned with the design of rolled-section beams and stanchions as structural elements. In Chapter 4 the treatment of structural connections is very comprehensive and reflects the degree of importance which the authors attach to this topic. Much of the work in this chapter is new and is based on recent research at the University of Aston and other institutions. Chapters 5 and 6 deal with complete, or major parts of structures, in which the interaction between members, and connections, is an important consideration. Finally an advanced treatment of the design of fabricated beams is given in Chapter 7.

The book is primarily intended for students in universities and polytechnics, but it is expected that practising engineers may find it a useful addition to their library. The design methods are illustrated by fully worked examples, which are preceded by the necessary analysis and theoretical work. With increasing reliance on computer programs it is hoped that this part of the book will help to provide the conceptual understanding essential to a good designer. In the examples 'looking back' is avoided by writing the appropriate equations in algebraic form before inserting numerical values. Extracts from BS 449 ('The use of structural steel in buildings') and from section tables are given in the appendix, making the book completely self-contained. References to research publications and other books are appended to each chapter, for the benefit of readers wishing to enquire further.

The decision to base the book on design to BS 449 was taken after much thought, bearing in mind that when the proposed new limit-state code is eventually issued, it will be some considerable time before it is universally adopted by practising engineers. Nevertheless the book contains a number of pointers to conditions at ultimate load. Plastic methods of analysis are described in detail, starting from first principles.

P.C.L. Croxton
L.H. Martin
J.A. Purkiss

Acknowledgements

The authors gratefully acknowledge the services of Mrs. G. Jones for her work on the word processor, and Miss G. Thomas and Mrs. S. Lancaster for their work in tracing the diagrams.

Extracts from BS 449 are reproduced by permission of the British Standards Institution, 2 Park Street, London W1A 2BS, from whom complete copies of the standard can be obtained. Extracts from the BCSA and Constrado publications are reproduced by permission of BCSA Ltd., Silverton House, 1 Vincent Square, London SW1P 2PJ, and Constrado, NLA Tower, 12 Addiscombe Road, Croydon CR9 3JH.

P.C.L. Croxton
L.H. Martin
J.A. Purkiss
University of Aston in Birmingham

Notation

A	cross sectional area,
	or area enclosed by the profile of a section,
	or length of leg of an angle or tee section,
	or maximum cantilever extension for a base plate,
	or lateral torsional buckling stress for plate girders where the flanges have equal second moments of area about the y-y axis
A_b	cross sectional area of a bolt
A_c	cross sectional area of a column
A_g	cross sectional area of a strip of gusset plate
A_p	cross sectional area of a steel base plate
A_{st}	cross sectional area of a steel stiffener
a_c	edge distance for a hole in a steel column flange
a_p	edge distance for a hole in a steel end plate
a_w	throat thickness of a fillet weld
a_1	net cross sectional area of the attached leg of an angle
a_2	cross sectional area of the unattached leg of an angle
B	breadth of a section of rectangular aspect,
	or minimum cantilever extension for a base plate
	or a factor used in the lateral torsional buckling of plate girders
B_b	breadth of flange for a steel I section beam
B_c	breadth of flange for a steel I section column
B_p	breadth of steel plate
b	breadth
b_c	cantilever distance to the centre line of a bolt hole for a column flange
b_g	maximum unsupported length for a gusset plate in a truss
b_p	cantilever distance to the centre line of a bolt hole for a steel end plate
b_{st}	width of steel stiffener
b_w	breadth of weld
b_{wc}	effective breadth of weld
C_o	Euler critical stress
C_s	allowable critical bending stress for lateral torsional buckling
C_x, C_y	distance from the heel of an angle to the x-x and y-y axes respectively
C_1, C_2, C_3	components of web bearing and buckling resistance
D	overall depth of a section
D_{av}	average depth of a tapered beam
D_c	depth of a castellated I section
D_p	depth of a plate
D_s	depth of hole in the web of a castellated girder
d	distance between the centroids of flanges for a beam,
	or depth of beam,
	or nominal diameter of a bolt
d_h	diameter of a hole

d_r	distance from the theoretical hinge to the resultant force
d_w	depth of weld,
	or depth of a web between flanges
E	Young's modulus of elasticity
e	eccentricity of an applied load
e'	equivalent eccentricity of an applied load
F_{be}	shear force on a bolt due to the applied load
F_{bs}	preload in a bolt
F_{bt}	axial force on a bolt due to the applied load
F_{bv}	shear force on a bolt due to the applied load
F_c	compressive force due to the applied load
F_{cw}	external force applied to local crushing of a web of a steel column
F_e	external applied force to a bolt in the elastic stage of behaviour
F_g	force applied to a gusset plate
F_p	force in an end plate or column associated with a prying force
F_t	tensile force related to an applied load
F_{wr}	resultant force per unit length of weld
F_{wx}	force per unit length of weld in the x direction transverse to the length of the weld
F_{wy}	force per unit length of weld in the y direction transverse to the length of the weld
F_{wz}	force per unit length of weld in the z direction parallel to the length of the weld
f	stress due to applied loads
f_b	stress due to an applied bending moment, or bearing stress due to an applied load
f_c	average axial stress due to an applied compressive load
f_{bc}	compressive stress due to an applied bending moment
f_{bt}	tensile stress due to an applied bending moment, or tensile stress in a bolt due to an applied load
f_{bw}	shear stress in a bolt due to an applied load
f_{cu}	characteristic 28 day cube crushing strength of concrete
f_e	equivalent stress due to applied loads
f_{gw}	axial buckling stress in a gusset plate at a distance w from the theoretical hinge
f_{gy}	specified guaranteed minimum yield stress for a gusset plate
f_q	shear stress due to an applied shear force
f_q'	average shear stress
f_{qp}	average shear stress in a web at factored load, or shear stress at point P
f_t	average axial stress due to an applied tensile load
f_u	ultimate tensile strength
G	elastic shear modulus
g	total grip length for a bolt, or the gauge dimension of a bolt group
H	horizontal applied load, or reaction

	or height of a gusset plate
h	height,
	or distance between the centres of areas of flanges
	or outstand of stiffener
h_1	height to eaves
I	second moment of area
I_g	second moment of area of a strip of gusset plate,
	or gross second moment of area of a section
I_o	second moment of area of a weld group about an axis O-O
I_{st}	second moment of area of a stiffener
I_u	major principal second moment of area
I_v	minor principal second moment of area
I_w	warping constant
I_{wG}	second moment of area of a weld group about an axis G-G
I_x	second moment of area of a weld group about an axis x-x,
	or second moment of area about the x-x axis
I_{xy}	product moment of area about the x-x and y-y axes
I_y	second moment of area of a weld group about an axis y-y,
	or second moment of area about the y-y axis
J	torsional constant
K_1	flange thickness modification factor in lateral torsional buckling
K_2	correction factor in lateral torsional buckling
k	interaction constant for combined bending and axial stress
k_b	factor for a bolt relating shear strength to tensile strength
k_c	factor associated with the prying force for a column flange
k_e	factor associated with the prying force in the elastic stage of behaviour
L	length,
	or span
L_b	length of stiff bearing
L_j	joint length in mm on one side of a bolted connection
L_w	total length of a weld
l	effective length
l_b	effective length of a beam web in buckling
l_w	length of a weld
l_o	horizontal length of a hole in a castellated beam
M	bending moment
M_c	critical bending moment in lateral torsional buckling
M_{ex}, M_{ey}	effective bending moments about the x-x and y-y axes respectively
M_p	plastic moment of resistance of a section,
	or plastic collapse moment
M_y	Yield moment or moment about the y-y axis
M_z	torque about the z axis
m	modular ratio of steel to concrete i.e. E_s/E_c,
	or small element of moment
N	applied axial force,
	or ratio of flange area at maximum bending moment to

	flange area at minimum bending moment
n	number, or stress index (p/Y_s)
n_b	number of bolts
n_{bt}	number of bolts in tension
n_l	number of levels of bolts in a friction grip joint
n_s	number of samples
P_b	web bearing or buckling resistance
P_{bs}	pre-load force on a HSFG bolt
P_q	shear resistance
P_{st}	resistance of a stiffener to buckling based on design stresses
p	an allowable design stress, or an axial stress at collapse
p_b	allowable bearing stress
p_{bc}	allowable bending compressive stress
p_{bt}	allowable bending tensile stress
p_c	allowable average axial compressive stress
p_{cr}	critical bending stress in lateral torsional buckling
p_e	allowable equivalent stress
p_p	an allowable design bending stress for a plate
p_q	an allowable design shear stress for a plate
p_q'	allowable average shear stress
p_t	allowable average axial tensile stress
p_w	allowable design shear stress for a weld
Q	shear force
Q_{be}	prying force acting on a bolt in the elastic stage of behaviour
Q_{ex}, Q_{ey}	effective shear forces in the x and y directions respectively
q	wind pressure
q_s	shear flow, i.e. shear force per unit length (at point S)
R	reaction force or resultant force, or applied transverse load, or outside radius, or radius of curvature
R_m	mean radius of a tube
R_w	resultant force of a weld group
r	radius, or fillet radius for an I section, or radius of gyration
r_a	fillet radius for an angle section
r_b	fillet radius for a beam section
r_c	fillet radius for a column section
r_x	radius of gyration about the x-x (major rectangular) axis
r_y	radius of gyration about the y-y (minor rectangular) axis
r_v	radius of gyration about the v-v (minor principal) axis
s	spacing of stiffeners, or distance around the profile of a section, or staggered pitch of a bolt group

s_g distance of the resultant force on a gusset plate from the theoretical hinge

s_h horizontal spacing between the centre lines of holes in an end plate

s_v vertical spacing between the centre lines of holes in an end plate

S_1, S_2, S_3 factors affecting the design wind speed

T applied tensile force,
or applied torsional moment,
or torsional parameter for instability,
or thickness of a flange

t thickness of a plate (web of an I section)

t_g thickness of a gusset plate

t_p thickness of a plate

t_w leg length of a fillet weld

u coordinate parallel to the major principal axis,
or lateral displacement

V actual wind speed,
or vertical reaction,
or shear force

V_s design wind speed

v coordinate parallel to the minor principal axis

W total load

W_g width of gusset plate perpendicular to the free edge

W_{sw} self weight of a member

w loading per unit length

w_c width of column flange associated with one bolt

w_{ex}, w_{ey} effective load intensities in the x and y directions

w_p width of plate associated with one bolt

x arithmetic mean,
or position of the centre of rotation from the centroid of the fastener group along the x-x axis,
or coordinate of a point in a section

x_g distance to the centroid of the weld group

x_n distance along the x-x axis to the nth fastener

Y_e effective yield stress

Y_s guaranteed minimum yield stress

y (or y_c)   position of the centre of rotation from the centroid of the fastener group along the y-y axis,
or distance from the neutral axis to the centroid of an area,
or distance from the neutral axis to the extreme compression fibre,
or coordinate of a point in a section

y_t distance from the neutral axis to the extreme tension fibre

Z_e elastic section modulus

Z_p plastic section modulus

Z_{pw} plastic modulus of a web

a linear coefficient of expansion,
or sway correction coefficient,

	or angle,
	or angle between x-x and u-u axes of a section,
	or a proportion of the length of a beam
β	angle,
	or ratio of end bending moments
γ	load factor at collapse
Δ	deflection,
	or displacement
δ	deflection
ε_{bs}	strain in a bolt due to preload
η	imperfection coefficient in lateral torsional buckling and strut buckling
θ	angle,
	or angular rotation in sway equations,
	or angle of twist
τ_1, τ_2	shear stress on a 45° plane for a fillet weld
τ_{crit}	elastic critical shear stress for buckling
λ	slenderness ratio,
	or ratio of the second moment of area of the compression flange to the second moment of area of both flanges about the y-y axis
μ	coefficient of friction for steel
ρ	density of a material
σ	standard deviation
σ_{cr}	critical bending stress in lateral torsional buckling
ϕ	angle of rotation,
	or angle of fracture plane for a fillet weld
ν	Poissons ratio (0.25 in BS 449)

Contents

5   Frames and Framing   237

List of Tables

1 General

1.1 DEFINITION OF A STRUCTURE AND STRUCTURAL DESIGN

A structure is an assemblage of members arranged in a regular geometrical pattern in such a way that they interact through structural connections to support loads and maintain them in equilibrium without excessive deformation. Structural members have the capacity to carry loads in a variety of ways, and may act in tension, compression, flexure, shear, torsion, or in combinations of these.

Structural design is the determination of the disposition and size of load bearing members and other components in a structure. The process consists of the application of theoretical analysis and design methods which have been validated by experimental work and practical experience.

1.2 BRIEF OUTLINE OF DESIGN DEVELOPMENTS IN FERROUS METALS

Prior to 1779 when the Ironbridge at Coalbrooke-dale on the Severn was completed, the main materials for load bearing structures were masonry or timber. Ferrous metals were only used for fastenings, armaments and chains.

The earliest use of cast iron columns in factory buildings was about 1780 which enabled relatively large span floors to be adopted. Due to a large number of disastrous fires around 1795, cast iron was also being used for beams with the floors carried on brick jack arches between the beams. This mode of construction was pioneered by Strutt in an effort to attain a fire proof construction technique. Cast iron, however, is weak in tension and this necessitates a tension flange larger than a compression flange.

Cast iron was mainly used for compression members, but for large span beams it was impracticable, and was occasionally disastrous as in the collapse of the Dee bridge designed by Robert Stephenson in 1874. The last probable use of cast iron in bridge works was in the piers for the Tay bridge which suffered an unsatisfactory end in 1879. It collapsed in high winds due to poor design and unsatisfactory supervision during construction.

In an effort to overcome the tensile weakness of cast iron, wrought iron was introduced in 1784 by Henry Cort. Wrought iron enabled the Victorian engineers to produce the following classic structures. Robert Stephenson's Brittania Bridge was the first box girder bridge and represented the first major collaboration between engineer, fabricator (Fairburn) and scientist (Hodgkinson). I.K.Brunel's Royal Albert Bridge at Saltash combined an arch and suspension bridge. Telford's Menai suspension bridge used wrought iron chains which have since been replaced by steel chains. Telford's Pont Cysyllte is a canal aqueduct near Llangollen. The first of the four structures has been replaced after a fire in 1970. The introduction of wrought iron also revolutionised ship building to enable Brunel to produce S.S. Great Britain.

Steel was first produced in about 1740, but was not available in large quantities until Bessemer invented his converter in 1856. The first major structure to use the new steel exclusively was Fowler and Baker's railway bridge at the Firth of Forth. The first steel rail was rolled in 1857 and installed at Derby where it was still in use 10 years later. Cast iron rails in the same position lasted about 3 months. Steel rails were in regular production at Crewe under Ramsbottom from 1866.

By 1840 standard shapes in wrought iron, mainly rolled flats, tees and angles, from which compound girders could be fabricated by riveting, were in regular production; and were appearing in structures about 10 years later. Wrought iron remained in use until around the end of the nineteenth century.

By 1880 the rolling of steel I sections had become widespread under the influence of companies such as Dorman Long. Riveting continued in use as a fastening method until around 1950 when it was superseded by welding.

1.3 PRODUCTION OF STEEL

The manufacture of standard steel sections may be conveniently divided into three stages:

    (a) Iron production.
    (b) Steel production.
    (c) Rolling.

    Iron production is continuous and consists of chemically reducing iron ore in a blast furnace, using coke and crushed limestone. The resulting material, called cast iron, is high in carbon, sulphur and phosphorus.

Steel production is a batch process and consists in reducing the carbon, sulphur and phosphorus levels and adding where necessary manganese, chromium, nickel, vanadium, etc. This process is now carried out using a Basic Oxygen Converter, which consists of a vessel charged with molten cast iron, steel scrap and limestone through which oxygen is passed under pressure to reduce the carbon content by oxidation. This is a batch process which typically produces about 250 - 300 tonnes every forty minutes. The electric arc furnace is only in limited use (around 5% of the U.K. steel production), and is generally used only for special steels such as stainless steel. The open hearth furnace and the Bessemer Converter are no longer used in the U.K.

From the converter the steel is "teemed" into ingots which are then passed to the rolling mills for successive reduction until the finished standard section is produced. The greater the reduction in size the greater the work hardening, which produces varying properties in a section. The variation in cooling rates of varying thickness introduces residual stresses which may be relieved by the subsequent straightening process. However, steel plate is now produced using a continuous casting procedure which eliminates ingot casting and subsequent processes before rolling, enabling economies to be made and better quality to be obtained. It is likely that by the end of this decade rolled sections will also be produced using the continous casting procedure.

1.4 TYPES OF STEEL

The steel used in structural engineering is a compound of approximately 98% iron and small percentages of carbon, silicon, manganese, phosphorus, sulphur, niobium and vanadium as specified in BS 4360. Carbon content is restricted to between 0.25 to 0.20 percent, to produce a steel that is weldable and not brittle. The niobium and vanadium are introduced to raise the yield strength of the steel; the manganese is introduced to improve corrosion resistance; and the phosphorus and sulphur are impurities.

There are three grades of steel that are specified for use in BS 449, i.e. grades 43, 50 and 55, with approximate yield strengths of 250, 350 and 430 N/mm^2 respectively. These three grades are subdivided depending on the chemical composition, heat treatment, yield strength, and thickness as given in BS 4360.

1.5 STANDARD STEEL SECTIONS AND PLATES

The optimisation of costs in the steelwork industry favours the production of steel members with standard cross sections and standard bar lengths of 12 or 15 m. The billets of steel are hot rolled to form bars, flats, plates, angles, tees, channels, I sections and tubes as shown in Fig. 1.1. The detailed dimensions of these sections are given in BS 4 Pt 1 and BS 4848 Pt 4. Details of some of these sections are given in the Appendix. Further details and safe load tables are given in the Structural Steelwork Handbook published by BCSA and Constrado.

Where thickness varies standard sections are identified by nominal size, i.e. 'depth x breadth x mass per unit length x shape', e.g. Universal beams, columns and channels. Where thickness is constant in a section the identification is 'breadth x depth x thickness x shape', e.g. tees and angle sections. A section is identified further by the grade of steel.

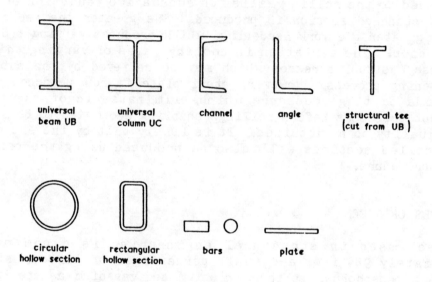

universal beam UB   universal column UC   channel   angle   structural tee (cut from UB)

circular hollow section   rectangular hollow section   bars   plate

Fig. 1.1. Standard steel sections.

To reduce costs steel plates should be selected from available stock sizes. Thicknesses are generally in the range 6, 8, 10, 12.5, 15 mm and then in 5 mm increments. Thicknesses of less than 6 mm are available but because of lower strength and poorer corrosion resistance they are not often used. Plate widths which are of standard stock sizes are in the range 1, 1.25, 1.5, 2, 2.5, and 3 metres, but narrow width plates are also available. The adoption of standard widths avoids extra work in cutting to size. Standard plate lengths are in the range 2, 2.5, 3, 4, 5, 6, 10, and 12 metres.

The application of some types of section are obvious. When a member is in tension a round or flat bar appears to be the obvious choice, but often the same member may be in compression under alternative loading and an angle, tee or tube is therefore more appropriate. The connection at the end of the member for the round bar or the tube, however, may be more difficult to make.

If a member is in bending about one axis then the 'I' section is the most efficient because a large proportion of the material is in the flanges at the extreme fibres. Alternatively, if a member is in bending about two axes at right angles and it also supports an axial load then a tube of rectangular cross section would be more appropriate. Connections to tubes are however generally more complicated.

Other steel sections available are cold formed from steel plate but sections are not standardised because of the infinite variety of possible shapes. They are used, for example, to form standard lattice beams, or glazing bars, or racks of shelving for a particular manufacturer. The use in building of cold formed sections in light gauge plate, sheet and strip steel 6 mm thick and under is dealt with in Addendum No. 1 to BS 449 Pt 2.

1.6 THE STEEL SKELETON

The standard steel sections are connected together as shown in Fig. 1.2, to form a steel skeleton for load bearing structures such as bridges, buildings, warehouses, sports stadia, schools, offices, etc. In some cases, e.g. a bridge, the steel skeleton forms most of the visible structure and only needs a deck to support traffic. In other situations the steel skeleton is almost completely hidden by cladding, e.g. an office building, and may rely on some of the walls and floors for stability.

Individually the members of a structure act as ties, struts, or beams but connected together they behave as a complete structure. If the structural connections are 'rigid', i.e. there is only a small relative rotation between connected members, and the flexural stiffness of the member is adequate, the structure may be designed to resist the forces and control the deflections without extra bracing. If the connections and/or members are less stiff then bracing is usually needed to reduce the deflections, redistribute the forces and achieve an economic structure. The bracing may be in the form of additional diagonal members, or infill brick wall, or concrete.

The bracing is introduced to stiffen the structure but not to interfere with its use by the client and is therefore generally introduced into the walls, floors or roofs. The optimisation of a structure is complicated and whilst a braced structure may use less steel than an unbraced structure, the unbraced structure may be more economic because of reduced labour costs.

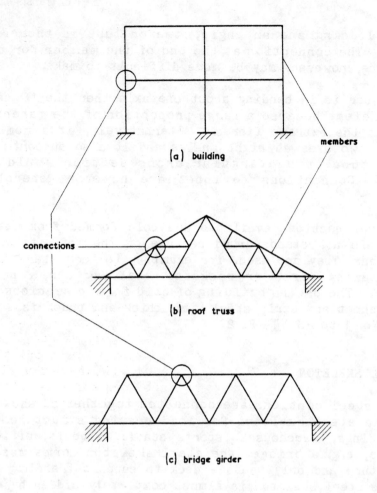

Fig. 1·2. Typical steel skeletons.

1.7   TYPES OF JOINT

1.7.1   <u>Structural joints</u>   A structural connection may be defined as
an assembly of components which are arranged to transmit forces from
one member to another.   A connection may be subjected to any
combination of axial force, shear force and bending moment in relation
to three perpendicular axes, but for theoretical purposes the
situation is generally reduced to forces in one plane.   The transfer
of forces through the components of a joint is often complex and
Chapter 4 is devoted to a detailed study of selected connections.

1.7.2   <u>Movement joints</u>   Movement joints are introduced into a
structure to take up the free expansion and contraction that may occur

on either side of the joint due to changes in temperature, shrinkage, expansion, creep, settlement, etc. These joints may be detailed to be watertight but do not generally transmit forces and are therefore not considered in this book. Details of movement joints may be obtained from Alexander and Lawson.

1.7.3 Construction joints Construction joints are introduced into a structure because components are manufactured to a convenient size, for transport and a joint therefore occurs between components. In some cases these joints transmit forces, e.g. a beam-to-column connection, but in other cases the only requirement is that the joints should be waterproof.

1.8 TOLERANCES

Tolerances are limits placed on unintentional inaccuracies that occur in dimensions and must be allowed for in design if elements and components are to fit together. In steelwork variations occur in the rolling process, in marking out for cutting during fabrication, and in setting out during erection. In the rolling process the allowable tolerances for length, width, thickness and flatness for plates are given in BS 4360. Length and width tolerances are positive while those for thickness and flatness are negative and positive. The dimensional and weight tolerances for sections are given in BS 4 Pt 1 and BS 4848 Pt 4 as appropriate. Further recommendations on tolerances are given in ECSS.

In the fabrication and erection processes a tolerance of $\pm$ 2 mm is generally acceptable. During fabrication there is a tendency for components to 'grow' rather than reduce, and the tolerance is therefore often specified as negative; it is often cheaper and simpler to insert packing rather than shorten a member, provided that the packing is not excessive. Where concrete work is associated with steelwork variations in dimensions are likely to be greater. When casting, for example, errors in dimensions may arise from shrinkage or from warping of the shuttering, especially when it is re-used. Larger tolerances are therefore accepted for work involving concrete.

Generally little attention is given to tolerances in structural engineering because the main members involving large dimensions can be made to fit by forcing them into position. Unfortunately this method is often used with smaller components, e.g. connections, and the 'built in' forces can then be large. Fortunately, these forces are often redistributed because of local yielding, without serious detrimental effects.

1.9 CLEARANCES (Cl. 57 BS 449)

A clearance dimension is an intentional allowance made on a theoretical dimension to accommodate tolerances and thus ensure that the components or members fit together. To facilitate erection all members and connections should be provided with the maximum clearance that is acceptable from structural and architectural considerations. A typical example is a connection between a steel column and a reinforced concrete base. As it could be disastrous if the base is set too high, a deliberate clearance of 50 mm or so is included in the design, with provisions for grouting under the base. Similar clearances are provided to allow lateral adjustment of the foundation bolts. Clearances between steelwork and concrete are particularly important since two different contractors may be involved. Some recommendations for clearances are given in the BCSA Structural Steel Handbook.

1.10 THE INITIATION OF A DESIGN

The demand for a structure originates with the client. The client may be a private person, private or public firm, local or national government, or a nationalised industry.

In the first stage preliminary drawings and estimates of costs are produced, followed by consideration as to which structural materials to use, i.e. reinforced concrete, steel, timber, brickwork, etc. If the structure is a building an architect only may be involved at this stage, but if the structure is a bridge or industrial building then a civil or structural engineer may prepare the documents.

If the client is satisfied with the layout and estimated costs then detailed design calculations, drawings and costs are prepared and incorporated into a legal contract document.

The contract document is usually prepared by a consultant engineer and the work is carried out by a contractor who is supervised by the consultant engineer. However, the larger firms, local and national government, and nationalised industries, generally employ their own consultant engineer. The work is still usually carried out by a contractor, but alternatively may be carried out by directly employed labour. A recent development is for the contractor to produce a design and construct package, where the contractor is responsible for all parts and stages of the work. General recommendations for steelwork tenders and contracts are given in Appendix E BS 449.

1.11 STEELWORK DRAWINGS

Detailed design calculations are essential for any steel work design but the sizes, dimensions and geometrical arrangement are usually presented as drawings. General arrangement drawings are often drawn to a scale of 1:100, while drawings of details may be to a scale of 1:20 or 1:10. Special details are drawn to a larger scale where necessary.

The drawings should be easy to read and should not include superfluous detail. Some important notes are:

(a) Members and components should be identified by logically related mark numbers.
(b) The main members should be represented by a bold outline (0.4 mm wide) and dimension lines should be unobtrusive (0.1 mm wide).
(c) Dimensions should be related to centre lines, or from one end; strings of dimensions should be avoided. Dimensions should appear once only so that ambiguity cannot arise when revisions occur. Fabricators should not be put in the position of having to do arithmetic in order to obtain an essential dimension.
(d) Clearances for erection purposes should be clearly shown.
(e) The grade of steel to be used should be clearly indicated.
(f) The size, weight and type of section to be used should be clearly stated.

1.12 FABRICATION OF STEELWORK

The drawings produced by the structural designer are used first by the steel fabricator and later by the contractor on site.

The steel fabricator obtains the material either direct from the rolling mills or from a steel stockist, and then cuts, drill and welds the steel components to form the members as shown on the drawings. In general, for British practice, the welding is confined to the workshop and the connections on site are made using bolts. In America, however, site welding is common practice.

When marking out, the measurements of lengths for overall size, position of holes etc. can be done by hand, but if there are several identical components then wooden or cardboard templates can be made to the exact dimensions. More recently automatic machines, controlled by punched paper tape have been introduced to cut and drill standard sections.

When fabricated, parts of the structure are delivered to site in the largest pieces that can be transported and erected. For example a lattice girder may be sent fully assembled to a site in this country, but sent in pieces to fit a standard transport container for erection abroad.

On site the general contractor may be responsible for the
assembly erection, connections, alignment and levelling of the
completed structure. Alternatively the erection work may be done by
the steel fabricator, or sublet to a specialist erector.

1.13   THE OBJECT OF STRUCTURAL DESIGN

The object of structural design is to produce a structure that will
not become unserviceable or collapse in its lifetime and which
fulfils the requirements of the client at reasonable cost.

These requirements may include any or all of the following:

(a)   The structure should not collapse locally or overall.
(b)   It should not be so flexible that deformations under load
      are unsightly or alarming, or cause damage to internal
      partitions and fixtures; neither should any movement due to
      live loads such as wind cause discomfort or alarm to the
      occupants/users.
(c)   It should not require excessive repair or maintenance due to
      accidental overload, or because of the action of the
      weather.
(d)   In the case of a building, the structure should be
      sufficiently fire resistant to give the occupants time to
      escape and to enable the fire brigade to fight the fire in
      safety.

The designer should be conscious of the costs involved which include:

(a)   The initial cost which includes fees, site preparation, cost
      of materials, and construction.
(b)   Maintenance costs - decoration and structural repair.
(c)   Insurance - chiefly against fire damage.
(d)   Eventual demolition.

It is the responsibility of the structural designer to design a
structure that is safe and which conforms to the requirements of the
local bye-laws and building regulations. Information and methods of
design are obtained from current British Standards and Codes of
Practice and in general these are 'deemed to satisfy' the local bye-
laws and building regulations. In exceptional circumstances, e.g. the
use of forms or methods validated by research or testing, a waiver to
the building regulations may be obtained.

A structural engineer is expected to keep up to date with the
latest research information. In the event of a collapse or malfunction
where it can be shown that within reason the engineer has failed to
foresee the cause or action leading to such a collapse, or has failed

to apply properly the information at his disposal, i.e. codes of practice, British standards, Building regulations, research or information supplied by manufacturers, then he may be sued for professional negligence. Consultancies and contractors carry professional liability insurance to mitigate the effects of such legal action.

1.14 FORCES INDUCED IN STRUCTURES

Forces are induced into members of a structure from self-weight, erection forces including lack of fit, imposed loads, wind, inertia forces due to changes in velocity of vehicles, earthquake forces, restrained temperature movements, differential settlement and prestressing.

The magnitude of the working loads on a particular structure varies, and if sufficient information were available the variation would probably be a Gaussian curve as shown in Fig. 1.3. In this graph the frequency of the loads, i.e. the number of times a particular load occurs, is plotted against the magnitude of the load. If the mean load were used to design a structure, then 50% of the loads would be greater than this and the risk is unacceptable. Instead the design load is chosen so that only 5% of loads are in excess of the value. It would be uneconomical to design for zero percentage to be in excess of the design value.

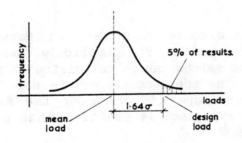

Fig. 1·3. Relationship between frequency and loads.

The forces induced into a structure from the gravitational self-weight of the members are significant for example in multi-storey buildings and large span beams. Self weight forces should always be considered but may be insignificant compared with other forms of loading. Self-weight is often described as a dead load and information on the density of materials is obtained from BS 648.

The forces induced into a structure during the erection stage of construction may also be significant, but often tend to be <u>ignored by</u> the <u>designer</u>. For example; reverse bending moments may be induced in a beam when lifted at centre span; or members that are too short, or too long, are made to fit; or supports not set at the correct level for redundant structures; or a column left unsupported during erection, which induces premature buckling. BS 449 does not given advice on erection forces and the designer therefore is expected to anticipate the conditions that are likely to arise.

A structural member or connection may not fit for a number of reasons, e.g. it may not have been manufactured correctly to the dimensions specified, or insufficient allowance may have been made for tolerances, or members or connections may be distorted before or after fabrication, e.g. by welding. The significance of lack of fit can only be assessed in many cases by engineering judgement. In some situations however the effect is obvious, e.g. forcing short or long members to fit will induce forces into the structure, or distorted faying surfaces for high strength friction–grip bolts reduce the frictional resistance, or inaccurate positioning of holding down bolts necessitates a new base plate. Further information on lack of fit in steelwork may be obtained from Mann and Morris.

There are gravitational forces other than self–weight, which are applied to structures, e.g., people, furniture, filing cabinets, vehicles, stored materials, fluids, etc. These are classified as imposed loads, and generally have the characteristics that they can be moved or changed in magnitude. Often these forces are expressed as uniformly distributed loads and values acceptable for design purposes are given in CP 3 Ch V Pt 1. Loading for bridges is given in BS 5400 Pt 2.

Wind loads may also be described as imposed loads but they are obviously not gravitational. For simplicity these are expressed as a uniform pressure and values can be calculated as shown in CP 3 Ch V Pt 2 and Example 6.1. The pressure values are dependent on geographical locality, shelter, slope and shape of the face of a structure. Pressure values are expressed as positive when pushing and negative when suction occurs.

Forces are induced into structures due to inertia forces from the change in velocity of vehicles or loads relative to the stationary structure, or the change in velocity of the earth relative to a stationary structure, i.e. an earthquake. The change in velocity of vehicles is generally not a problem but the change in velocity of loads when lifting using a crane can increase the static load considerably. Guidance for increases for static loads for light cranes is given in BS 449 Cl. 7 and for heavy cranes in BS 2573. Earthquake forces can be devastating and are a major consideration in design for countries in earthquake zones. Generally in the British Isles this is not a problem and therefore no guidance is available from British sources. Where information is required American codes of practice should be consulted.

Loads that require special consideration are oscillating loads which produce increases and decreases in forces in members. The fluctuation in forces and consequent changes in stress produce fatigue conditions which result in failure loads less than those occurring in static conditions of loading. Further information on the effects of fatigue is given in Section 1.21.

All structures should be designed to accommodate the strains due to changes in temperature. BS 449 gives a temperature range of $-70^{\circ}C$ to $+50^{\circ}C$ for exposed structures in the U.K. If temperature strains are restrained large stresses, additional to those produced by the loading, can be introduced into a structure. For example, a temperature change of $10^{\circ}C$ can produce an increase in stress of $E\alpha t = 210 \times 10^3 \times 12 \times 10^{-6} \times 10 = 25$ N/mm^2 in a member which is not allowed to expand freely. If ignored local yielding and/or buckling can occur.

Differential settlement of the foundations results in large forces being induced into stiff hyperstatic structures. Where differential settlement is expected, and cannot be easily or economically prevented, then methods of construction shall be used which adapt to the site conditions. For example, the hyperstatic structure should be built on a raft, or jacking pockets should be included so that the desired levels can be maintained, or the structure should be made statically determinate so that forces are not induced into the structure.

Prestressing is a means of preloading a member in compression when subsequent loading is known to produce tension. Practically this is accomplished by extending reinforcing rods using jacks and anchoring the rods at the ends of the member. The anchor is effected by nuts at the ends of threaded rods or by wedges.

Prestressing of steelwork is not common except for high strength friction-grip bolts used in connections. In contrast prestressing is used for reinforced concrete to overcome the weakness of concrete in tension as compared with its great strength in compression. Steel however does not exhibit this weakness and prestressing is only used on rare occasions, e.g. to provide upward camber for a beam or to strengthen the tension flange of a beam which has corroded. Prestressing forces must be allowed for in design calculations.

Forces may be induced in a structure by any one of the loads or situations described above. Some of the loads or situations occur simultaneously and the forces must be a summation of those involved. A designer must however use his judgement as to the most likely combination for the worst possible situation that is likely to occur. Further guidance on this is given in Chapter 2.

1.15   DESIGN STRESSES

If a structure is to be designed so that the stress does not exceed a specified value it is necessary to examine the stress-strain relationship for steel.   The tensile stress-strain relationship for a low strength steel, e.g. grade 43, is of the form shown in Fig. 1.4a. From 'a' to the 'limit of proportionality' the steel behaves 'elastically', i.e. there is no permanent deformation when the load is removed.   Up to this point the slope of the curve is approximately constant for all grades of structural steel, and the value for the slope given in BS 449 Appendix B is $E = 210$ kN/mm^2.   The stress/strain curve can be idealised by the tri-linear graph abcd.   From 'b' to 'c' 'yielding' takes place, i.e. there is a relatively large deformation with little or no increase in stress.   From 'c' to 'd' 'work hardening' occurs and the stress increases until fracture occurs. The yield stress is the critical value for design purposes, not the fracture stress, because after yielding structures form collapse mechanisms which result in large unacceptable deformations.

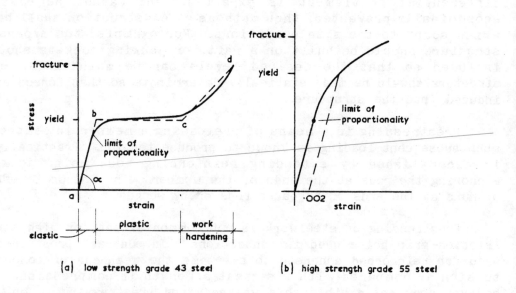

(a) low strength grade 43 steel   (b) high strength grade 55 steel

Fig. 1·4.   Tensile stress-strain relationships for steel.

If a large number of tensile specimens are tested then there is a variation in strength and the graph relating the number of samples with the same strength (frequency) against the yield stress will be a normal Gaussian distribution as shown in Fig. 1.5.   If the mean 'yield stress' is used for design at the ultimate load, then 50% of the samples would be at, or below, this value and this risk is not acceptable.   The stress used for design purposes is therefore lower than the mean stress, and is described as the 'minimum guaranteed yield stress'. This stress is defined as the value for which only approximately 5% of the samples are less than the guaranteed yield stress.

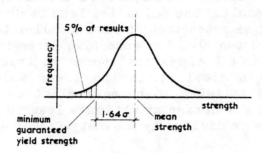

Fig. 1·5. Relationship between frequency and yield strength.

The small risk of collapse associated with the minimum guaranteed yield stress is considered acceptable because the probability that a weak member would be associated with the maximum load is very small. The 'minimum guaranteed yield stress' is controlled in practice by the steel manufacturer who takes samples during the rolling process. Ideally a large number of samples should be taken at frequent intervals but this is impractical and uneconomical. Small numbers of samples are therefore taken and the minimum guaranteed yield stress calculated from the equation for a normal Gaussian distribution.

$$Y_S = f_{mean} - 1.64\ \sigma \qquad\qquad (1.1)$$

where the standard deviation for n samples

$$\sigma = [\ \Sigma(f_{mean} - f)^2/(n-1)]^{1/2}$$

The value of the yield stress obtained from Equation (1.1) must be greater than or equal to the minimum guaranteed yield stress as specified in BS 4360.

Currently the design of steelwork in buildings is to BS 449 which is based on stresses at working loads and common sense requires, that for safety reasons, the allowable stress should be less than the minimum guaranteed yield stress. The minimum guaranteed yield stress is therefore divided by a safety factor greater than unity, for example in BS 449 the value for tension is 1.63, for axial compression is 1.7, for maximum bending stress is 1.52, and for bolts is 1.96. These factors ensure that the 'allowable stress' does not exceed the 'elastic limit of proportionality' shown in Fig. 1.4 and allows for accidental overloading, minor dimensional errors, etc.

The tensile stress-strain relationship for high strength steel, e.g. grade 55, is similar to that shown in Fig. 1.4b. There is no definite yield plateau, as with grade 43 steel, and a comparable limit must therefore be defined. The value specified is 0.2% proof stress, i.e. the stress at which, if the specimen were unloaded, a permanent strain of 0.002 would result. The unloading line is shown dotted in Fig. 1.4b, and it should be noted that it runs parallel to the loading line. This stress may be too high for some high strength steels, and a further alternative is 0.7 times the stress at fracture. For a particular high strength steel it is the lower value which is appropriate, and which is taken to be equivalent to the minimum guaranteed yield stress for design at ultimate load conditions. At working load this value is divided by a safety factor as previously explained.

BS 449 was conceived in 1932 primarily as a specification for a working load design method and the design stresses given are therefore the allowable stresses. The allowable stresses given in tables in the Appendix are taken from BS 449.

1.16   FAILURE CRITERIA FOR STEEL

The structural behaviour of materials at or close to failure may be described as ductile or brittle. A typical brittle material is cast iron which exhibits an approximate linear load-displacement relationship until fracture occurs suddenly at a relatively small displacement. In contrast mild steel is a ductile material which initially yields by sliding on a shear plane exhibiting large plastic deformations followed by fracture after work hardening.

The ductile yield condition is critical in design situations for mild steel. The yield condition can be produced in various stress situations e.g. tension, compression, shear or combinations of these stresses, and for convenience they are related to the yield stress in tension.

There are four generally acceptable theoretical yield criteria which are as follows:

(a)   The maximum stress theory, which states that yield occurs when the maximum principal stress reaches the uniaxial tensile yield stress.

(b)   The maximum strain theory, which states that yield occurs when the maximum principal strain reaches the uniaxial tensile strain at yield.

(c)   The maximum shear stress theory, which states that yield occurs when the maximum shear stress reaches half of the yield stress in uniaxial tension.

(d)   The distortion strain energy theory, or shear strain energy theory, which states that yielding occurs when the shear strain energy reaches the shear strain energy in simple tension. For a material subject to principal stresses f_1, f_2 and f_3 it is shown by Timoshenko that this occurs when

$$(f_1 - f_2)^2 + (f_2 - f_3)^2 + (f_3 - f_1)^2 = 2Y_s^2 \qquad (1.2)$$

This theory was developed by Huber, Von-Mises and Hencky.

Equation (1.2) is not expressed in terms of principal stresses in BS 449 Cl. 14c and 14d, but alternatively in terms of direct stresses f_b, f_{bc} or f_{bt}, and shear stress f_q on two mutually perpendicular planes. It can be shown from Mohr's circle of stress that the principal stresses

$$f_1 = (f_b + f_{bc})/2 - [(f_b - f_{bc})^2/4 + f_q^2]^{1/2} \qquad (1.3)$$

and

$$f_2 = (f_b + f_{bc})/2 + [(f_b - f_{bc})^2/4 + f_q^2]^{1/2} \qquad (1.4)$$

If Equations (1.3) and (1.4) are inserted in Equation (1.2) with $f_3 = 0$ and Y_s is equal to the allowable equivalent stress p_e then

$$p_e^2 = f_{bc}^2 + f_b^2 - f_{bc}f_b + 3f_q^2 \qquad (1.5)$$

This equation is given in Cl. 14d BS 449, and if f_{bc} is replaced by f_{bt} with a change of sign then this also agrees with BS 449. If $f_b = 0$ then

$$p_e^2 = f_{bc}^2 + 3f_q^2 \qquad (1.6)$$

This equation is given in Cl. 14c BS 449.

These equations are yield criteria and the values of the allowable equivalent stresses p_e shown in Table 1 BS 449 Cl. 14d are approximately $Y_s/(\text{materials factor})$. The materials factor is approximately 1.1. These values are relatively high when compared with the individual allowable stresses in bending, bearing and shear shown in Tables 2,9 and 10 BS 449 (see Appendix) respectively. Table 1 must however only be used in combined stress conditions.

1.17   OVERALL STABILITY

Some structures, e.g. tall slender buildings and retaining walls, are susceptible to being overturned by wind or other forces. BS 449 Cl. 10 states that 'the stability of the structure as a whole or any part of it shall be investigated, and weight or anchorage shall be provided so that the least restraining moment, including anchorage shall not be less than the sum of 1.2 times the maximum overturning moment due to dead loads and 1.4 times the maximum overturning moment due to imposed loads and wind'. Some structures, e.g. low rise wide buildings, are obviously stable and these calculations are not considered necessary. Other structures such as a retaining wall are very susceptible to overturning forces from imposed loads and retained material.

1.18   DEFLECTION LIMITS

Under working loads elastic deformations in members produce deflections in the structure as a whole. If these deflections are excessive the structure does not give confidence to the user. It is also difficult to fit components such as facing slabs and brickwork and finishes, e.g. plaster develops cracks which are unsightly. Empirical deflection limits are therefore specified and the value given in BS 449, Cl. 15, for beams under imposed loading is that the deflection limit should not exceed span/360. It should be noted that this limit is related to the imposed load only because deflections due to dead load are produced progressively during construction and generally do not result in cracks. The deflection limit used for single storey buildings is that the maximum deflection should not exceed height/325 as given in Cl 31b BS 449.

1.19   FIRE RESISTANCE AND PROTECTION OF STEELWORK

1.19.1   _Fire resistance_   In a fire a temperature of 1500°C may be reached, and this temperature is above the melting point of steel. Lower temperatures however will cause reductions in both the yield stress and the ultimate tensile strengths. Young's modulus will only show a slight decrease up to 400°C then a dramatic decrease above 400°C. On the other hand at about 600°C Young's modulus reduces by approximately 50%. Typically, for both grade 43 and grade 50, the yield stress is reduced by 50% at 600°C according to Smith and Stirling.

In design situations the 'critical temperature' is important, and is defined as the temperature at which the strength of the steel is reduced to a level such that the structure just collapses. The critical 'temperature' phenomenon will not tend to occur with composite construction because of the presence of the concrete. Behaviour in this case is similar to reinforced concrete. For elastic designs, in accordance with BS 449, the critical temperature is around 450°C, but for plastic designs to BS 449, this will be slightly lower thus giving a smaller margin of safety under fire conditions. If the steelwork fails to reach the critical temperature during a fire, any loss in strength in grade 43 steel will be recoverable. This may not apply to grade 50. If the temperatures reached are in excess of the critical temperature, full or partial collapse will occur, and the structure rendered fit only for demolition.

Malhotra considers in detail the methods that may be adopted for the design, under fire conditions, for steel structural elements (excluding joints). The methods delineated by Smith and Malhotra have been adopted by BS 5950 part 8. The ECCS have published a set of European recommendations for the fire safety of steel structures which give methods of design for beam and column elements together with framed structures.

1.19.2 <u>Fire protection</u> In general bare steelwork will need protection to give adequate fire resistance to the structure by limiting the temperature rise of the steelwork. Traditionally this has been done by encasing the steelwork in concrete, and is specified in BS 449 Cl 30b. This concrete casing may also assist, to a limited extent, in resisting the forces applied to the structure, but it is expensive due to the costs of formwork and of placing the concrete.

Lightweight cladding (gypsum plaster board), or intumescent paints or sprayed mineral fibres have come into use, and for a description of this treatment, Elliot should be consulted.

It may be possible to demonstrate in certain cases, especially for external steelwork, by consideration of the likely fire load, that any protection is unneccessary, and that a waiver to Part E of the Building Regulations be obtained. (see Elliot, and also Law and O'Brien).

The required fire resistance for a building is usually expressed in hours and this concept is related to the time required to evacuate the building and the time required to control the fire. The smaller the area in which the fire can be contained the shorter the escape and fire fighting times, although these will be affected by the materials in store. It is therefore advantageous to divide buildings into small sections by fire resisting walls and floors connected by fire resisting self-closing doors. These doors prevent the spread of flames and smoke which hinder escape and fire fighting.

1.20   STRESS CONCENTRATIONS

Structural members and connections often have abrupt changes in geometry and often contain holes for bolts. These features produce 'stress concentrations', which are localised stresses greater than the average stress in the member, e.g. tensile stresses adjacent to a hole are approximately three times the average tensile stress. If the average stresses in the components are low then stress concentrations may be ignored, but if the average stresses are high then appropriate methods of analysis must be used to cater for these effects. The effect of stress concentrations has been shown to be critical in plate web girders in recent history.   Stress concentrations are also associated with fatigue as described in Section 1.21.

1.21 FATIGUE

The term 'fatigue' is generally associated with metals and is the reduction in strength that occurs due to the progressive development of existing small pits, grooves or cracks.   The rate of development of these cracks depends on the size of the crack and on the magnitude of the stress variation in the material.   The number of stress variations, or cycles of stress, that a material will sustain before failure is called the 'fatigue life'.   There is a linear experimental relationship between the log of the stress range and the log of the number of cycles.   Research into the strength of welded structures is described by Munse.

All structures are subject to varying loads but the variation in stress depends on the ratio of dead load to imposed load, or whether the load is cyclic in nature as occurs where machinery is involved. In bridges and cranes fatigue effects are more likely to occur because of the cyclic nature of the loading which causes reversals of stress from tension to compression and vice versa, and an estimate must be made of the number of load cycles the structure is likely to experience during the life of the structure.   Guidance on this is given in BS 2573 for cranes and BS 5400 Pt 3 and 10 for bridges.   In these documents design stresses are related to the stress range and the number of load cycles.

1.22 RESIDUAL STRESSES IN STEEL

Residual stresses are induced into steel during rolling, welding constraining the structure to a particular geometry, force fitting of individual components, lifting and transportation.

Welding for instance raises the local temperature of the steel which then expands relative to the surrounding metal.   When it cools it contracts inducing tensile stresses in the weld and the immediately adjacent metal.   These tensile stresses are balanced by compressive stresses in the metal on either side.

During rolling, the whole of the steel section is initially at a uniform temperature, but as the rolling progresses some parts of the cross-section become thinner than others and consequently cool more quickly.   Thus, as in a welded joint, the parts which cool last have a residual tensile stress and the parts which cool first may be in compression.   Since the cooling rate also affects the yield strength of the steel, the thinner sections tend to have a higher yield stress than the thicker sections, and a tensile test piece cut from the thin web of a Universal beam will probably have a higher yield stress than one cut from the thicker flange. The residual stress and the yield stress in rolled sections is also affected by the cold straightening which is necessary for many rolled sections before leaving the mills.

The designer must realise that residual stresses are present in steelwork but many investigators believe that when the steel reaches the plastic stage of behaviour the residual stresses are relieved by the large strains that occur, and therefore do not significantly affect the ultimate strength of the structure unless elastic buckling occurs. Information on residual stresses in a steel box girder is given by Ogle.

1.23 BRITTLE FRACTURE OF STEEL

Brittle fracture of steel occurs at high speed with little visible signs of plastic deformation, and it is more likely to occur in welded structures as shown by Stout, Tor and Ruzek. The essential conditions leading to brittle fracture are that:

(a) There must be a tensile stress in the material but it need not be very high, and may be a residual stress from welding.
(b) There must be a notch or defect or hole in the material which produces a stress concentration.
(c) The temperature of the material must be below the transition temperature.

The mechanism of failure is that the notch, defect, or hole raises the local tensile stress to values as high as three times the average tensile stress. The material which generally fails by a shearing mechanism now tends to fail by a brittle cleavage mechanism which exhibits considerably less plastic deformation. A drop in temperature encourages the cleavage failure. A ductile material, which has an extensive plastic range is more likely to resist brittle fracture and the ability to absorb energy is called 'toughness'. 'Toughness' is often measured in practice by the Charpy V-notch impact test as specified in BS 4360, but this test is only a guide to resistance to brittle fracture.

1.24 CORROSION AND PROTECTION OF STEELWORK

Corrosion of steel reduces the cross section of members and an understanding of the corrosion process is therefore of interest to the structural engineer. Corrosion is a chemical reaction between iron, water and oxygen, which produces a hydrated iron oxide called rust. Electrons are liberated in the reaction and a small electrical current flows from the corroded area to an uncorroded area.

The elimination of water, oxygen or the electrical current reduces the rate of corrosion. In contrast pollutants in the air e.g. sulphur dioxides from industrial atmospheres and salt from marine atmospheres, increase the electrical conductivity of water and accelerate the corrosion reaction.

Steel is particularly susceptible to atmospheric corrosion which is often severe in coastal or industrial environments and the corrosion may reduce the section size due to pitting or flaking of the surface. Modern rolling techniques and higher strength steels result in less material being used, and for example, the webs of I beam sections may be only 6 mm thick. As a general rule Cl. 12 BS 449 specifies a minimum thickness of 8 mm for steel exposed to the weather or other corrosive influences, and 6 mm for steel not so exposed. For sealed hollow sections these limits are reduced to 4 mm and 3 mm respectively.

Corrosion in steel usually takes the form of rust which is a complex oxide of iron. The rust builds up as a deposit on the surface and may eventually flake off. The coating of rust does not inhibit corrosion except in special steels and the corrosion progresses beneath the rust forming conical pits, and the general thickness of the metal is reduced. The pits are conical in shape and can act as 'stress raisers', i.e. centres of high local stress, and in cases where there are cyclic reversals of load, may become the initiating points of fatigue cracks or brittle fracture. The corrosion properties of any unprotected steel, are dependent on its chemical composition, the degree of pollution in the atmosphere, and the frequency of wetting and drying of the steel.

Low strength carbon steels are inexpensive but particularly susceptible to atmospheric corrosion which is often greatest in industrial or coastal environments. High strength low alloy steels (Cr-Si-Cu-P) do not pit as severely as carbon steels and the rust that forms becomes a protective coating against further deterioration. These steels therefore have several times the corrosion resistance of carbon steels.

The longer steel remains wet the greater the corrosion and therefore the detailing of steelwork should include drainage holes, avoid pockets, and allow the the free flow of air for rapid drying.

The most common and cheapest form of protection process is to clean the surface by sand or shot blasting, and then to paint with a red lead primer, generally in the workshop prior to delivery on site. Joint contact surfaces need not be protected unless specified. On site the steel is erected and protection is completed with an undercoat and a finishing coat or coats of paint.

In the case of surfaces to be welded steel should not be painted, or metal coated, within a suitable distance of any edges to be welded, if the paint specified or the metal coating is likely to be harmful to welders or impair the quality of the welds. Welds and adjacent parent metal should not be painted prior to de-slagging, inspection and approval.

Encasing steel in concrete provides an alkaline environment and no corrosion will take place unless water diffuses through the concrete carrying with it SO_2 and CO_2 gases from the air in the form of weak acids. The resulting corrosion of the steel and the increase in pressure spalls the concrete. Parts to be encased in concrete should not be painted or oiled, and where friction grip fasteners are used protective treatment should not be applied to the faying surfaces.

A more expensive protection is zinc, or aluminium spray coating which is sometimes specified in more corrosive atmospheres. A further improvement is hot dip zinc galvanising, or finally the use of stainless steels. These and other forms of protection are described in BS 5493.

1.25 TESTING OF STEELWORK

1.25.1 Destructive testing   Testing of small sample pieces of steel is generally used as a means of controlling the structural properties of the steel during the manufacturing process. If, however, a structure collapses, then tests may be carried out on selected parts of the structure in an attempt to isolate the causes of collapse. Testing of large scale structures is expensive and is generally only carried out for research purposes, or when sample testing of the material in a collapse does not provide sufficient information on the mechanism of collapse.

The size, shape, position of gauges and method of testing of small sample pieces of steel is given in BS 4360 and BS 18. The tensile test is most frequently employed, and yields the values of Youngs modulus, limit of proportionality, yield stress or proof stress, percentage elongation and ultimate stress. The Charpy V-notch test for impact resistance, is used to measure toughness, i.e. the total energy, elastic and plastic, which can be absorbed by a specimen before fracture. The test specimen is a small span beam of rectangular cross section with a V notch at mid length. The beam is fractured by a blow from a swinging pendulum, and the amount of energy absorbed is calculated from the loss in height of the pendulum swing after fracture. Details of the test specimen and procedure are given in BS 4360 and BS 131. The Charpy V-notch test is often used to determine the transition temperature from brittle to ductile behaviour.

1.25.2 <u>Non-destructive testing</u> Structures which are unconventional, and/or methods of design which are unusual or not fully validated by research, should be subject to acceptance tests. Essentially this consists of loading the structure to ensure that it has adequate strength to sustain a total load of 2(deadload) + (imposed load), and adequate stiffness to resist a total load of (dead load) + 1.5(imposed load). Details of the tests are given in Appendix A of BS 449.

Where welds are of vital importance, e.g. in pressure vessels, they should be subject to non-destructive tests. The defects that can occur in welded connections are: slag inclusions; porosity; lack of penetration and sidewall fusion; liquation, solidification, hydrogen cracking, lamellar tearing and brittle fracture.

A surface crack in a weld may be detected visibly but alternatively a dye may be sprayed onto the joint which seeps into the cracks. After removing any surplus dye the weld is sprayed with a fine chalk suspension and the crack then shows as a coloured line on the white chalk background. A variant of this technique is to use fluorescent dye and a crack then shows as a bright green line in ultra violet light. A surface crack may also be detected if the weld joint area is magnetised and sprayed with iron powder. The powder congregates along a crack, which shows as a black line.

Other weld defects cannot be detected on the surface and alternative methods must be used. Radiographic methods use an X-ray, or gamma-ray, source on one side of the weld and a photographic film on the other. The rays are absorbed by the weld metal, but if there is a hole or crack there is less absorption which shows as a dark area on the film. Not all defects are detected by radiography since the method is sensitive to the orientation of the flaw, e.g. cracks at right angles to the X-ray beam are not detected. Radiography also requires access to both sides of the joint and the method is therefore most suitable for in line butt welds for plates.

An alternative method to detect hidden defects in welds uses ultra-sonics. If a weld contains a flaw then high frequency vibrations are reflected. The presence of a flaw can therefore be indicated by monitoring the reduction of transmission of ultra-sonic vibrations, or by monitoring the reflections. The reflection method is extremely useful for welds where access is only possible from one side. Further detailed information may be obtained from Gourd.

1.26 STRUCTURAL SAFETY

It is self evident that a structure should be 'safe' during its lifetime, i.e. free from risk of collapse. There are however other risks associated with a structure and the term 'safe' is now being

replaced by the broader term 'serviceable'. A structure shoul
during its lifetime become 'unserviceable', i.e. it should be
from risk of collapse, deterioration, fire, cracking, deflection, etc.

Ideally it should be possible to calculate mathematically the
risk involved in structural safety but this area of research has been
neglected. Recently reports such as the CIRIA report 63, have
introduced the designer to the elegant and powerful concept of
'structural reliability'. Methods have been devised whereby
engineering judgement and experience can be combined with statistical
analysis for the rational computation of partial safety factors in
codes of practice. However, in the absence of complete understanding
and data concerning aspects of structural behaviour, absolute values
of reliability cannot be determined.

It is not practical, nor is it economically possible, to design a
structure that will never fail. It is always possible that the
structure will contain material that is less than the required
strength or that it will be subject to loads greater than the design
loads.

It is therefore accepted that 5% of the material in a structure
is below the design strength and 5% of the applied loads will be
greater than the design load. This does not mean therefore that
failure is inevitable, because it is extremely unlikely that the weak
material and excessive loading will combine to produce collapse.

Fortunately there are now relatively few structural failures and
when they do occur they are generally associated with human error, or
in design calculations, or during construction or use of the
structure.

Errors that occur in structural design calculations and affect
structural safety are:

(1) Ignorance of the physical behaviour of the structure under
    load, which introduces errors in the basic assumptions used
    in the analysis.
(2) Errors in estimating the loads, especially the erection
    forces.
(3) Numerical errors in the calculations. These should be
    eliminated by checking.
(4) Ignorance of the significance of certain effects, e.g.
    residual stresses, fatigue, etc.
(5) Introduction of new materials or methods which have not been
    proved by tests or experiments.
(6) Insufficient allowance for clearance or temperature strains.
(7) Neglecting the tolerance allowance.

Errors that can occur on construction sites or in workshops are:

(1) Using the wrong grade of steel. This is also important in welding where the wrong type of electrode may be inadvertently used.
(2) Using the wrong weight of section. A number of sections are of the same nominal size but differ in thickness of web, flange, etc.
(3) Errors in manufacture, e.g. holes in the wrong position.

Errors that occur in the life of the structure and also affect safety are:

(1) Overloading due to change of use.
(2) Removal of structural material.
(3) Poor maintenance.

REFERENCES

Alexander, S.J. and Lawson, R.M. (1981) - Movement design in buildings, Technical Note 107, CIRIA London.

British Constructional Steelwork Association (1978) - Structural steel handbook, properties and safe load tables.

British Standards Institution (1972) - CP 110 The structural use of concrete, Pt. 1.

British Standards Institution (1969) - BS 449 The use of structural steel in building, Pt. 2.

British Standards Institution (1982) - BS 5400 Steel, concrete and composite bridges, Pt. 3.

British Standards Institution (1980) - BS 5400 Code of practice for fatigue, Pt. 10.

British Standards Institution (1983) - BS 5950 Structural steelwork for buildings, Pt. 8.

British Standards Institution (1979) - BS 4360 Specification for weldable structural steels.

British Standards Institution (1949) - BS 648 Schedule of weights of building materials.

British Standards Institution (1967) - CP 3 Dead loads and imposed loads, Ch V, Pt. 1.

British Standards Institution (1972) - CP 3 Wind loads and dynamic loads for light cranes, Ch V, Pt. 2.

British Standards Institution (1977) - BS 2573 Permissible stresses in cranes and design rules, Pt. 1.

British Standards Institution (1978) - BS 5400 Steel concrete and composite bridges, specification for loads, Pt. 2.

British Standards Institution (1974) - BS 5135 Specification for metal arc welding of carbon and carbon manganese steels.

British Standards Institution (1977) - BS 5493 Code of practice for protection of iron and steel structures from corrosion.

British Standards Institution (1980) - BS 4 Specification for hot rolled sections, Pt. 1.

British Standards Institution (1972) - BS 4848 Equal and unequal angles, Pt. 4.

British Standards Institution (1971) - BS 18 Methods for tensile testing of metals - steel (general), Pt. 2.

British Standards Institution (1972) - BS 131 The Charpy V-notch impact test on metals, Pt. 2.

Buchanan, R.A. (1972) - Industrial Archeology in Britain, Penguin.

Construction Industry Research and Information Association (1977) - Rationalisation of safety and serviceability factors in structural codes, Report 63.

Cossons, N. (1975) - The BP Book of Industrial Archeology, David and Charles.

Derry, T.K. and William, T.I. (1960) - A Short History of Technology, Oxford University Press.

Elliott, D.A. (1981) - Fire and Steel Construction - Protection of Structural Steelwork (second edition), CONSTRADO.

Elliott, D.A. (1980) - Fire and Steel Construction - The Building Regulations 1976 - Application to steelwork of Part E: section I: structural fire precautions, CONSTRADO.

European Convention for Constructional Steelwork (1981) - European Recommendations for Steel Construction, Construction Press.

European Convention for Constructional Steelwork (1983) - European Recommendations for the Fire Safety of Steel Structures, Construction Press.

Gourd, L.M. (1980) - Principles of Welding Technology, Edward Arnold.

Low, M. and O'Brien, T. (1981) - Fire and Steel Construction - Fire Safety of Bare External Structural Steel, CONSTRADO.

Malhotra, H.L. (1982) - Design of Fire-resisting Structures, Surrey University Press.

Mann, A.P. and Morris, L.J. (1981) - Lack of fit in steel structures, Report 87, CIRIA, London.

Munse, W.H. (1964) - Fatigue of Welded Steel Structures, Welding Research Council.

Ogle, M.H. (1982) - Residual stresses in a steel box-girder bridge, CIRIA Tech Note 110.

Pannel, J.P.M. (1946) - An Illustrated History of Civil Engineering, Thames and Hudson.

Rolt, L.T.C. (1970) - Victorian Engineering, Penguin.

Smith, C.I. and Stirling, C. (1983) - Analytical methods and design of firesafe steel structures, three decades of fire safety, BRE.

Stout, R.D., Tor, S.S. and Ruzek, J.M. (1951) - The effect of fabrication procedures on steels used in pressure vehicles, Welding Journal, 30.

Timoshenko, S. (1946) - Strength of Materials, part II, Pub. D. Van Nostrand Co.

2   Rolled-Section Beams

2.1   INTRODUCTION

The design of members subject to bending, which is covered by BS 449 Chapter 4, Section B, divides conveniently into two parts: the design of beams consisting of standard rolled sections, and the design of fabricated beams such as plate girders. With fabricated beams the proportions and layout of the webs and flanges are selected by the designer to provide properties of flexural strength and stiffness economically and without local instability. Rolled beam sections, on the other hand, are mass produced in nominal sizes, each of which covers a range of sections, thus enabling the designer to choose the combination of flexural strength, transverse and lateral stiffness suitable for his purpose. The required properties may not be provided as efficiently as by a carefully designed fabricated section but the comparative cheapness of rolled sections makes them more economical for most purposes. In general, fabricated beams are designed when, by reason of size, shape, or some other special requirement, suitable rolled beams are not available.

In this chapter the basic theoretical work needed for the design of beams in general is presented first. Only the results of theorems are quoted, with examples of their application. For proofs and more detailed discussion of the theorems the reader is directed to a standard text book on Strength of Materials. The works by Megson and Ryder which are given in the bibliography at the end of this chapter, are examples. Section 2.10 covers the design of beams by plastic theory, which is one of the alternatives allowed by BS 449. A number of complete worked solutions of design problems are given in Section 2.11. The design of fabricated beams is discussed and illustrated by examples in Chapter 7.

2.2   ELASTIC BENDING

2.2.1   <u>Sectional axes and sign conventions</u>   For all standard sections rectangular centroidal axes x-x and y-y are defined parallel to the main faces of the section, as shown in Fig. 2.1. The position of these axes is given in the tables of steel sections issued by Constrado (see Appendix). In angles, and other sections where the rectangular and principal axes do not coincide, the principal axes are denoted by u-u and v-v. The major axis u-u is conventionally inclined to the x-x axis by an angle α , as shown in Fig. 2.1(e) and (f). For equal angles, α = 45 degrees.

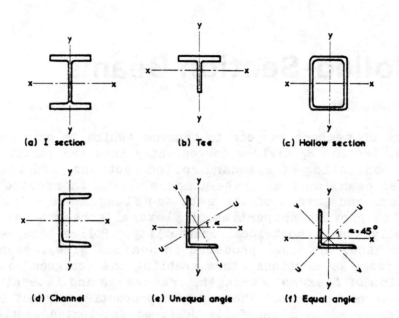

Fig. 2·1. Sectional axes.

(a) I section   (b) Tee   (c) Hollow section

(d) Channel   (e) Unequal angle   (f) Equal angle

In problems involving simple uniaxial or biaxial bending of symmetrical sections a strict sign convention is not necessary, but for the solution of complex problems it is desirable. In this book the positive conventions of sagging curvature and downward deflections are adopted; and the direction of the angle α is anti-clockwise, consistent with the Constrado tables. Fig. 2.2(a) shows the coordinates of a point P in the positive quadrant of a section and the positive directions for externally applied forces and couples. The positive direction of the corresponding stress resultants (shear forces and bending moments) are shown for the horizontal and vertical planes in Fig. 2.2(b). Positive directions relative to the u-u and v-v axes can be inferred. The convention for moments has been chosen so that positive moments give tensile stresses in the positive quadrant of the section.

Normally the coordinates of points in a section relative to the rectangular axes are known, or can easily be obtained. The coordinates relative to the principal axes are given by

$$u = x \cos\alpha + y \sin\alpha$$

$$v = y \cos\alpha - x \sin\alpha$$

(2.1)

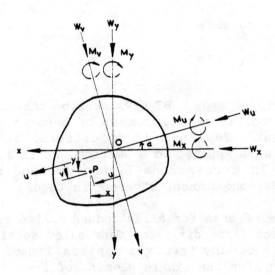

(a) Coordinates and external loads

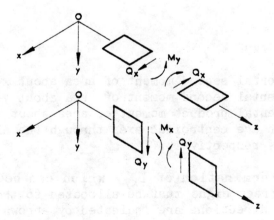

(b) Stress resultants

Fig. 2·2. Sign conventions.

External and shear forces transform in exactly the same way, thus

$$W_u = W_x \cos \alpha + W_y \sin \alpha$$

$$W_v = W_y \cos \alpha - W_x \sin \alpha$$

(2.2)

However the directions chosen for M_y and M_v are inconsistent with the rules for a right hand set of axes, which gives rise to changes in sign, thus

$$M_u = M_x \cos \alpha - M_y \sin \alpha$$

$$M_v = M_y \cos \alpha + M_x \sin \alpha$$

(2.3)

2.2.2   <u>Second moments of area</u>   BS 449, section tables, and some older literature, refer to the second moment of area of a section as the 'moment of inertia'.   The term is traditional in design, but is strictly incorrect when applied to a section, which is a plane surface having no mass.   In more recent literature and modern codes of practice the term 'second moment of area' is used.

Second moments of area for all standard rolled sections are given in the section tables (Appendix); for fabricated sections they must be calculated.   The procedure involves application of the theorems of parallel axes which, for the single element of area A in Fig. 2.3, can be stated as follows.

$$I_x = I_a + Ay^2$$

$$I_y = I_b + Ax^2$$

(2.4)

$$I_{xy} = I_{ab} + Axy$$

where I_x = elemental second moment of area about x-x
 I_y = elemental second moment of area about y-y
 I_{xy} = elemental product moment of area about x-x and y-y
 a-a and b-b are centroidal axes through the element, parallel to
 x-x and y-y respectively.

For the determination of I_{xy}, which can be either positive or negative, the correct signs must be allocated to the coordinates x and y.   The positive directions are indicated by arrows in Fig. 2.3.

When second moments of area about the rectangular axes have been computed, the direction of the principal axes can be obtained from

$$\tan 2\alpha = 2I_{xy}/(I_y - I_x)$$

(2.5)

The principal second moments of area are then given by

$$I_u = I_x\cos^2 a + I_y\sin^2 a - I_{xy}\sin 2a$$

$$I_v = I_x\sin^2 a + I_y\cos^2 a + I_{xy}\sin 2a$$

(2.6)

If I_x is arranged to be greater than I_y, then a will be less than 45 degrees and I_u will be the major principal second moment of area. A negative result for Equation (2.5) indicates that a is to be measured clockwise from x-x.

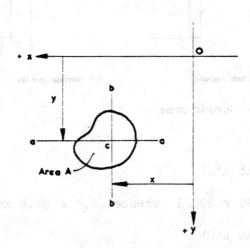

Fig. 2·3. Parallel axes for an element.

Example 2.1   Find the directions of the principal axes and the values of the principal second moments of area for the angle section in Fig. 2.4(a). Dimensions are in mm.

For the calculation of section properties the work is simplified considerably, with insignificant loss of accuracy, by using the dimensions of the section profile, i.e. the shape formed by the centre-line of the elements, as shown in Fig. 2.4(b).

The position of the centroid O is found by taking moments of area about the centre lines of each leg in turn, thus

Areas		mm^2
A'B'	140 x 20 =	2800
A'C'	290 x 20 =	5800

		8600

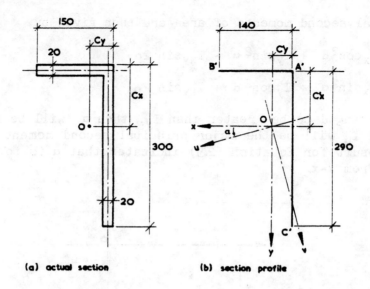

(a) actual section   (b) section profile

Fig. 2·4. Unequal angle.

Taking moments about A'B'

$$8600\ C_x' = 5800 \times 290/2 \quad \text{whence} \quad C_x' = 97.8 \text{ mm}$$

Taking moments about A'C'

$$8600\ C_y' = 2800 \times 140/2 \quad \text{whence} \quad C_y' = 22.8 \text{ mm}$$

Hence for the full section

$$C_x = 107.8 \text{ mm} \quad \text{and} \quad C_y = 32.8 \text{ mm}$$

The coordinates of the centroids of the legs AB and AC are therefore given by

Leg A'B' $x = 140/2 - C_y' = 47.2$ mm

$y = -C_x' = -97.8$ mm

Leg A'C' $x = -C_y' = -22.8$ mm

$y = 290/2 - C_x' = 47.2$ mm

The second moments of area about the rectangular axes are obtained in the usual way by applying the parallel axes formula to each leg. Thus for I_x:

$$I_x \text{ (leg)} = bd^3/12 + A(\text{leg}) \, y^2$$

where b and d are the dimensions of the leg parallel to the x-x and y-y axes respectively.

$$10^6 \text{ mm}^4$$

Leg A'B'	$140 \times 20^3/12 + 2800 \times (-97.8)^2 =$	26.87
Leg A'C'	$20 \times 290^3/12 + 5800 \times 47.2^2 \quad =$	53.57
	$I_x \quad =$	80.44

Similarly for I_y:

$$I_y \text{ (leg)} = db^3/12 + A(\text{leg}) \, x^2$$

$$10^6 \text{ mm}^4$$

Leg A'B'	$20 \times 140^3/12 + 2800 \times 47.2^2 \quad =$	10.81
Leg A'C'	$290 \times 20^3/12 + 5800 \times (-22.8)^2 =$	3.21
	$I_y =$	14.02

The product moment of area I_{xy} is obtained by applying the parallel axes formula

$$I_{xy} \text{ (leg)} = I_{ab} + A(\text{leg})xy$$

to each leg. For each leg the term I_{ab} is equal to zero, because the parallel axes through the centroid of the leg are principal axes. Hence for I_{xy} of the whole angle:

$$10^6 \text{ mm}^4$$

Leg A'B'	$2800 \times 47.2 \times (-97.8) \quad =$	-12.93
Leg A'C'	$5800 \times (-22.8) \times 47.2 \quad =$	-6.24
	I_{xy}	-19.17

The direction of the principal axes is given by Equation (2.5), i.e.

$$\tan 2\alpha = 2I_{xy}/(I_y - I_x)$$

i.e. $2a$ = arctan [2 x (-19.17)/(14.02 - 80.44)]

i.e. $2a$ = 30 degrees, a = 15 degrees.

The principal second moments of area are given by Equation (2.6), i.e.

$$I_u = I_x \cos^2 a + I_y \sin^2 a - I_{xy} \sin 2a$$

$$I_v = I_x \sin^2 a + I_y \cos^2 a + I_{xy} \sin 2a$$

Whence, substituting for I_x, I_y, I_{xy} and a

$$I_u = 85.58 \times 10^6 mm^4 \qquad I_v = 8.88 \times 10^6 \ mm^4$$

As a check on the transformation, the invariant properties of the sum of second moments of area can be invoked, thus

$$I_x + I_y = I_u + I_v$$

which is found to be correct in this case.

2.2.3 **Bending of a symmetrical section** When either of the rectangular axes is an axis of symmetry the normal bending stress at any point in the section is given by

$$f = M_x y / I_x + M_y x / I_y \qquad \qquad (2.7)$$

If the directions of the bending moments and the coordinates are in accordance with the sign convention of Fig. 2.2, a positive result indicates that the stress is tensile.

For simple bending about the x-x axis the stress in the extreme fibres of the section is given by

$$f_{max} = M_x / Z_{ex} \qquad \qquad (2.8)$$

where $Z_{ex} = I_x / y_{max}$, the elastic section modulus relative to the x-x axis.

Similarly, for simple bending about the y-y axis

$$f_{max} = M_y / Z_{ey} \qquad \qquad (2.9)$$

where $Z_{ey} = I_y / x_{max}$

Values of Z_{ex} and Z_{ey} are quoted for Universal beam and column sections, joists and channels, in the section tables (Appendix); Z_{ey} for channels refers to the outer ends of the flanges. For structural tees two values of Z_{ex} are given, referring to the extreme fibres in the table and the stalk.

Example 2.2   Calculate the maximum extreme fibre stresses in a
standard 292 x 419 x 113 kg structural tee cut from a Universal beam.
The tee is loaded by two bending moments as shown in Fig. 2.5.

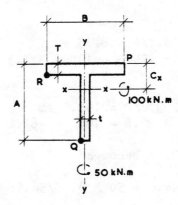

Fig. 2.5. Structural tee in biaxial bending.

From the section tables in the Appendix, the section has the following
dimensions and properties.

    Width (B) = 293.8 mm
    Depth (A) = 425.5 mm
    Thickness of web (t) = 16.1 mm
    Thickness of flange (T) = 26.8 mm
    Depth to centroid (C_x) = 10.8 cm = 108 mm
    Second moments of area
        I_x = 24660 cm^4 = 246.6 x 10^6 mm^4
        I_y = 5676 cm^4 = 56.76 x 10^6 mm^4
    Elastic moduli
        x-x axis
            Flange 2277.0 cm^3 = 2.2770 x 10^6 mm^3
              Toe   777.6 cm^3 = 0.7776 x 10^6 mm^3
        y-y axis   386.5 cm^3 = 0.3865 x 10^6 mm^3

     By inspection the maximum compressive stress occurs at point P
because the stresses from both moments are compressive in the quadrant
containing P and the extreme fibres with respect to both axes
intersect at P.   The stress can therefore be obtained directly, using
the elastic moduli for the flange and the y-y axis.   Hence, combining
Equations (2.8) and (2.9) and cancelling 10^6 in numerator and
denominator,

$$f_p = 100/2.277 + 50/0.3865 = 173 \text{ N/mm}^2$$

The maximum tensile stress can occur either at point Q or point R, depending upon the relative values of the bending moments. It is necessary to check both points.

Using the sign convention of Fig. 2.2, both bending moments are positive, and the coordinates of the points are given by

Point Q

$$x = t/2 = 16.1/2 = 8.05 \text{ mm}$$
$$y = A - C_x = 425.5 - 108 = 317.5 \text{ mm}$$

Point R

$$x = B/2 = 293.8/2 = 146.9 \text{ mm}$$
$$y = T - C_x = 26.8 - 108 = -81.2 \text{ mm}$$

For point Q, Equation (2.7) becomes

$$f_Q = 100 \times 317.5/246.6 + 50 \times 8.05/56.76 = 135.8 \text{ N/mm}^2$$

Similarly, for point R

$$f_R = 100 \times (-81.2)/246.6 + 50 \times 146.9/56.76 = 96.5 \text{ N/mm}^2$$

The maximum tensile and compressive stresses are therefore:

$$f_{bt} = 136 \text{ N/mm}^2 \text{ (tension)}$$

$$f_{bc} = 173 \text{ N/mm}^2 \text{ (compression)}$$

where the subscripts are those prescribed by BS 449 for bending stresses.

As a general comment on the evaluation of expressions such as Equation (2.7), where an accurate result depends upon the correct interpretation of a sign convention, it is the writer's experience that mistakes in signs can easily occur. Checking of the results by inspection cannot be too strongly advised.

2.2.4 __Unsymmetrical bending__   When a section is subject to a bending moment about an axis which is not a principal axis, the effect is the same as if the section were subject to the components of the bending moment acting about the principal axes. In other words the bending is effectively biaxial.

For standard rolled angles the principal second moments of area and the directions of the principal axes are given in the section tables (Appendix).   It is therefore a fairly simple matter to transform bending moments and coordinates to the principal axes by means of Equations (2.3) and (2.1).   The bending stress is then given by

$$f = M_u v/I_u + M_v u/I_v \qquad\qquad (2.10)$$

This is the same as Equation (2.7), but with all the terms related to the principal axes.   If the sign convention of Fig. 2.2 is observed, a positive result indicates tension.

In other cases the additional calculations required for the solution of the problem by principal axes can be avoided by the use of 'effective bending moments'.   These are modified moments which can be considered to act about the rectangular axes of the section.   The bending stress is then given by an expression having exactly the same form as Equation (2.7), thus

$$f = M_{ex} y/I_x + M_{ey} x/I_y \qquad\qquad (2.11)$$

where M_{ex} and M_{ey} are the effective moments about the x-x and the y-y axes respectively and are given by

$$M_{ex} = (M_x - M_y I_{xy}/I_y)/D$$

$$M_{ey} = (M_y - M_x I_{xy}/I_x)/D \qquad\qquad (2.12)$$

where $D = 1 - I_{xy}^2/(I_x I_y)$

These expressions are derived from the application of conventional bending theory to curvature in both the xz and yz planes. One such derivation is given by Megson.

By successive differentation with respect to z (the longitudinal dimension), similar expressions for the effective shear force and effective load intensity can be obtained, thus

$$Q_{ex} = (Q_x - Q_y I_{xy}/I_x)/D$$

$$\qquad\qquad (2.13)$$

$$Q_{ey} = (Q_y - Q_x I_{xy}/I_y)/D$$

and

$$w_{ex} = (w_x - w_y I_{xy}/I_x)/D$$

$$\qquad\qquad (2.14)$$

$$w_{ey} = (w_y - w_x I_{xy}/I_y)/D$$

It should be noted that the quantities I_x and I_y in Equations (2.12) are interchanged in Equations (2.13) and (2.14).   This is because the expressions for shear forces and load intensities in the x direction are obtained by successive differentiation of bending moments about the y-y axis and vice versa.

All bending problems with unsymmetrical sections can be solved simply by replacing ordinary loads, shears, and bending moments by their effective counterparts. Note however that these effective counterparts have values relating to both the x-x and y-y axes, even if the section is only loaded in the direction of one of the rectangular axes.

Example 2.3   Calculate the bending stresses in the angle section of Example 2.1 when it is subjected to the bending moments shown in Fig. 2.6(a). The positive sign convention for bending moments is shown for reference in Fig. 2.6(b).

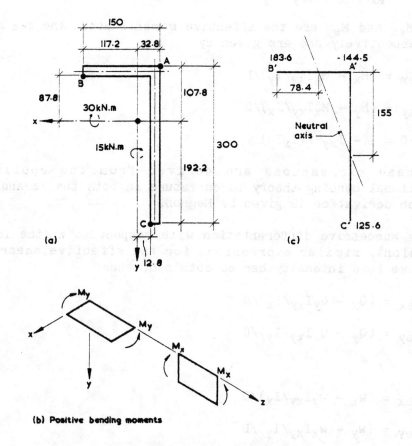

(a)

(b) Positive bending moments

(c)

Fig. 2·6. Bending stresses in an unequal angle.

Both bending moments are positive, i.e.

$$M_x = 30 \text{ kN.m} \qquad\qquad M_y = 15 \text{ kN.m}$$

From Example 2.1 the second moments of area in 10^6 mm^4 are

$$I_x = 80.44 \qquad I_y = 14.02 \qquad I_{xy} = -19.17$$

The effective bending moments are given by Equations (2.12), i.e.

$$M_{ex} = (M_x - M_y I_{xy}/I_y)/D$$

$$M_{ey} = (M_y - M_x I_{xy}/I_x)/D$$

where $D = 1 - I_{xy}^2/(I_x I_y)$

whence $M_{ex} = 74.92$ kN.m and $M_{ey} = 32.86$ kN.m

The bending stress due to the combined effective moments is given by Equation (2.11), i.e.

$$f = M_{ex}y/I_x + M_{ey}x/I_y \tag{1}$$

The maximum compressive stress occurs at point A, which is at an extreme fibre with respect to both axes and receives compression from both moments.

$$y_A = -107.8 \text{ mm and } x_A = -32.8 \text{ mm}$$

Substitution into Equation (1) gives

$$f_A = -177.3 \text{ N/mm}^2 \text{ (i.e. compression)}$$

There is no point on the section which receives tensile stress from both moments, so by inspection the maximum tensile stress could occur either at point B or point C.

At point B:

$$y_B = -87.8 \text{ and } x_B = +117.2$$

Substitution into (1) gives

$$f_B = 192.9 \text{ N/mm}^2 \text{ (i.e. tension)}$$

Similarly at point C:

$$y_C = +192.2 \text{ and } x_C = -12.8$$

and hence

$f_C = 149.0$ N/mm^2 (i.e. tension)

The maximum bending stresses are therefore

$f_{bt} = 192.9$ N/mm^2 and $f_{bc} = 177.3$ N/mm^2

Note that although the position of the centroid and the values of second moments of area can be calculated without significant error from the profile dimensions of the section, the same is not true of the stresses. For comparison the stresses at corresponding points on the profile are shown in Fig. 2.6(c).

These values can be used to find where the neutral axis intersects the profile. Thus the distance from B' is given by

$x_0 = 140 \times 183.6/(183.6 + 144.5) = 78.4$ mm

and the distance from A' by

$y_0 = 290 \times 144.5/(144.5 + 125.6) = 155$ mm

Example 2.4 Re-calculate the bending moments at points A, B and C in the angle of the previous example by considering bending about the principal axes.

The coordinates of the points are transformed in accordance with Equations (2.1), i.e.

$u = x \cos a + y \sin a$

$v = y \cos a - x \sin a$

From the previous example $\sin a = 0.2588$, $\cos a = 0.9659$, giving the coordinates in mm:

Point	x/y axes		u/v axes	
	x	y	u	v
A	-32.8	-107.8	-59.6	-95.6
B	117.2	-87.8	90.5	-115.1
C	-12.8	192.2	37.5	189.0

The bending moments M_u and M_v transform in accordance with Equations (2.3), i.e.

$M_u = M_x \cos a - M_y \sin a$

$$M_v = M_y \cos a + M_x \sin a$$

From the previous example $M_x = 30$ kN.m, $M_y = 15$ kN.m, and hence $M_u =$ 25.10 kN and $M_v = 22.25$ kN. All moments are positive in accordance with the convention.

The bending stress is given by Equation (2.10), i.e.

$$f = M_u v/I_u + M_v u/I_v$$

which, with the coordinates just calculated, gives the following stresses in N/mm^2

Point	A	B	C
Stress	-177.3	193.0	149.1

These, allowing for rounding errors in calculation, are the same as were obtained by the use of effective moments.

2.3   SHEAR EFFECTS

When a beam is bent elastically by a system of transverse loads, plane sections no longer remain plane after bending, but are warped by shear strains. In most cases the effect is small and the errors introduced by the use of conventional bending theory are negligible. Important exceptions are discussed briefly in Section 2.3.2. Formulae for the calculation of shear stresses in an elastic beam are derived by considering the variation in bending stresses along a short length of beam, using conventional theory.

2.3.1   Shear stresses in a symmetrical section   Consider the very short length of beam in Fig. 2.7(a). At a point S in the web the shear stresses on the vertical and longitudinal section are complementary and are given by the established formula

$$f_{qs} = QAy_c/(It_s) \tag{2.15}$$

where Q = the vertical shear force on the section
   A = the hatched area, i.e. the part of the section between
       the point S and the extreme fibres
   y_c = the distance from the centroid of area A to the
        neutral axis
   I = the second moment of area of the whole section about the
       neutral axis
   t_s = the thickness of the section at the point S

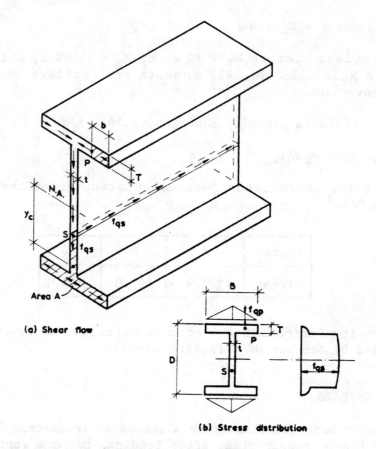

(a) Shear flow

(b) Stress distribution

Fig. 2·7. Shear stresses in an I beam.

The formula cannot be used to obtain the vertical shear stress in the outstanding parts of the flange. However as this must be equal to zero at the top and bottom faces, it must be very small. In fact the resistance of the section to vertical shear is provided almost entirely by the web.

The resultant of the longitudinal shear stress in the web is in equilibrium with the change in the normal tensile force on area A due to the variation in bending moment along the beam. Similar longitudinal stresses exist in the flanges and give rise to horizontal complementary shear stresses in the directions shown. For example, at point P in the top flange: $A = bT$, $t_s = T$, $y_c = (D - T)/2$, and Equation (2.15) becomes

$$f_{qp} = Qb(D - T)/(2I) \qquad (2.16)$$

This expression is linear with respect to the variable b, and f_{qp} has a maximum value at the centre of the flange where b = B/2, i.e.

$$f_{qp}(max) = QB(D - T)/(4I) \qquad (2.17)$$

The complete distribution of shear stress on the cross section is shown in Fig. 2.7(b).

Equation (2.15) can be expressed in terms of the shear flow, which is the product of the shear stress and the thickness of the section, thus

$$q_s = tf_{qs} = QAy_c/I \qquad (2.18)$$

In the longitudinal sense the shear flow is equal to the shear force per unit length of beam, and is a convenient quantity for the calculation of the shear force to be resisted by bolts or welds in a fabricated or compound section. Examples of its application are given in Chapter 7.

2.3.2 <u>Shear lag</u>   The simple theory of bending is based on the assumption that plane sections remain plane after bending. In reality shear strains cause the section to warp. The effect in the flanges is to modify the bending stresses obtained by the simple theory, producing higher stresses near the junction of a web and lower stresses at points remote from it. This effect is described as 'shear lag'. The discrepancies produced by shear lag are minimal in rolled sections, which have relatively narrow and thick flanges. However in plate girders or box sections having wide thin flanges the effects can be significant when they are subjected to high shear forces, especially in the vicinity of concentrated loads where the sudden change in shear force produces highly incompatible warping distortions.

2.3.3 <u>Shear stresses in thin walled open sections</u>   Steel sections are usually composed of relatively thin elements, for which the analysis can be simplified:

    (a) by referring all dimensions to the profile of the section;
    (b) by assuming that the shear stress does not vary across
        the thickness;
    (c) by ignoring any shear stresses acting at right angles to the
        section profile. As these are equal to zero at each outside
        surface they must always be very small in a thin walled
        section.

If it is further assumed that the load is applied in such a way that no twisting of the beam occurs, the shear flow at a point S on the profile of the section is given by

$$q_s = q_0 - (Q_{ey}/I_x) \int_0^s ty \, ds - (Q_{ex}/I_y) \int_0^s tx \, ds \qquad (2.19)$$

where Q_{ex} and Q_{ey} are the effective shear forces obtained either from Equation (2.13) or by applying the effective loads obtained from Equation (2.14). The variable s is the distance around the profile to the point of interest, starting from any point at which the shear flow q_0 is known. At any open end, such as the end of a flange, the value of q_0 is zero. The direction of s can be chosen arbitrarily and, provided the sign convention of Fig. 2.2 is adopted, a positive sign for q_s indicates that the shear flow is in the direction chosen for s. The application of Equation (2.19) is demonstrated in the next example.

Example 2.5   Calculate the shear stresses in the angle section from Example 2.1, which is arranged as a simply supported beam carrying two central point loads, as shown in Fig. 2.8(a). The dimensions of the profile are given in Fig. 2.8(b), and the sign convention for the shear stresses is shown for reference in Fig. 2.8(c). Both external loads act in the positive direction.

To the left of mid-span the shear forces (kN) are

$$Q_x = 5 \quad Q_y = 10$$

From Example 2.1 the second moments of area (10^6 mm^4) are:

$$I_x = 80.44 \quad I_y = 14.02 \quad I_{xy} = -19.17$$

The effective shear forces are given by equation (2.13), thus

$$Q_{ex} = (Q_x - Q_y I_{xy}/I_x)/D$$

$$Q_{ey} = (Q_y - Q_x I_{xy}/I_y)/D$$

where $D = 1 - I_{xy}^2/(I_x I_y)$

whence $Q_{ex} = 10.95$ kN and $Q_{ey} = 24.97$ kN

The shear flow is given by Equation (2.19), i.e.

$$q_s = q_0 - (Q_{ey}/I_x) \int_0^s ty \, ds - (Q_{ex}/I_y) \int_0^s tx \, ds \qquad (1)$$

For the horizontal leg, starting from the left hand end:

$$q_o = 0 \quad s = s_1 \quad x = (117.2 - s_1) \text{ mm} \quad y = -97.8 \text{ mm}$$

Substitution of these values into (1) and integrating gives

$$q_{s1} = 0.00781 \, s_1{}^2 - 1.224 \, s_1 \tag{2}$$

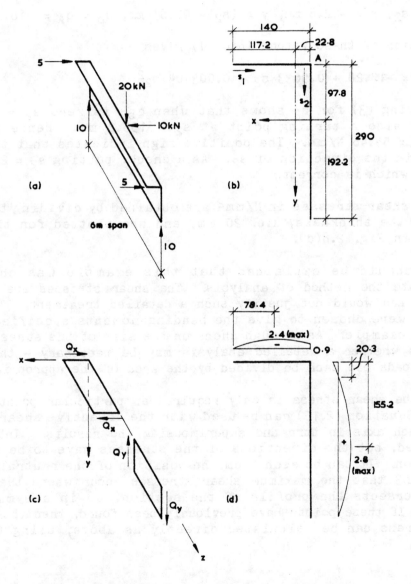

Fig. 2·8. Distribution of shear stress on an unsymmetrical
section.

This equation shows that $q_{s1} = 0$ only when $s_1 = 0$. Differentiating with respect to s_1 and equating to zero gives a turning point at $s_1 = 78.4$ mm. Hence, from (2) $q_{s1}(max) = -47.96$ N/mm, and at $s_1 = 140$ mm, $q_{s1} = -18.28$ N/mm.

The negative signs indicate that the shear flow is in the opposite direction to s_1.

For the vertical leg

$$s = s_2, \quad x = -22.8 \text{ mm}, \quad y = (s_2 - 97.8) \text{ mm}, \quad q_o = q_A = -18.28 \text{ N/mm}$$

Substitution of these values into (1) gives

$$q_{s2} = -18.28 + 0.9633 s_2 - 0.003104 s_2^2 \tag{3}$$

Solving (3) for s_2 shows that when q_{s2} is zero $s_2 = 20.3$ mm. There is also a turning point at $s_2 = 155.2$ mm. Hence from (3) $q_{s2}(max) = 56.46$ N/mm. The positive sign indicates that the shear flow is in the direction of s_2. As a check, putting $s_2 = 290$ gives $q_{s2} = 0$, which is correct.

The shear stresses in N/mm^2 are obtained by dividing the shear flows by the thickness, i.e. 20 mm, and are plotted for the whole section in Fig. 2.8(d).

It should be explained that this example was chosen to demonstrate the method of analysis. The shear stresses are actually very low and would not justify such a detailed treatment. The loads and span were chosen to give the bending moments specified in the previous example. As a rough check on the size of the stresses - to determine whether a detailed analysis may be necessary - the actual applied loads can each be divided by the area of the appropriate leg.

If the shear stress is only required at particular points in the section Equation (2.15) can be used with the effective shear forces, taking each axis in turn and superimposing the results. Integration is avoided, but the directions of the stresses have to be found by inspection. It can be seen from the position of the neutral axis in Example 2.3 that the maximum shear stresses occur where the neutral axis intersects the profile of the section, as in a symmetrical section. If these points have previously been found, then the maximum shear stress can be calculated directly as above, using Equation (2.15).

2.3.4 <u>Shear centre</u>   Equation (2.19) is only valid if no twisting of the beam occurs at the section considered. Torsion in a section can be generated by a transverse load if the resultants of the shear stresses in the elements of the section produce a torque. To

counteract this, the line of action of the applied load must pass
through the shear centre.   In a symmetrical section the shear centre
lies on an axis of symmetry, and loads applied along such an axis do
not cause twisting.   In some sections the position of the shear centre
can be inferred directly from the direction of the shear flow;
examples are given in Fig. 2.9.   In (a) the shear centre lies at the
intersection of two axes of symmetry and is coincident with the
centroid; in (b) and (c) it lies at the intersection of lines of shear
flow; in (d), if the flanges are of the same size, the shear stresses
in them set up opposing torques about the centroid, which is therefore
the shear centre.

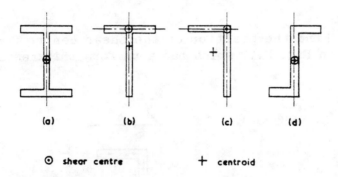

(a)   (b)   (c)   (d)

⊙ shear centre   + centroid

Fig. 2.9.   Position of shear centre.

For the general case of an unsymmetrical thin walled open section
subject to biaxial bending, with shear forces Q_x and Q_y, the position
of the shear centre can be found by determination of the shear flow
from Equation (2.18) or (2.19), applying Q_x and Q_y in turn, assuming
that they pass through the shear centre. Consider, for example, the
section profile in Fig. 2.10.   If point B is chosen as the fulcrum it
is only necessary to find the resultant shear forces in the leg CD due
to Q_x and Q_y in turn.   These forces produce torques equal to $Q_x y'$ and
$Q_y x'$ respectively.   By taking moments about B the values of y' and x'
can be obtained.   There is no need to calculate shear stresses in AB
or BC because their lines of action pass through the point B, and
generate no moment.   The resultant shear forces in CD are obtained by
integrating the shear stresses obtained by Equation (2.19) along the
leg.

The above process is tedious since, for each of Q_x and Q_y, the
corresponding effective shear forces Q_{ex} and Q_{ey} must be calculated
and applied.   If there is an axis of symmetry Equation (2.18) can be
used and the analysis is simplified considerably, as in the next
example.

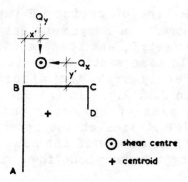

Fig. 2·10. Shear centre - unsymmetrical design.

Example 2.6   Find the position of the shear centre for the channel section shown in Fig. 2.11 which has a uniform thickness.

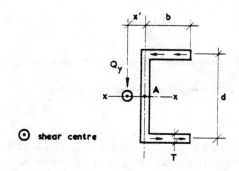

Fig. 2·11. Shear centre of a channel.

As the shear centre lies on the axis of symmetry x-x, there is no need to consider Q_x. If point A at the intersection of axis x-x with the centre line of the web is taken as the fulcrum, then only the shear force in the flanges need be considered since the resultant shear force in the web produces no moment about A. Equation (2.18) is

$$q_s = QAy_c/I$$

The distribution of shear flow in the flanges is linear with zero at the ends and the maximum at the web centre-line. Hence, for maximum shear flow: A = bT, Q = Q_y, y_c = d/2, and I = I_x, which gives

$$q_s(max) = Q_ybTd/(2I_x)$$

The resultant shear force is equal to half the maximum shear flow multiplied by the flange width, i.e.

$$Force = Q_yb^2Td/(4I_x)$$

The torque about A from both flanges is equivalent to the torque produced by the applied shear force when it passes through the shear centre, thus

$$Q_yx' = Q_yb^2Td^2/(4I_x)$$

from which

$$x' = b^2d^2T/(4I_x) \qquad\qquad (2.20)$$

2.3.5   Shear stresses in thin walled closed sections

The shear stress and shear flow in a symmetrical closed section can be obtained directly from Equations (2.15) and (2.18) respectively. For an unsymmetrical section Equation (2.19) can be used, but the analysis is complicated by the fact that q_o is not known at any point. The problem can be solved by first cutting the section at some point and finding the position of the shear centre in the resulting open section. The shear flow in the closed section then results from the combined action of the applied shear loads transferred to the shear centre of the cut (open) section, and the torque on the closed section due to the transference of the loads. A similar approach is used to find the position of the shear centre of the closed section. (Torsion of thin walled closed sections is discussed in Section 2.4.3).

2.4   TORSION

Torsion in structural engineering arises from a variety of causes including, for example, beams cranked or curved in plan, transverse loads applied through brackets, and distributed loads whose line of action does not pass through the shear centre of the section. It is important to realise that in torsion problems, the lever arm for the torque is measured from the shear centre of the section, not the centroid. In particular, distributed loads on unsymmetrical sections such as angles or channels usually act through the centroid of the section, as in the case of the self-weight, or through some other line which is offset from the shear centre axis. For example in a 381 x 102 channel section the lever arm for loads acting through the centroid is approximately 50 mm. The effect of the self-weight acting alone is usually negligible.

2.4.1   <u>Uniform and non-uniform torsion</u>   In general, the cross sections of members subject to torsion do not remain plane, but tend to warp. The degree of warping that takes place depends on the shape of the section, and is most pronounced in thin walled channels. In some sections, such as angles and tees, solid and hollow circular sections, and square box sections of uniform thickness, warping is virtually non-existent, while in others, such as closed box sections of general shape, its effect is small. In I sections most of the warping takes place in the flanges; its effect on the web is very small and can be ignored. If the torque is applied only at the ends of the member and warping is not restrained, the flanges remain virtually straight and maintain their original shape, as shown in Fig. 2.12(a). The result is that the sectional planes of the flanges rotate in opposite directions, producing warping displacements which are constant along the whole length of the member. Under these circumstances the member is said to be in a state of uniform or St. Venant torsion. The analysis of members in uniform torsion is covered by Sections 2.4.2 to 2.4.4, which follow.

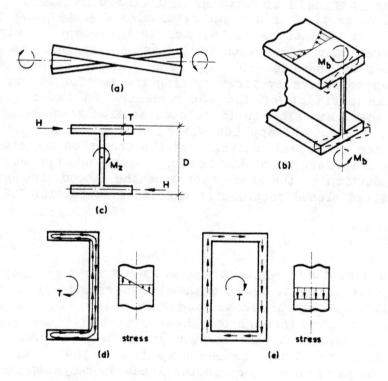

Fig. 2·12. Torsion of thin walled sections.

In practice warping is frequently restrained, particularly in beams, where the torsional effects are due to the action of transverse loads. Restraints arise from the action of structural connections, or from the incompatibilities in warping displacements that occur when the torque is not uniform along the length of the member. Warping restraint increases the torsional stiffness of a member, and at any point along its length the applied torque is resisted by two components, one due to St. Venant torsion, and the other to warping torsion from the effects of the restraints. The proportions of the two components depend on the type of loading and the distance from a restraint.

Both components of torsion produce shear stresses parallel to the walls of the section, and their combined effects can locally be greater than the effect of St. Venant torsion alone. However in members other than channels with very thin walls the increase in shear stress can usually be ignored in design. In beams the maximum shear stress can be obtained approximately by combining the effects of transverse shears, using Equation (2.15), with the shear stresses from torsion, assuming that the whole of the applied torque results in St. Venant torsion, as in Example 2.7.

A more significant effect of warping restraint in the design of beams is the introduction of longitudinal stresses. The effect is illustrated for an I beam in Fig. 2.12(b). In this case warping displacements are confined to the flanges, whose positions if warping were allowed to occur freely are shown by the dotted outlines. Bimoments M_b are induced in the planes of the flanges when warping is restrained, and these give rise to tensile and compressive stresses, as shown.

A full treatment of the analysis of members subjected to warping torsion is beyond the scope of this introductory text, and the reader is referred to the excellent book by Zbirohowski-Koscia. The analysis is tedious, but for I beams a conservative estimate of the longitudinal stresses due to warping torsion can be obtained by assuming that each flange acts independently and is bent in its own plane by an analogous system of lateral loads which replace the applied torques, as in Fig. 2.12(c). The value of the equal and opposite lateral loads H, analogous to the applied torque M_z is given by

$$H = M_z/(D - T) \tag{2.21}$$

The ends of the flanges can be assumed to be either fixed or simply supported, depending on whether or not warping is restrained by the structural connections. The results obtained by this method are conservative because in reality the warping stresses are produced only by the warping component of the applied torque, not the whole torque as assumed. However, Equation (2.21) can be useful in preliminary designs where it is necessary to assess whether the effects of torsion are likely to be significant, or not.

2.4.2 <u>Circular sections</u>   The elastic (St. Venant) theory of torsion of prismatic members with solid and hollow circular sections can be expressed by the single well established formula:

$$T/J = f_q/r = G\theta/L \tag{2.22}$$

where T = the applied torque
      J = the torsion constant which, for a circular section, is equal to the polar second moment of area
     f_q = the shear stress at radius r
     G = the shear modulus
     θ = the angle of twist
     L = the length of the bar

The polar second moment of area for a solid section of radius R is given by

$$J = \pi R^4/2 \tag{2.23}$$

For a thin walled tube of mean radius R_m and wall thickness t, the approximate formula

$$J = 2\pi R_m^3 t \tag{2.24}$$

can be used. The error is below 3 percent for t/R_m ratios of one third or less, and is on the safe side.

The polar second moment of area is twice the second moment of area about a diameter, which is quoted in the section tables for solid circular sections (BCSA-Constrado Handbook).

The distribution of shear stress along the radius of a solid circular section is linear, with zero at the centre and the maximum at the outside surface. For a thin walled tube, the stress also varies linearly across the wall thickness and in the range of standard structural tubes unsafe errors in the shear stress of up to about 18 percent are introduced by the use of R_m in Equation (2.22) instead of the outside radius.

2.4.3 <u>Thin walled open sections</u>   The torsional constant J for a thin rectangle of width b and thickness t is given approximately by

$$J = bt^3/3 \tag{2.25}$$

This formula is accurate when b/t is infinite and gives unsafe errors of 6 percent when b/t = 10, and 10 percent when b/t = 6. These b/t ratios are typical of the webs and flanges of Universal column and beam sections.

Most sections used in steelwork design are composed of thin rectangles, and for a complete section the torsional constant can be obtained by summing the torsional constants for each rectangular element, i.e.

$$J = \sum bt^3/3 \qquad\qquad (2.26)$$

In standard rolled sections the root fillets at the junctions of the web and the flanges give additional torsional stiffness.

The shear stresses in an open section under St. Venant torsion vary from zero on the centre-line of the wall to a maximum on the outside surface, as shown in Fig. 2.12(d), and their direction is reversed on each side of the the centre-line. The shear flow constitutes a closed loop. The maximum stress in any element of thickness t is given by

$$f_q = Tt/J \qquad\qquad (2.27)$$

The maximum shear stress in the section therefore occurs in the thickest element. At re-entrant corners the flow lines are crowded together, giving rise to very high stress concentrations. The effect is reduced by the provision of fillet radii. The shear stress is zero at outside corners.

The angle of twist is given by

$$\theta = TL/(GJ) \qquad\qquad (2.28)$$

where J is from Equation (2.26).

2.4.4 **Thin walled closed sections**   The shear stress distribution for closed sections is shown in Fig. 2.12(e). The flow is unidirectional with respect to the profile, contrasting with open sections. Variations in stress across the thickness of the section are ignored. The shear flow is constant at all points on the profile and is given by

$$q = T/(2A) \qquad\qquad (2.29)$$

where A is the total area enclosed by the profile, as shown in Example 2.7. The shear stress is a maximum in the thinnest part of the section and is obtained by dividing the shear flow by the thickness. This is in direct contrast to an open section where the maximum shear stress occurs in the thickest part.

The angle of twist is given by

$$\theta = TL/(4A^2G) \oint (ds/t) \qquad (2.30)$$

The evaluation of this integral is demonstrated in the next example.

As in open sections, additional stresses are introduced when warping is restrained. Formulae are given for calculation of the additional longitudinal stresses in the design of box girders in Appendix B of BS 5400 Part 3 'Code of practice for the design of steel bridges'. Equations (2.29) and (2.30) are derived from the Bredt-Batho hypothesis in which it is assumed that the shape of the section remains unchanged. To ensure that this assumption remains valid it may be necessary to stiffen the section with internal diaphragms at intervals along the length, and at points where concentrated loads are applied.

<u>Example 2.7</u>   Find the maximum shear stress in the box section of Fig. 2.13 which is subject to a torque T = 200 kN.m and a shear force S = 500 kN, as shown. Assume that the section is adequately stiffened to prevent distortion of the profile and ignore the effects of warping restraints. Calculate the angle of twist per metre length.

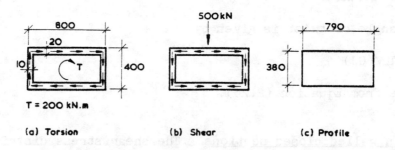

(a) Torsion   (b) Shear   (c) Profile

Fig. 2.13. Torsion and transverse shear in box section.

The torsion and shear may be considered separately and the resulting shear flows are shown in Figs. 2.13(a) and (b). For both calculations the profile dimensions can be used, as in Fig. 2.13(c). The shear modulus is 80 kN/mm^2.

First consider the torsion.

The area A enclosed by the profile is

$$790 \times 380 = 0.3002 \times 10^6 \text{ mm}^2$$

Hence from Equation (2.29), the shear flow

$$q = T/2A = 200/(2 \times 0.3002) = 333 \text{ N/mm}$$

The shear stress is a maximum in the web, where the section is thinnest, i.e.

$$f_q = q/t = 333/10 = 33.3 \text{ N/mm}^2$$

Now from Equation (2.30), the angle of twist

$$\theta = TL/(4A^2G) \oint (ds/t)$$

$$= \frac{200 \times 10^6 \times 1000}{4 \times (0.3002)^2 \times 10^{12} \times 80 \times 10^3} \times 2(790/20 + 380/10)$$

$$\theta = .00107 \text{ radians/metre}$$

As this angle is very small the box is clearly very stiff torsionally.

Now consider the direct shear force. The shear stress is given by Equation (2.15), thus

$$f_{qs} = QAy_c/It_s$$

The maximum shear stress is in the webs at the neutral axis and therefore

$$Ay_c(\text{flange}) = 20 \times 790 \times 380/2 = 3.002 \times 10^6$$
$$Ay_c(\text{webs}) = 2 \times 10 \times 380/2 \times 380/4 = 0.361 \times 10^6$$
$$\overline{}$$
$$3.363 \times 10^6$$
$$\overline{}$$

The second moment of area I is

$$2(10 \times 380^3/12 + 20 \times 790 \times 190^2) = 1232 \times 10^6 \text{ mm}^4$$

Hence, noting that for two webs $t_s = 2 \times 10 = 20$

$$f_{qs} = 500 \times 10^3 \times 3.363 \times 10^6/(1232 \times 10^6 \times 20) = 68.2 \text{ N/mm}^2$$

Combining the torsional and direct shear stresses, the maximum shear stress in the web is

$$33.3 + 68.2 = 101.5 \text{ N/mm}^2$$

2.5   BENDING STRESSES IN DESIGN

2.5.1   Buckling of compression flange

The most common section used for beams in steelwork design is the I section, which consists of two flanges connected by a relatively slender web. In bending, the longitudinal tensile and compressive stresses are mainly concentrated in the flanges, while the web provides the bulk of the resistance to shear.

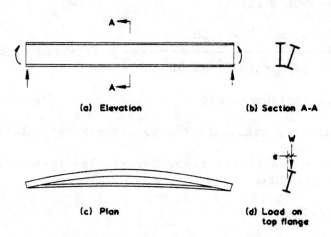

(a) Elevation   (b) Section A-A

(c) Plan   (d) Load on
                                       top flange

Fig. 2·14. Lateral torsional buckling of beam.

The behaviour of such a beam can be illustrated by considering the beam in Fig. 2.14, which rests on simple end supports and is subjected to bending moments at each end, as shown. If the moments are gradually increased from zero the beam initially deflects elastically in the plane of bending and for a short stocky beam this in-plane elastic deflection could continue until yielding of the extreme fibres occurred. In most practical beams however a critical bending moment is reached, at which the compression flange becomes elastically unstable and buckles laterally. The tension flange, on the other hand, tends to remain straight, so that twisting also takes place. These effects are shown in Fig. 2.14 (a), (b) and (c).

The analysis of lateral torsional buckling in beams is generally complex. However, if simplifying assumptions are made, it can be shown that, for an ideal beam, the maximum stress in the compression flange at the critical bending moment is dependent upon two parameters. These are: the ratio of the overall depth of the section to the flange thickness (D/T ratio), and the slenderness ratio (l/r_y), which is defined as the effective length of the compression flange divided by the radius of gyration of the section about its minor axis.

The meaning of the term 'effective length' with regard to beams, is the subject of Section 2.5.3. The radius of gyration is defined as

$$r_y = \sqrt{(I_y/A)} \tag{2.31}$$

where A = the gross area of the section. For standard rolled sections r_y is quoted in the section tables (Appendix).

By adopting an approach similar to that used in the derivation of the Perry-Robertson formula for compression members (Section 3.3.1), the critical stress in the compression flange of an ideal beam at ultimate load is limited to the yield stress, and modified to take account of the imperfections, such as residual stresses and lack of straightness, that exist in real beams.

When establishing design rules for a complex problem like the lateral torsional buckling of beams, where the design is influenced by a number of variable factors, there must be a compromise between simplicity and economy. For example, in BS 449 the variable effects of imperfections are combined by assuming an empirical initial curvature, as in the design of compression members. Other variable errors are introduced by ignoring the effects of different loading patterns and basing allowable stresses on the most severe case, which is that of Fig. 2.14, and by assuming simplified relationships between section properties. Other variations, such as the effect of end restraints and the height of the load above the shear centre, are discussed in the next section.

In BS 449, the upper limits at working load to the allowable stresses in tension or compression for all beams, are given in Table 2 (Appendix). These stresses give a factor of safety against yielding of 1.52.

BS 449, Table 3 (Appendix) gives the allowable stresses in the compression flange, based on lateral torsional buckling. These stresses are tabulated directly against the ratios l/r_y and D/T. There appears to be some doubt as to the derivation of Table 3, but it is clearly semi-empirical.

The allowable stresses in compression for plate girders, also based on lateral torsional buckling, are set out in BS 449, Tables 7 and 8 (Appendix), the derivation of which is described in Chapter 7.

2.5.2 _Criteria for bending stresses_   The design criteria for various categories of beams are specified in BS 449, Cl. 19. For all beams the maximum tensile stress should not exceed the appropriate allowable stress p_{bt} from Table 2. The maximum compressive stress for beams other than plate girders should not exceed the allowable stress p_{bc} in

compression as set out below.   The design of plate girders is
described in Chapter 7.

Category (a)

Rolled I beams, broad flange beams, Universal beams and columns,
castellated beams with lateral support (those without lateral support
are designed as plate girders), rolled channels, rolled Z beams, and
compound beams having equal flanges of uniform section throughout, but
where the second moment of area of the compression flange exceeds that
of the tension flange when taken about the y-y axis:

For these beams the allowable compressive stress p_{bc} is the lesser of
that obtained from Table 2 or from Table 3, as appropriate to the
grade of steel and the thickness of the section.

Category (b)

Compound girders as in (a) above, but with curtailed flanges:

The design of these beams is described in Chapter 7.

Category (c)

Angles and tees:

If the flange (table) is in compression the allowable stress p_{bc} is
obtained from Table 2, i.e. lateral torsional buckling is not
relevant. However, if the leg (stalk) is in compression p_{bc} is the
lesser of the appropriate stresses from Table 2 or Table 8 using the
following formula for C_s (N/mm^2)

$$C_s = (A + K_2B)y_c/y_t \qquad\qquad (2.32)$$

where $K_2 = -1.0$
        y_c and y_t are the distances from the neutral axis to the
        extreme fibres in compression and tension respectively.
        A and B are obtained from Table 7 for the appropriate l/r_y and
        D/T ratios.

For calculation of D/T, D is the overall section depth, and T is the
thickness of the leg.   The design of a tee subjected to combined
bending and axial loads is covered in Example 6.1 (Chapter 6).

Category (d)

Tubes and rectangular hollow sections:

Provided that the ratio of depth to breadth does not exceed 4, p_{bc} is
obtained from Table 2.   (All the standard sections comply)

2.5.3 <u>Effective lengths</u> The effective span of a beam is defined in Cl. 24 of BS 449 as the distance between the centres of the supports, except in two cases. These are: when the 'rigid' method of design is used (Chapter 5), and in the 'simple' design approach when beams span between stanchions (Chapter 3, Section 3.3.7). In both these cases the effective span is taken as the distance between the assumed points of application of the reactions. In order to take account of different end restraints, the slenderness ratio is based on an effective length l, obtained by multiplying the effective span by a factor, as specified in BS 449, Cl. 26. These factors, which are set out in Table 2.1 below are based on the assumption that the ends of the beam are restrained against torsion and that the method of loading does not increase the torsion once buckling has commenced. If these two conditions are not fulfilled the effective lengths must be increased by 20 percent.

Restraint against lateral bending	Factor
No restraint, e.g. Fig. 2.15(a)	1.0
Partial restraint, e.g. by cleated flange connections - Fig. 2.15(b)	0.85
Full restraint, e.g. Fig. 2.15(c)	0.7

Table 2.1 Effective length factors for simply supported beams

Practical examples of end connections which restrain the beam against torsion are given in Fig. 2.15(a), (b) and (c). Full restraint is also provided by building the end of the beam into a wall. A bottom cleat alone, as in Fig. 2.15(d), is not sufficient. The effect of a load applied above the shear centre, without adequate lateral support of the compression flange, is to produce an additional torque, as shown in Fig. 2.14(c). Practical examples are given in Fig. 2.15(e) and (f). In (e) the lack of a proper structural connection allows the lower beam to twist. If the flanges were bolted together the upper beam would provide additional lateral support to the lower beam and twisting would be prevented. In (f) the bottom of the column transmitting the load is free to move laterally, accentuating torsion in the lower beam. No torsion is possible when the load is applied through a secondary beam connected to the web, as in Fig. 2.15(g). Full restraint against torsion can be provided by other types of external support, provided that they are capable of exerting on the compression flange a reaction H, as shown in Fig.

2.15(b), equal to at least 2.5 percent of the maximum compressive
force occuring in the flange, i.e.

$$H \geq 0.025M_{max}/(D - T)$$ (2.33)

where M_{max} = the maximum bending moment in the beam.

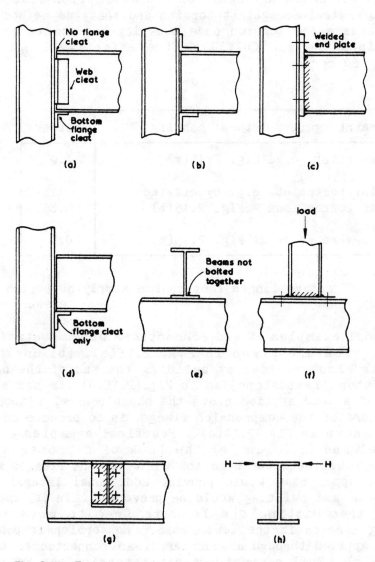

Fig. 2·15. End restraints and load conditions.

The effective length of a beam is reduced by structural elements such as secondary beams framing into it in such a way as to provide lateral support to the compression flange, as in Fig. 2.15(g). Where a number of lateral supports are provided, as in Fig. 2.16, the effective length is the greatest of L_1, L_2, or L_3, provided that the ends of the beam are restrained against torsion.

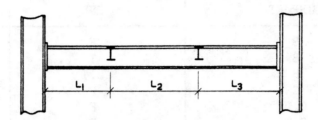

Fig. 2·16. Lateral supports along the span.

The conditions for effective lateral support are that lateral displacement of the compression flange must be completely prevented at the point of support; and the supporting member should be capable of providing a minimum reaction equal to H, as given by Equation (2.33).

Where there are a number of lateral supports the minimum reaction to be provided by each one is H/n, where n is the number of supports.

Where the same member, or members, provide lateral support to a series of parallel beams with solid webs, the minimum total reaction H should be calculated from the beam in the series which has the maximum compressive force. This rule is varied for a series of roof trusses or lattice girders where the supporting members should be capable of providing a total reaction of

$$H = 0.0125(n + 1)F_c \qquad (2.34)$$

where n = the number of trusses, up to a maximum of 5
      F_c = the maximuum force in the compression flange of one truss.

A similar rule applies to beams supporting slab construction, where the lateral support to the compression flange is provided either by friction with the concrete, or by a series of angles or studs or similar short connectors welded to the compression flange and embedded in the concrete. The total minimum lateral force to be provided is again equal to H (from Equation (2.33)), which is assumed, for frictional force, to be uniformly distributed along the interface between the compression flange and the concrete, or, for connectors, to be equally divided between them. The application of this rule is explained in Example 2.11.

CASE NO.	RESTRAINTS		EFFECTIVE LENGTH FACTOR
	T = Torsional restraint	B = Built in	
	P = Partial torsional restraint	C = Continuous	
	L = Lateral restraint	F = Free end	
	SUPPORT	END	
1	B	F	0.85
2	B	T	0.75
3	B	L, T	0.5
4	C	F	3.0
5	C, P	F	2.0
6	C, T	F	1.0

Table 2.2   Effective length factors for cantilevers

For cantilevers the effective length factors can be obtained from Table 2.2 above. It is normal in cantilevers for the reactive moment to be provided by continuous construction at the support. Building the support into masonry provides torsional restraint, but it is unlikely that sufficient reactive moment could be provided by the masonry itself. Continuous construction by itself does not necessarily give torsional restraint. Fig. 2.17 gives some examples of how restraints can be provided in practice.

Cases 4, 5 and 6 in Table 2.2 all have free ends. If however the ends are restrained, as in cases 2 and 3, the effective lengths should be reduced by the following factors.

For restraints as in case 2: a factor of 0.882, i.e. 0.75/0.85
For restraints as in case 3: a factor of 0.588, i.e. 0.5/0.85

2.5.4 **Maximum slenderness ratio**   Cl. 25, BS 449 specifies a maximum slenderness ratio of 300 for any beam. For cantilevers with an effective length factor of 3.0 (Case 4 in Table 2.2) this is sometimes difficult to achieve, and structures have been built with cantilevers having slenderness ratios in excess of 300. In such cases it is necessary to carry out loading tests, as specified by BS 449 (Cl. 9c and Appendix A), after the structure has been built.

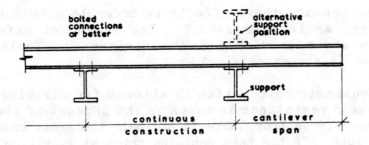

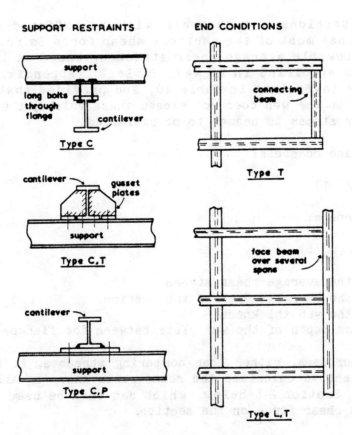

Fig. 2·17. Cantilever restraints for Table 2·2.

2.5.5 <u>Criteria for shear stresses</u>   Cl. 23, BS 449 specifies that the maximum shear stress should not exceed the allowable maximum shear stress p_q set out in Table 10 (Appendix). This clause states that the distribution of the shear stress, 'having regard to the elastic behaviour of the members in flexure', should be taken into account. The effects of torsion are not mentioned, but clearly if a member is subjected to torsion as well as to transverse loading, it would be necessary to combine these effects in order to obtain the maximum shear stress, as in Example 2.7.   The factor of safety against yielding in shear provided by the stresses in Table 10 varies slightly, but is approximately 1.26.

An approximate calculation is allowed for circular tubes, for which the shear resistance is taken as the product of the allowable maximum shear stress from Table 10 and half the gross cross-sectional area of the tube.   If the tube contains holes at a critical section an accurate analysis has to be made.

For I sections and channels with unstiffened webs, it is recognised that most of the vertical shear force is resisted by the web, and allowable average shear stresses, based on the web cross section, are specified in BS 449, Table 11 (Appendix).   Values in Table 11 are lower than in Table 10, and provided that the average shear stress in the web does not exceed these values, the criterion of maximum shear stress is deemed to be satisfied.

For I beams and channels:

$$f_q' = Q/(Dt) \tag{2.35}$$

For plate girders:

$$f_q' = Q/(d_W/t) \tag{2.36}$$

where f_q' = the average shear stress
  D = the overall depth of the section
  t = the web thickness
  d_W = the depth of the web plate between the flanges.

For design purposes, rather than comparing stresses, it is frequently more convenient to calculate the shear resistance, or allowable shear force, as in Section 2.6 below, which can then be used as the upper limit to the shear force on the section.

2.6   DESIGN OF UNSTIFFENED WEBS

The design of the unstiffened web of a rolled I beam or channel involves the determination of the resistance of the web when three alternative failure modes are considered, i.e.

1.   Resistance to shear
2.   Resistance to bearing
3.   Resistance to buckling.

2.6.1   <u>Shear resistance</u>   For I beams and channels BS 449 allows the shear resistance of the web to be based on the average stress.   Thus,

$$P_q = p_q' Dt/1000 \tag{2.37}$$

where P_q = the shear resistance in kN
$\quad\quad p_q'$ = the allowable average shear stress in N/mm^2, from Table 11
$\quad\quad\quad$ (Appendix)
$\quad\quad D$ = the overall depth of the section in mm
$\quad\quad t$ = the web thickness in mm

2.6.2   <u>Bearing resistance</u>   When a heavy concentrated load is applied to the flange of a beam it can cause crushing of the web by direct compression.   The load is dispersed through three main components, namely the stiff bearing, the flange plate if fitted, and the beam itself.   A stiff bearing is one which is sufficiently stiff to exert a uniform pressure along its whole length.   A dispersion angle of 30° is assumed (Cl. 28e, BS 449), and the most critical part of the web is at the base of the root fillets, i.e. at a distance of (T + r) from the outside face of the flange.   Typical situations where the bearing resistance of the web should be considered are shown in Fig. 2.18.   In case (a) the beam is continuous over the bearing and the load is dispersed on both sides.   At an end support, as in case (b), the dispersal is to one side only.   In case (c) the concentrated load acting on the column is the force exerted by the compression flange of the connecting beam, due to the bending moment at the support.

Considering first the end bearing in Fig. 2.18(b), the contributions to the direct bearing resistance made by the components are:

Beam component (kN)$\quad\quad\quad\quad\quad\quad C_1 = (T + r)\cot 30^\circ \times tp_b/1000$
Stiff bearing component (kN/mm)$\quad C_2 = tp_b/1000$$\quad\quad\quad\quad\quad\quad$(2.38)
Flange plate component (kN/mm)$\quad C_3 = tp_b \cot 30^\circ/1000$

where T, r and t are the flange thickness, the root radius, and the web thickness, in mm, as given in the section tables (Appendix)
$\quad\quad\cot 30^\circ = 1.732$
$\quad\quad p_b$ = the allowable bearing stress in N/mm^2 from BS 449, Table 9 (Appendix)

The components C_2 and C_3 are expressed in terms of force per unit length and force per unit thickness respectively, so the total direct bearing resistance of the whole assembly is given by

$$P_b = C_1 + L_b C_2 + t_p C_3 \quad (kN) \tag{2.39}$$

where L_b = the length of stiff bearing in mm
$\quad\quad t_p$ = the thickness of the flange plate in mm

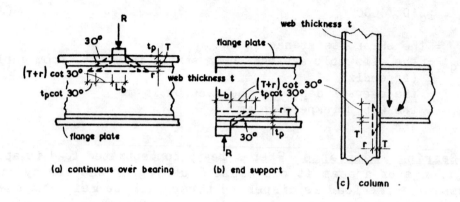

Fig. 2·18. Web bearing situations.

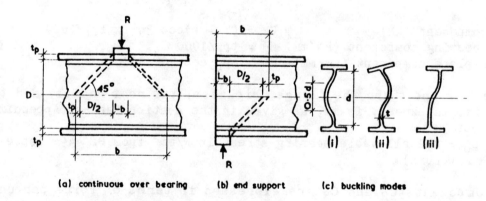

Fig. 2·19. Web buckling situations.

Where the beam is continuous over the bearing, as in Figs. 2.18(a) and (c), the values of C_1 and C_3 are doubled, but C_2 remains unaltered.

The values of the C factors for web bearing and web buckling are quoted for Universal beams and columns, joists and channels, in the 'Safe Load Tables' of the BCSA Constrado handbook.

2.6.3 <u>Buckling resistance</u> Buckling of the web can occur in exactly the same situations as failure by bearing. For design purposes the web is assumed to act as a strut whose width b is the intercept on the centre-line of the beam obtained by a dispersion angle of 45°, as shown in Fig. 2.19 (a) and (b).

For the reader who is unfamiliar with the theory of strut behaviour, the following reasoning leading to Equations (2.40) will be better understood after reading Chapter 3.

The allowable compressive stress p_c in the strut is obtained from BS 449, Table 17 (Appendix). The radius of gyration r_y of the strut about its minor axis is equal to $[(bt^3/12)/(bt)]^{1/2}$. The effective length l, assuming no rotation or relative lateral displacement of the flanges, is as shown in case (i) of Fig. 2.19 (c), i.e. 0.5d, where d is the clear depth of the web between fillets. The slenderness ratio l/r_y, for entry into Table 17, is therefore 1.732 d/t.

Using the same procedure as for web bearing, the three components of buckling resistance are given by

Beam component (kN) $C_1 = Dtp_c/2000$
Stiff bearing component (kN/mm) $C_2 = tp_c/1000$ (2.40)
Flange plate component (kN/mm) $C_3 = C_2$

The total buckling resistance of the assembly is given by Equation (2.39) using the above C values. The effect on the C values when the beam is continuous over the bearing, is exactly the same as for web bearing, i.e. C_1 and C_3 are doubled, but C_2 remains the same.

2.6.4 <u>Design procedure</u> In most cases the design of a beam is governed by bending stresses and it is only necessary to check that the resistance of the web is adequate, after the size of the section has been determined. The following simple procedure may be adopted

1. Determine the maximum shear force Q at any point along the beam and compare it with the shear resistance P_q obtained from Equation (2.37), or from the 'Safe Load Tables' in the BCSA-Constrado Handbook. If Q exceeds P_q then select a new section with a thicker web, or consider stiffening the web locally with side plates; otherwise proceed to stage 2 below.

2.   At connections where concentrated loads (or reactions) are applied compare the value of each load with P_b from Equation (2.39), considering web bearing and web buckling at the section concerned. If both values of P_b so obtained exceed the applied load then no further action is necessary. On the other hand if at any connection P_b is less than the applied load it will be necessary either to increase the length of stiff bearing, or to provide stiffeners, or both. The additional length of stiff bearing required can be obtained from

$$L_b(\text{extra}) = (R - P_b)/C_2 \ (\text{mm}) \qquad\qquad (2.41)$$

where R = the applied load at the connection.

It is important that the correct values of C factors are used, i.e. for an end bearing or for continuous over the bearing.

The maximum permissible length of effective stiff bearing is D/2 for an end bearing and D when the beam is continuous over the bearing. If the required length of stiff bearing is excessive, and the beam is satisfactory in other respects, the provision of load-bearing stiffeners can be considered. The design of these is described in Chapter 7.

2.6.5 Special considerations The design procedures outlined above for determining the strength of an unstiffened web in an I beam are based on the provisions of Cls. 27(e) and 28(a) of BS 449 and the recommendations given in the BCSA-Constrado Handbook. Experiments by Astill et al have shown that the strength of the web is reduced by an increase in bending and axial stresses in the web. These can be allowed for in bearing by applying Equation (2.43) as described in Section 2.7. Axial compressive stresses can be added directly to f_{bc} when evaluating Equation (2.43).

The experiments also showed that the strength of the web was significantly reduced when the line of action of the load was not in the plane of the web (causing rotation of the loaded flange), and when relative lateral displacement of the flanges was not restrained by lateral support at the loaded point. The change in effective length resulting from the modified buckling modes is illustrated in Fig. 2.19(c), cases (ii) and (iii). Astill et al found effective lengths varying from 0.7d to 1.5d, as compared with 0.5d implied by Cl. 28a, BS 449. As a result of the experiments two additional paragraphs were added in 1975 to the above clause, to the effect that the slenderness ratio of the web should be increased if flange rotation and lateral displacement are not restrained. Guidance is given by Astill et al, but it is recommended that, as the mechanism of failure is still not completely understood, circumstances liable to produce flange rotation or relative lateral displacement should be avoided wherever possible.

2.7 COMBINED STRESSES

Under certain loading conditions, for example when beams are subjected to heavy concentrated loads, it is possible that while the stresses due to bending, shear, and bearing do not separately exceed the appropriate allowable stress, yielding can still occur at some point in the section due to the combined effects of the stresses. A critical point is the junction of the web with the flange, at the base of the root fillets. Cl. 14, BS 449 gives formulae based on von Mises failure criterion (see Section 1.17), from which an equivalent stress f_e can be obtained, thus.

For bending and shear:

$$f_e^2 = (f_{bt}^2 + 3f_q^2) \qquad \text{or} \qquad f_e^2 = (f_{bc}^2 + 3f_q^2) \qquad (2.42)$$

For bearing, bending, and shear:

$$f_e^2 = (f_{bt}^2 + f_b^2 + f_{bt}f_b + 3f_q^2)$$

$$\text{or} \quad f_e^2 = (f_{bc}^2 + f_b^2 - f_{bc}f_b + 3f_q^2) \qquad (2.43)$$

where f_{bt}, f_{bc}, f_q and f_b are the bending stress (tensile or compressive), shear stress, and bearing stress, acting together at the point.

The calculated equivalent stress f_e must not exceed the appropriate allowable equivalent stress from BS 449, Table 1 (Appendix). The stresses in Table 1 give a factor of safety against yielding of approximately 1.1. This factor appears to be quite small, but it is reasonable to assume that maximum stresses due to the individual effects of bending, shear and bearing are only likely to coexist over a very small area of the section, so the initial yielding would only be local and would not cause the beam to collapse.

If the resistance of the beam to bearing has been obtained by Equation (2.39), the actual bearing stress f_b can be obtained from

$$f_b = p_b R / P_b \qquad (2.44)$$

where p_b = the allowable bearing stress
  R = the applied load or reaction
  P_b = the bearing resistance.

2.8 DEFLECTIONS

The reasons for limiting deflections have been explained in Chaper 1. For beams, the deflections due to all loads excluding the dead weight of the structure must not exceed 1/360 of the span. No specific limit is placed on deflections due to dead load because they take place progressively as the structure is being built and, unlike imposed loads, which are applied after the load-bearing structure has been completed, do not have such a damaging effect on partitions and finishes. Nevertheless there is a general requirement for the designer to ensure that no deflections should be so great as to impair the efficiency or strength of the structure, or make it unsightly. Even quite small deflections in long beams can be unsightly and it may be necessary to counteract the visual effects of dead-load deflections by rolling the beams to give an upwards camber, or by varying the thickness of the casing.

2.8.1 Deflection calculations For beams in general the methods of calculating deflections are well established and will not be described in this book. Such methods include the use of integration, area moments, strain energy, and slope-deflection equations, and can be found in any standard text book on structural analysis. For simple beams however the method of superimposing standard cases usually involves considerably less calculation. A large number of standard cases for simply supported and built-in beams, and cantilevers, is given in 'The Steel Designer's Manual'. Some of the most useful cases are given in Figs. 2.20 and 2.21 below.

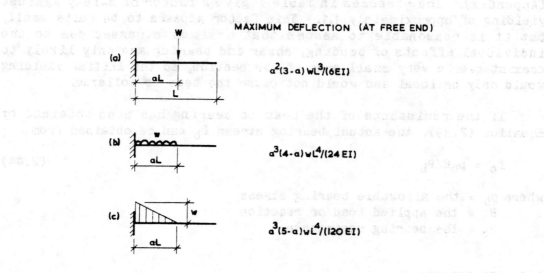

MAXIMUM DEFLECTION (AT FREE END)

(a) $a^2(3-a)WL^3/(6EI)$

(b) $a^3(4-a)wL^4/(24EI)$

(c) $a^3(5-a)wL^4/(120EI)$

Fig. 2·20 Deflection of cantilevers.

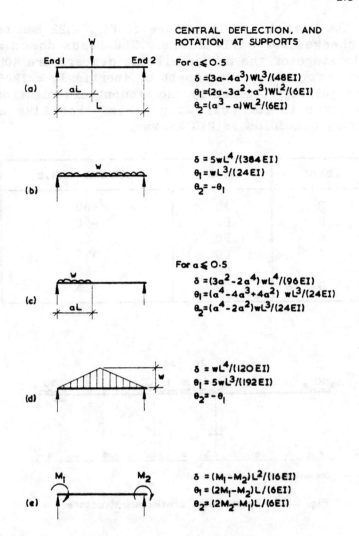

CENTRAL DEFLECTION, AND
ROTATION AT SUPPORTS

For a ≤ 0.5

$\delta = (3a - 4a^3)WL^3/(48EI)$

$\theta_1 = (2a - 3a^2 + a^3)WL^2/(6EI)$

$\theta_2 = (a^3 - a)WL^2/(6EI)$

$\delta = 5wL^4/(384EI)$

$\theta_1 = wL^3/(24EI)$

$\theta_2 = -\theta_1$

For a ≤ 0.5

$\delta = (3a^2 - 2a^4)wL^4/(96EI)$

$\theta_1 = (a^4 - 4a^3 + 4a^2) wL^3/(24EI)$

$\theta_2 = (a^4 - 2a^2)wL^3/(24EI)$

$\delta = wL^4/(120EI)$

$\theta_1 = 5wL^3/(192EI)$

$\theta_2 = -\theta_1$

$\delta = (M_1 - M_2)L^2/(16EI)$

$\theta_1 = (2M_1 - M_2)L/(6EI)$

$\theta_2 = (2M_2 - M_1)L/(6EI)$

Fig. 2·21. Displacements of simply supported beams.

For simply supported beams the central deflection rather than the maximum is given, so that deflections from individual load cases can be added. For most loading cases the central deflection only differs by a small percentage from the maximum. In case (a) of Fig. 2.21, for example, the difference is always within 2.5 percent. A notable exception is the case of equal end moments acting in the same direction, when the central deflection is zero. However in such a case the deflections at other points along the beam are likely to be small. A more accurate analysis should be performed if it is suspected that the maximum deflection could exceed span/360.

Example 2.8   The symmetrical structure of Fig. 2.22 has been designed and is to be checked for deflections.   The loads shown are imposed loads.   At this stage of the design all the members are 406 x 140 x 30 Universal beams, for which the moment of inertia I_x = 12452 cm^4.   The moments at the joints, obtained by the 'moment distribution' method of analysis are given in the following table.   Positive moments are clockwise.   Young's modulus is 210 kN/mm^2.

Joint	Span	Moment (kN.m)
B	AB	+60
B	BC	-60
C	BC	-6
C	CG	+72
C	CD	-66
G	CG	+36

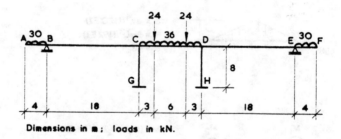

Dimensions in m; loads in kN.

Fig. 2.22. Symmetrical continuous structure.

For all beams EI = 210 x 10^3 x 12452 x 10^4 = 26.15 x 10^{12} N.mm^2

Span CD

1. Uniform load (Fig. 2.21(b))

   Total load = wL = 36 kN, span = 12 m

   δ_1 = 5wL^4/(384EI)

   = 5 x 36 x 10^3 x (12000)^3/(384 x 26.15 x 10^{12})

   = 30.9 mm

2. Concentrated loads (Fig. 2.21(a))

   For each load: W = 24 kN,   a = 3/12 = 0.25

$\delta_2 = 2(3\alpha - 4\alpha^3)WL^3/(48EI)$

$\quad = 2[3 \times 0.25 - 4 \times (0.25)^3] \times 24 \times 10^3 \times (12000)^3/...$

$\quad ...(48 \times 26.15 \times 10^{12})$

$\quad = 45.4$ mm

3. End moments (Fig. 2.21(e))

$M_1 = -66$ kN.m, $M_2 = +66$ kN.m (by symmetry)

$\delta_3 = L^2(M_1 - M_2)/(16EI)$

$\quad = (12000)^2 \times (-66 - 66) \times 10^6/(16 \times 26.15 \times 10^{12})$

$\quad = -45.4$ mm (i.e. upwards)

Total deflection $= \delta_1 + \delta_2 + \delta_3 = 30.9$ mm

Span/360 $= 12000/360 = 33.3$ mm

Span BC

$M_1 = -60$ kN.m, $M_2 = -6$ kN.m

$\delta = L^2(M_1 - M_2)/(16EI)$

$\quad = (18000)^2 \times (-60 + 6) \times 10^6/(16 \times 26.15 \times 10^{12})$

$\quad = -41.8$ mm (i.e. upwards)

In this case an accurate analysis gives - 43.4 mm at 7.35 m from B

Span/360 $= 50$ mm

Cantilever span AB

For this span the deflection is due to the flexure of the cantilever, assuming the beam is horizontal at B, plus the effect of the anti-clockwise rotation of the beam at B, i.e. $-\theta_1$ for span BC

Span BC (Fig. 2.21(e))

$\theta_1 = L(2M_1 - M_2)/(6EI)$

$\quad = 18000(-2 \times 60 + 6) \times 10^6/(6 \times 26.15 \times 10^{12})$

= −0.01308 radians

1.   Deflection at A due to rotation

   $\delta_1 = -L\ \theta_1$

      = 4000 x 0.01308

      = 52.32 mm

2.   Deflection due to load (Fig. 2.20(b))

   Total load wL = 30 kN

   Putting   a = 1

   $\delta_2 = wL^4/(8EI)$

      = 30 x 10^3 x $(4000)^3/(8$ x 26.15 x $10^{12})$

      = 9.18 mm

   Total deflection = $\delta_1 + \delta_2$

                  = 52.3 + 9.2 = 61.5 mm

   Span/360 = 4000/360 = 11.1 mm

Conclusion:

The deflection of the cantilevers exceeds span/360, so stiffening
would be required. Note that it would not be sufficient just to
stiffen the cantilever span AB, because the deflection solely due to
rotation in BC is excessive. It would therefore be necessary to
stiffen both spans, either by providing a heavier section for the
spans, or by plating the flanges of the existing beam over the support
B and for part of the spans on either side. The deflections in the
other spans are satisfactory. The limit of span/360 is severe for
cantilevers and may be reduced in a future code of practice.

2.8.2 Span/depth ratios for simply supported beams   An initial
estimate for the depth of a trial section can be obtained for a
uniformly loaded simply supported beam.

   Letting f = the maximum bending stress,

   f = My/I = (WL/8 x D/2)/I

from which W, the total load on the span is given by

W = 16fI/(LD)

Now assuming the deflection to be the maximum permissible

$$L/360 = 5WL^3/(384EI)$$

Substituting for W, putting $E = 210 \times 10^3$, and re-arranging, the span/depth ratio is given by

L/D = 2800/f   (2.45)

Setting f to the maximum allowable bending stress, the span/depth ratio (L/D) for each steel grade is given in Table 2.3 below.

Grade	span/depth ratio
43	17
50	12
55	10

Table 2.3   Span/depth ratios for simply supported beams

Note that since Young's modulus is a constant for all grades of steel, the stiffness of a beam does not increase with a higher steel grade. Consequently the ratio in Table 2.3 is reduced in the higher grades to allow for higher stresses. In other words, if the design is governed by deflection there is no advantage to be gained by using a higher grade of steel.

2.9   CONTINUOUS BEAM LOADING PATTERNS

In Chapter 1 it was explained that there is a necessity for the designer to predict the combinations of loads likely to produce the most severe effects on a structure. When continuous beams are designed it is important to realise that the same load combination may not be carried simultaneously by all the spans. For example on warehouse floors where goods are regularly being stacked and removed it is quite common for the loads on individual spans to alternate between dead load only and dead plus imposed loads. A similar situation can occur on bridge decks where moving traffic can form a variety of patterns.

In general the load pattern giving the greatest bending moment may not be the same as that giving the greatest shear, or deflections. It is usually necessary therefore to consider several different patterns in order to obtain the most severe effects. For beams with more than two spans none of the maximum effects is obtained when all the spans are fully loaded.   Consider the three span beam in Fig. 2.23.   The structural dead loads are always carried simultaneously by all spans, but the imposed loads and possibly some of the non-structural dead loads can be varied.

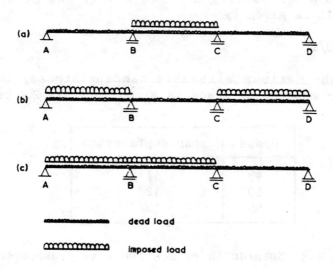

Fig. 2·23. Load patterns on a 3 span beam.

Case (a), with the imposed load on the centre span, gives the maximum sagging bending moment and the maximum deflection in the centre span.   In some cases it could cause uplift on the end supports, although this is not common.   The minimum sagging moment occurs in the end spans with this pattern.   It is possible that stress reversals could occur, which could be important from the point of view of fatigue calculations, and would also require lateral support of the bottom flange to be considered.

Case (b), with the imposed load on the outer spans, gives the maximum sagging moment and deflection in the outer spans, and the minimum sagging moment in the inner span.   It also gives the maximum reactions from the outer supports, and hence the maximum positive shear in AB, and the maximum negative shear in CD.

Case (c), with imposed loads on spans AB and BC, gives the greatest hogging moment over the support B and the greatest reaction at B.   It also gives the greatest positive shear in BC.

Similar conclusions can be drawn for beams with a greater number of spans.   In general the effect on any given span, of loads on another span which is not adjacent to it, is very small and can be neglected without serious error.   For example, if a fourth span DE carrying imposed load were added in Case (c), all the effects at the support B would be increased, but only slightly.

It follows that for a multispan beam it is only necessary to examine two kinds of loading pattern, making a total of five possible loading patterns, as follows.

1. Alternate spans with maximum and minimum load:- There are two possible patterns, i.e. 101010 ... etc. or 010101 ... etc., where a zero indicates a span with minimum load.

2. Every third span unloaded:- There are three possible patterns, i.e. 01101101 ... etc, 10110110 ... etc. or 11011011 ... etc.

Some designers ignore the second set of loading patterns, replacing it with the single load-case of all spans fully loaded. This approximation gives errors on the unsafe side, but they are quite small.

When the positions of the loads on individual spans are variable, as in the case of moving wheel loads from travelling cranes, vehicles, etc. influence lines may be necessary. Some influence lines are given in 'The Steel Designer's Manual' for three and four-span beams with equal spans, or spans in which the ratio I/L is constant. For other cases, influence lines must be constructed, either by analysis, or by the use of models.

2.10   PLASTIC DESIGN OF BEAMS

When a beam is designed elastically the allowable stresses are set at values which give pre-determined factors of safety against yielding under working load conditions. However, if the loads on the structure are increased until collapse occurs, it does not follow that the load factor, which is the ratio of collapse load to working load, is necessarily the same as the factor of safety. In fact, it will be shown that the load factor is dependent on section shape, on the action of redundancies, and on the type of loading. It can be reasoned therefore that an ultimate load method of design, employing a constant load factor against collapse, is a more logical approach when the strength of a beam is being considered. For many structures it also gives a more economical design.

In 1948 a clause was added to BS 449 allowing designs based on 'the principles of plastic design' to be used, provided that deflections at working load do not exceed those specified for elastic design. This clause implies that no economies in design time can be obtained from the use of plastic design methods, although for beams they tend to be somewhat easier than elastic methods, because elastic analyses have to be carried out also to obtain deflections. Plastic methods can only be justified therefore where an economy in the weight of steel can be achieved. A particular example of a type of structure where economies have been achieved by plastic design is the pitched roof portal frame which is manufactured in quantity for a wide variety of low-rise buildings including factories, warehouses, agricultural buildings, etc.

This Chapter deals with the design of single span and continuous beams. The design of frames is discussed in Chapter 5.

2.10.1 <u>Development of a plastic hinge</u>   The stress-strain curve resulting from a tensile test on a specimen of mild steel (grade 43) has the form shown in Fig. 1.4(a). The curve can be idealised for design purposes as the tri-linear diagram a-b-c representing the elastic, plastic, and work hardening stages of deformation. The strain during plastic deformation is in the order of ten times the maximum elastic strain.

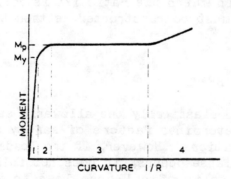

(a) Moment / curvature relationship

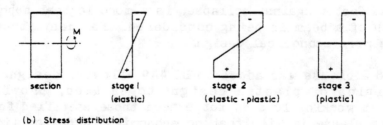

(b) Stress distribution

Fig. 2.24. Stages of elastic - plastic bending.

Fig. 2.24(a) shows the moment/curvature relationship for a beam made from the same material. The corresponding distribution of stress on a section symmetrical about the neutral axis is shown in Fig. 2.24(b). During stage 1 bending is elastic until the bending moment reaches the yield moment M_y, when the stress in the extreme fibres first reaches the yield point. On further bending (stage 2) the stress in the outer fibres remains approximately constant and yielding progresses towards the neutral axis until the bending moment M_p is reached. At this point virtually the whole section has yielded and a plastic hinge forms in the beam. M_p is the 'plastic moment of resistance' of the beam. Further rotation at the hinge takes place at constant moment until the end of stage 3, at which point the stress in the extreme fibres starts to increase again due to work hardening, and the bending moment at the plastic hinge also increases.

The amount of rotation at the plastic hinge that can take place at constant moment is determined by the length of the yield plateau in the stress–strain relationship for the material. In mild steels such as grade 43 it is considerable and the work hardening stage is frequently ignored. For the higher grades of steel the stress/strain curve is similar to that shown in Fig. 1.4(b). Yielding and work hardening occur simultaneously and there is no yield plateau. A plastic hinge can form, but rotation does not take place at constant moment. For the design of simple structures this increase in bending moment is ignored and the plastic moment is calculated on the assumption that the stress in the fully plastic section is equal to the design yield stress Y_s. For the higher grades of steel, Y_s is in reality a proof stress (see Chapter 1).

The longitudinal spread of plasticity in the region of a plastic hinge is equal to the length of beam over which the bending moment exceeds the yield moment M_y. This is shown for a point load in Fig. 2.25.

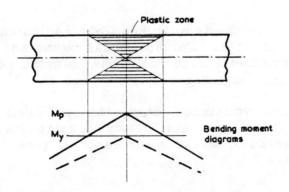

Fig. 2·25. Spread of plasticity in a plastic hinge.

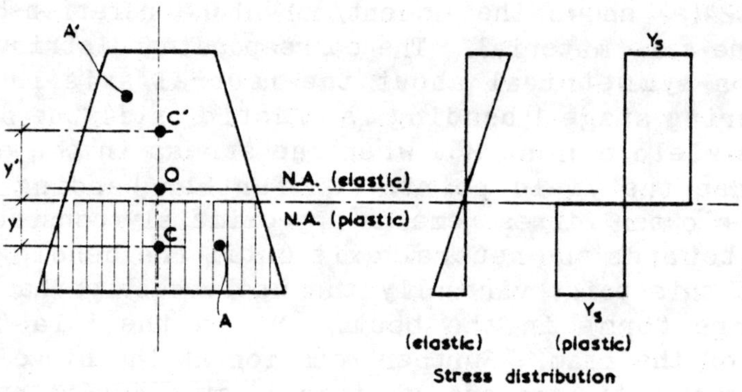

Fig. 2.26. Movement of neutral axis.

2.10.2 Plastic modulus   For the general case of a mild steel section symmetrical about the plane of bending, the stress distributions in the elastic and fully plastic states are shown in Fig. 2.26. For equilibrium of normal forces, the tensile and compressive resultants of the bending stresses must be equal. In the elastic state this condition is achieved when the neutral axis passes through the centroid (point O); but in the fully plastic state, since the stress is uniformly equal to the yield stress, equilibrium is obtained when the neutral axis divides the section into two equal areas, namely A (hatched) and A' (clear), as shown. If y and y' are the distances between the centroids C and C' of these areas and the neutral axis, then the plastic moment is given by

$$M_p = Y_s(Ay + A'y') \tag{2.46}$$

which can be written as

$$M_p = Y_s Z_p \tag{2.47}$$

where Z_p is equal to the term in brackets and is called the 'plastic modulus' of the section. For Universal beams and columns, joists, channels, and structural hollow sections, plastic moduli are given in the section tables (Appendix).

For a section symmetrical about its centroidal axis, the neutral axis is common to both the elastic and plastic states. Consider for example a rectangle of width b and depth d. Area A = A' = bd/2 and y = y' = d/4. Hence

$$Z_p = bd^2/4 \tag{2.48}$$

Example 2.9 Find the plastic moduli and the plastic moment of resistance for the channel in Fig. 2.27(a) assuming grade 50 steel with a yield stress of 345 N/mm². Consider both axes of the section.

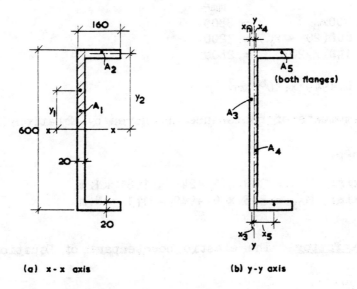

(a) x-x axis   (b) y-y axis

Fig. 2.27. Plastic moduli of a channel.

x-x axis

Since the section is symmetrical about the x-x axis, the plastic modulus is twice the moment of area of half the section. Dividing each half into areas A_1 (hatched) and A_2 (clear), as shown

$$Z_{px} = 2(A_1 y_1 + A_2 y_2)$$

$A_1 = 300 \times 20 = 6000 \text{ mm}^2 \qquad y_1 = 600/4 = 150 \text{ mm}$
$A_2 = 140 \times 20 = 2800 \text{ mm}^2 \qquad y_2 = 300-10 = 290 \text{ mm}$

$$Z_{px} = 2(6000 \times 150 + 2800 \times 290) = 3.424 \times 10^6 \text{ mm}^3$$

y-y axis

Since, for the fully plastic section, the neutral axis divides the section into two equal areas and the total web area is greater than the total area of the flange outstands, the y-y axis must lie within the web. Referring to Fig. 2.27(b), its distance from the left hand face is x_n, such that

$$x_n \times 600 = \text{half total area}$$

i.e. x_n = (6000 + 2800)/600 = 14.67 mm
and, taking first moments of area about y-y

$$Z_{py} = A_3 x_n/2 + A_4(20 - x_n)/2 + 2A_5(140/2 + 20 - x_n)$$

where $A_3 = 600 x_n$ = 8800
$\qquad A_4 = 600(20 - x_n)$ = 3200 mm^2
$\qquad A_5 = 140 \times 20$ = 2800

Hence Z_{py} = 0.4949 x 10^6 mm^3

The plastic moments of resistance are given by Equation (2.47). i.e.

$$M_p = Y_s Z_p$$

i.e.   x-x axis: M_{px} = 345 x 3.424 = 1181 kN.m
$\qquad$ y-y axis: M_{py} = 345 x 0.4949 = 171 kN.m

2.10.3 Shape factor   The elastic counterpart of Equation (2.47) is

$$M_y = Y_s Z_e \qquad\qquad (2.49)$$

where M_y is the yield moment, and Z_e is the elastic section modulus.

The ratio Z_p/Z_e is a geometric property of the section called the 'shape factor'. From Equations (2.47) and (2.49) it is also equal to the ratio M_p/M_y and therefore represents the reserve moment capacity of the section between the onset of yielding and the development of full plasticity. The shape factor is always greater than unity. For example, a rectangle has $Z_e = bd^2/6$ and $Z_p = bd^2/4$ and hence the shape factor = 1.5. For standard rolled sections it varies from section to section within the range, and with the axis about which bending takes place.

Typical values of the shape factor for various sections are given below in Table 2.4.

Section	x-x	y-y
Rectangle (solid)	1.5	1.5
Circle (solid)	1.7	1.7
Thin walled tube	1.27	1.27
Universal beam or column	1.15	1.5
Channel	1.2	1.9
Rectangular hollow section	1.24	1.16

Table 2.4   Typical shape factors

The shape factor is less when a considerable part of the section is concentrated at some distance from the neutral axis, for example in I sections about the x-x axis and hollow sections.

2.10.4 <u>Effect of axial force</u>   The effect of axial force on a fully plastic section can be demonstrated by considering the rectangular section in Fig. 2.28.

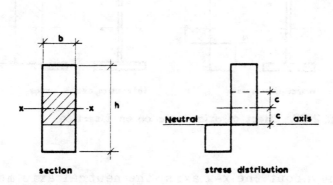

Fig. 2·28. Effect of axial force.

The axial force is resisted by the hatched area, and for equilibrium of normal forces

$$pbh = 2bcY_s$$

where p = the mean axial stress, obtained by dividing the axial force by the area of the whole section.

Hence, defining $n = p/Y_s$

$$c = nh/2$$

The reduced plastic modulus is obtained by subtracting the plastic modulus of the hatched area from the plastic modulus of the whole section about x-x, thus

$$Z_{pr} = bh^2/4 - bc^2$$

which, on substitution for c, gives

$$z_{pr} = bh^2/4(1 - n^2) \qquad (2.50)$$

The same procedure can be adopted with an I section except that the neutral axis can either be in the web or the flange depending upon whether the axial force is small or large respectively. Defining n as the ratio of mean axial stress to yield stress, as before, and using the section dimensions shown in Fig. 2.29, formulae can be obtained for the reduced plastic moduli.

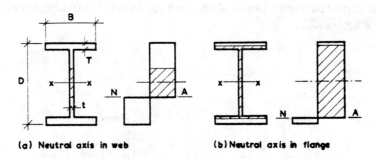

(a) Neutral axis in web   (b) Neutral axis in flange

Fig. 2·29. Effect of axial force on an I section.

For bending about the x-x axis, the neutral axis moves from the web into the flange when n exceeds n_c, given by

$$n_c = t(D - 2T)/A \qquad (2.51)$$

where A = the area of the whole section.

When n is less than or equal to n_c

$$Z_{pr} = Z_{px} - an^2 \qquad (2.52)$$

When n exceeds n_c

$$Z_{pr} = b(1 - n)(c + n) \qquad (2.53)$$

The constants a, b and c are given by

$$a = A^2/(4t)$$
$$b = A^2/(4B) \qquad (2.54)$$
$$c = 2BD/A - 1$$

For bending about the y-y axis the change point for n is given by

$$n_c = tD/A \qquad (2.55)$$

The reduced plastic modulus can be obtained by Equations (2.52) and (2.53), using the following constants

$$a = A^2/(4D)$$
$$b = A^2/(8T)$$
$$c = 4TB/A - 1$$

(2.56)

The values of a, b, c and n_c are quoted for Universal beam and column sections in the section tables (Appendix).

2.10.5 <u>Instability problems</u> The full rotation required at a plastic hinge in a beam may not be realised unless a lateral support is provided at the hinge position. It is also necessary to provide a lateral support at other points along the span to ensure that premature failure by lateral torsional buckling does not occur. Within the plastic region, where the bending moment exceeds the yield moment M_y, Morris and Randall give maximum allowable slenderness ratios of 70, 60, and 53 for steel grades 43, 50, and 55 respectively. In the elastic parts of the span the maximum slenderness ratios given by BS 449 can be used, based on the stresses obtained from the bending moment diagram at collapse, divided by the load factor. The spacing of lateral supports is calculated in Example 2.13 (Section 2.11).

When a section contains slender elements, the formation of a plastic hinge may be inhibited by local buckling. There are two kinds of local instability to be considered when selecting a section for a structure designed by plastic theory. The following limits to section dimensions and axial stress are quoted by Morris and Randall.

1. Flange instability, which occurs when the width to thickness ratio B/T exceeds the values given in column 2 of Table 2.5. Plastic hinge action is not permissible when these values are exceeded, either when the section is to be used as a beam or as a stanchion. All Universal beams are suitable in grade 43 steel, but the B/T ratios of Universal beams in the higher grades of steel, and of Universal columns in all grades should be checked before the section is selected.

2. Web instability, which occurs in the form of premature buckling when the depth to thickness ratio (D - 2T)/t exceeds the values given in column 3 of Table 2.5, unless the axial stress on the section is limited. All Universal column sections are satisfactory in this respect, but some universal beam sections are liable to be affected if used as stanchions.

The limiting axial stress can be obtained from the appropriate formula in column 4 of Table 2.5. It is not permissible for the depth to thickness ratio of the web to be so high that the axial stress given by these formulae is reduced below zero. These upper limits are given in column 5 of the table.

Steel grade	Flange Limits B/T	Web Limits		
		d_w/t	Limiting axial stress (N/mm^2)	Max d_w/t
43	18	53	$t(328t - 3.86d_w)/A$	85
50	15.25	45	$t(432t - 6.00d_w)/A$	72
55	13.75	40	$t(448t - 7.00d_w)/A$	64
$d_w = D - 2T$			A = area of whole section	

Table 2.5 Section limitations in plastic design

In a compression member local instability cannot be completely isolated from the general instability of the whole member. The interaction is briefly explained for elastic members in the next chapter and for plastic design in Chapter 5. The subject is extremely complex and a full treatment is beyond the scope of an introductory text. At this stage however, it is necessary to point out that the limiting stresses described above are upper bounds which can only be achieved when the general stability of the member is ensured by adequate lateral bracing at or near the position of the plastic hinge, and at suitable intervals along the member.

2.10.6 **Effect of holes**   The presence of holes in the flanges of a beam has no effect on the plastic design unless they occur at a plastic hinge, in which case due allowance should be made; thus for an I beam

$$\text{reduction in } Z_{px} = ndT(D - T) \tag{2.57}$$

where n = the number of holes (assumed to be equal) in each flange
      d = the diameter of the holes

If the holes occur in one flange only, the reduction can be taken as half of that given by Equation (2.57), ignoring the loss of symmetry.

2.10.7 **Effect of shear**   The effect of shear on the plastic moment of a solid rectangular section is usually negligible, but for an I section, in which most of the shear force is resisted by the web, it can be significant and, under conditions of high shear stress, cause an appreciable reduction in the plastic moment of resistance.

A practical method of calculating the reduction in plastic moment is due to Hayman and Dutton and has been verified experimentally by

Baker, Horne and Heyman. This, and an alternative approach, are fully described by Horne. The method is based on the assumption that when the section becomes fully plastic under the action of bending and shear, the yield stress in the flanges is unchanged and remains at the design yield stress Y_s. The effective yield stress in the web is reduced however by the presence of shear stresses.

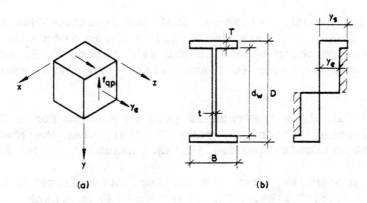

Fig. 2·30. Effect of shear on plastic moment.

The stresses on an element of the web are shown in Fig. 2.30(a) and the corresponding distribution of bending stresses on the fully plastic web is shown in Fig. 2.30(b). The reduction in plastic moment is due to the hatched area, i.e.

$$\text{Reduction in } M_p = Z_{pw} (Y_s - Y_e) \tag{2.58}$$

where Y_e = the effective yield stress in the web
Z_{pw} = the plastic modulus of the web, given by

$$Z_{pw} = td_w^2/4 \tag{2.59}$$

where d_w = the depth of the web, i.e. D - 2T

The value of the effective yield stress in the web is derived from Von Mises criterion for failure, by application of Equation (2.42), thus

$$Y_e^2 = Y_s^2 - 3f_{qp}^2 \tag{2.60}$$

where f_{qp} is the average shear stress in the web given by

$$f_{qp} = Q/td_w \tag{2.61}$$

where Q = the shear force at the plastic hinge.

Note that when $f_{qp} = Y_s/\sqrt{3}$, $Y_e = 0$ and the web yields due to the action of the shear stress alone.

In very slender webs the shear stresses can cause buckling, and limits are prescribed for the ratio of depth to thickness. These are the limits given in column 5 of Table 2.5.

Morris and Randall recommend that the reduction due to shear in the plastic modulus of standard rolled I sections can be ignored unless the average shear stress in the web exceeds $Y_s/3$, or $Y_s/4$ when the ratio of overall depth to flange width (D/B ratio) exceeds 2.5.

Example 2.10   Calculate the reduced plastic moment for a 762 x 267 x 197 Universal beam section, in grade 50 steel, when the shear force on the section at ultimate load is 1500 kN. Assume $Y_s = 345$ N/mm^2.

The relevant properties, from the section tables (Appendix), are: D = 769.6, t = 15.6, T = 25.4, $Z_p = 7167$ cm^3, from which

$$d_w = D - 2T = 769.6 - 2 \times 25.4 = 718.8 \text{ mm}$$

Depth to thickness ratio $d_w/t = 46$. This is less than the upper limit prescribed in column 5 of Table 2.5, but is greater than the limit given in column 3; so that if an axial force were also present it would need to be limited in accordance with the formula given in column 4.

The plastic modulus of the web is given by Equation (2.59), thus

$$Z_{pw} = td_w^2/4 = 15.6 \times (718.8)^2/4 = 2.015 \times 10^6 \text{mm}^3$$

The average shear stress from Equation (2.61) is

$$f_{qp} = Q/td_w = 1500 \times 10^3/(718.8 \times 15.6) = 134 \text{ N/mm}^2$$

The effective yield stress in the web is obtained from Equation (2.60)

$$Y_e^2 = Y_s^2 - 3f_{qp}^2 = 345^2 - 3 \times 134^2 = 65342$$

Hence $Y_e = 256$ N/mm^2

From Equation (2.58), reduction in $M_p = Z_{pw}(Y_s - Y_e)$

$$= 2.015(345 - 256) = 180 \text{ kN.m}$$

$$M_p = Z_p Y_s = 7167 \times 345/10^3 = 2473 \text{ kN.m}$$

Reduced M_p = 2473 - 180 = <u>2293 kN.m</u>, (i.e a reduction of 7 percent).

The maximum possible reduction in M_p occurs when $Y_e = 0$, i.e. when $f_{qp} = Y_s/\sqrt{3}$, and the web makes no contribution at all to the plastic moment of resistance. In this example:

$$f_{qp}(\text{yield}) = 345/\sqrt{3} = 199 \text{ N/mm}^2$$

giving a reduction in M_p of

$$Z_{pw}Y_s = 2.015 \times 345 = \underline{695 \text{ kN.m}}$$

which is a reduction of 28 percent.

2.10.8 <u>General principles of design</u>   The principles underlying the formation of a plastic hinge in a beam have been described in Section 2.10.1. When sufficient hinges have formed the beam becomes a mechanism and collapse occurs.

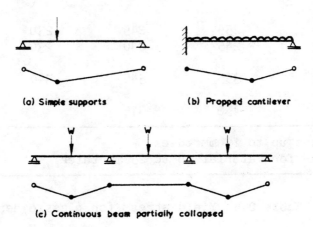

(a) Simple supports   (b) Propped cantilever

(c) Continuous beam partially collapsed

o Real hinge

• Plastic hinge

Fig. 2.31. Collapse mechanisms.

Three hinges, either real or plastic, are sufficient to constitute a mechanism in any span, as shown in Fig. 2.31(a), (b) and (c). A real hinge is a point where there is no resistance to rotation, for example a simply supported end or a pinned connection between members. In a

continuous beam collapse may only be partial, plastic hinges occurring
in one or more spans, the other spans remaining elastic as in Fig.
2.31(c).   The load carried by the beam at collapse, either partial or
complete, is the ultimate load capacity of the beam.

The method of design is to multiply the working loads by a load
factor γ , and then determine the minimum value of plastic modulus for
the beam such that collapse will just occur at the factored load.   Any
other forces, such as shear or axial force, which may affect the
plastic moment capacity of the beam, are also obtained from the
factored loads.   A section is then selected with a plastic modulus
equal to or as near as possible above the value calculated.

The values of yield stress used in plastic design are specified
in BS 4360 and are reproduced below in Table 2.6.   Grade 55 steel is
not generally used for plastic design.

Yield stress for hot-rolled I section and plates (N/mm^2)				
Steel grade	Thickness (mm)			
	Up to 16	16+ to 40	40+ to 63	63+ to 100
43A I-sections Plates	255 * 245	245 240	230 230	225 220
50B	355	345	340	325
55C	450	** 415		
* For plates up to 19 mm thick ** 430 N/mm^2 for material up to 25 mm thick				

Table 2.6   Yield stress for plastic design

For beams and small framed structures plastic design is based on
the assumption that elastic deformations prior to collapse have no
bearing on the ultimate load capacity of the structure.   It is
therefore possible to use the simplified rigid-plastic idealisation
shown in Fig. 2.32 in which a member is assumed to be infinitely stiff
until a plastic hinge forms.   The lengths of members between the
hinges are therefore assumed to be straight.

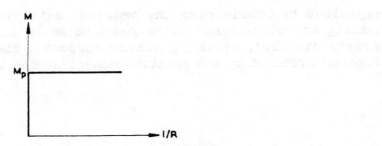

Fig. 2.32. Rigid - plastic idealisation.

2.10.9 <u>Statically determinate structures</u> Consider the simply supported beam in Fig. 2.33. At working load the beam is subjected to a uniformly distributed load w. If Y is the load factor against collapse, the ultimate, or collapse load, is Yw. As the beam is simply supported, a single plastic hinge is all that is required for a mechanism, and at collapse this occurs at mid-span where the bending moment is maximal. Taking moments about the plastic hinge for one side of the beam

$$M_p = YwL^2/8 \qquad\qquad (2.62)$$

i.e. $M_p = Y M_w$, where M_w is the elastic bending moment at mid-span due to the working load w. Any economy to be attributed to plastic design depends therefore on the relative values of Y and the shape factor.

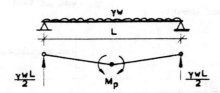

Fig. 2.33. Statically determinate beam.

Economy can only result when the shape factor of the section exceeds $Y/1.52$. For example, with a load factor of 1.75 plastic design would only be more economical for sections having shape factors greater than 1.15. This result applies to all statically determinate structures.

2.10.10 <u>Statically indeterminate single span beams</u> The development of a collapse mechanism in a statically indeterminate beam can be

explained by considering the propped cantilever in Fig. 2.34.   The
bending moment diagram can be constructed by superimposing the 'free
moment diagram', assuming simple supports, and the 'fixing moment
diagram' produced by the reactive moment at the built-in support.

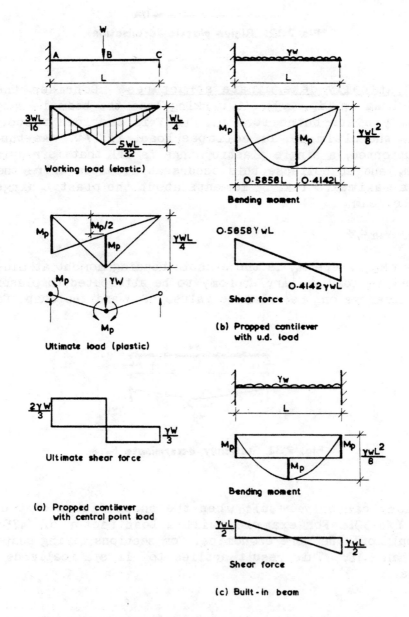

Fig. 2.34. Collapse mechanisms in single span beams.

The bending moments in the elastic state at working load can be obtained by any of the standard methods of elastic analysis. The maximum free moment under the load is WL/4 and the fixing moment at the support is 3WL/16, giving a net bending moment of 5WL/32 at mid-span, as shown.

If the load is increased, the bending moments increase in proportion until a plastic hinge forms at the built in support, where the bending moment is greatest. On further increase, the moment at this plastic hinge remains constant at M_p. The moment at mid-span continues to increase until it also reaches M_p, with the formation of a second plastic hinge. There is now a mechanism consisting of two plastic hinges at A and B and a real hinge at C, and collapse occurs.

At this stage the load has increased to γW, where γ is the load factor. The fixing moment at the support and the net moment at mid-span are both equal to M_p. The value of M_p in terms of γW can be obtained from the geometry of the diagram, remembering that at mid-span the free bending moment is $\gamma WL/4$, and hence

$$3M_p/2 = \gamma WL/4$$

whence

$$M_p = \gamma WL/6 \tag{2.63}$$

The required section will have a plastic modulus equal to or greater than

$$Z_p = M_p/Y_s \tag{2.64}$$

The shear forces can be obtained from the mechanism diagram, by taking moments about each end, thus:

Shear force at A = $\gamma W/2 + M_p/L$
Shear force at C = $\gamma W/2 - M_p/L$

which, on substituting for M_p from Equation (2.63), give $2\gamma W/3$ at A and $\gamma W/3$ at C, as shown in the bottom diagram.

Similar reasoning can be applied to the propped cantilever with a uniformly distributed load, in Fig. 2.34(b). The position of the plastic hinge on the span coincides with the position of zero shear and divides the span into the lengths 0.5858L and 0.4142L, as shown. From the geometry of the diagram it can be shown that the plastic moment is given by

$$M_p = \gamma wL^2/11.66 \tag{2.65}$$

For the built-in beam in Fig. 2.34(c) the plastic moment is

exactly half of the free bending moment, i.e.

$$M_p = \gamma wL^2/16 \qquad\qquad (2.66)$$

The shear forces are equal to half the applied load, by symmetry.

If a shape factor of 1.15 and the allowable bending stresses of BS 449, Table 2 are assumed, the load factor against collapse for a beam designed elastically can be expressed as 1.75 k_1/k_2, where k_1 and k_2 are the factors relating design loads and bending moments for plastic and elastic designs respectively. The results for the above beams are tabulated in Fig. 2.35, from which it can be seen that load factors varying from 1.75 to 2.55 are obtained, in spite of the fact that the safety factor against yielding applied to the bending stresses is constant. For plastic design it is current practice to use a load factor of 1.7.

BEAM	k_1	k_2	$1.75\, k_1/k_2$
 Any loading			1.75
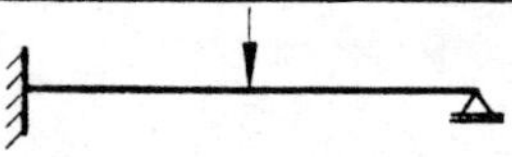	6	16/3	1.97
	11.66	8	2.55
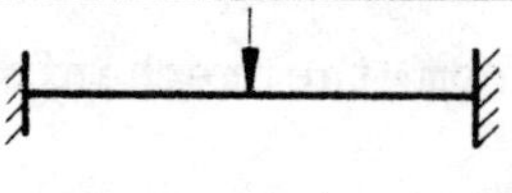	8	8	1.75
	16	12	2.33

Fig. 2.35. Load factors with elastic design.

2.10.11 Continuous beams   In a continuous beam there are negative bending moments over the internal supports, while the positive bending moments occur within the spans. As in the case of single span beams, the net bending moment diagram can be produced by combining the fixing moment diagram, consisting of straight lines, with the free bending moment diagram for each span. In a collapsed span the maximum positive and negative bending moments are equal to the plastic moment of resistance, so each span can be treated separately and the loading patterns described in Section 2.9 are not necessary for plastic design.

For beams of uniform section, simultaneous collapse in all the spans only occurs in special circumstances. It is possible to design for simultaneous collapse by varying the section as required, but this does not necessarily lead to economy.

Consider the beam in Fig. 2.36(a) which has two equal spans, equally loaded, and is of uniform section.

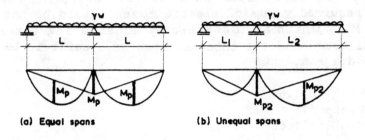

(a) Equal spans   (b) Unequal spans

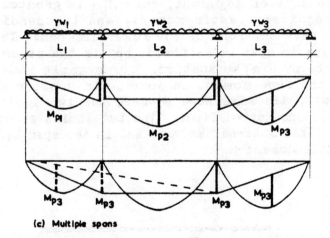

(c) Multiple spans

Fig. 2·36. Continuous beams with uniform section.

In this case collapse occurs in both spans, which behave exactly as propped cantilevers, and the required plastic moment of resistance is given by Equation (2.65), i.e. $M_p = \gamma wL^2/11.66$. If the loads, or spans, are not equal, then the most economical design would normally be achieved with a uniform section, but a plastic hinge would only form in one span. In Fig. 2.36(b) for example, a hinge forms in the longer span and the required plastic moment is $\gamma wL_2^2/11.66$. The moments and shears in the shorter span can easily be obtained by statics if required.

When there are more than two spans, the beam may be designed with a uniform section capable of restricting collapse to the span with the largest moments. Alternatively a basic section may be selected to produce a collapse mechanism in the span with the smallest moments. The beam is then reinforced with flange plates to provide the necessary plastic moments of resistance in the other spans. Either solution may produce the more economical design, depending upon a variety of circumstances; but it should be borne in mind that mere weight of steel is not a sufficient criterion of economy.

Consider the three-span beam in Fig. 2.36(c). For a uniform section the required value of plastic moment would be the greatest of M_{p1}, M_{p2} or M_{p3}, which are obtained by considering the end spans as propped cantilevers and the internal span as a built-in beam, as shown in the upper diagram, thus

$$M_{p1} = \gamma w_1 L_1^2/11.66$$
$$M_{p2} = \gamma w_2 L_2^2/16 \qquad\qquad (2.67)$$
$$M_{p3} = \gamma w_3 L_3^2/11.66$$

Assume, for the sake of argument, that M_{p3} is greatest. This then would be the required plastic moment, and the bending moment at collapse would have the form of the lower diagram. The position of the reactant line in span L_3 is fixed, but in the other two spans it is indeterminate by simple statics. For example if M_{p1} is greater than M_{p2}, and the free moment in span L_1 is greater than M_{p3}, the reactant line will lie somewhere between the two limiting positions shown by the full and dotted lines. The two limits are defined by the moments at the first internal support and in the span L_1, which cannot exceed the plastic moment M_{p3}.

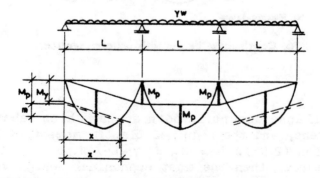

Fig. 2·37. Non-uniform section.

For a design with a non-uniform section consider the beam in Fig. 2.37, which has uniform loading on three equal spans. In this case the centre span requires the smaller plastic moment, i.e. $Mp = \gamma wL^2/16$. The two outer spans would need to be reinforced with flange plates capable of providing an additional plastic moment m, as shown by the dotted line drawn parallel to the reactant diagram.

The theoretical points for curtailment of the flange plates are where this dotted line intersects the free moment diagram. In practice the plates would probably be continued right to the ends of the beam and also extended beyond the second intersection point (denoted by the dimension x). Extension beyond the theoretical curtailment points is necessary to ensure that a plastic hinge does not form in the basic section. The extension should be at least long enough for the welds to transmit sufficient force into the flange plate, as in the design of compound beams (Chapter 7). Furthermore, at a plastic hinge yielding extends over a length of beam on either side of the theoretical hinge location, and if it was required to prevent yielding altogether in the basic section it would be necessary to extend the flange for the distance x' to the intersection of the chain-dotted line, representing the yield moment M_y, with the bending moment diagram.

2.10.12 Reductions in M_p   For the sake of economy, the lightest section consistent with the required plastic moment of resistance is usually selected. Even so, with standard rolled sections there is usually some reserve capacity above what is actually required, and this can often be used to accommodate any reduction in moment of resistance due to high shear stresses or holes drilled in flanges. Since the reductions do not necessarily occur at all the hinge positions, some redistribution of plastic moments may be possible.

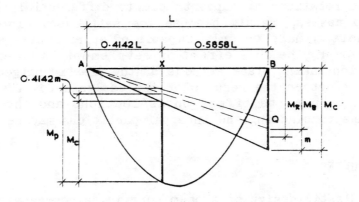

Fig. 2.38. Redistribution of plastic moments.

The method of re-distributing the plastic moments can be explained by considering the end span AB of a continuous beam, shown in Fig. 2.38. Assume that the required moment obtained from the analysis is M_c and that the chosen section has a plastic moment of resistance M_p. Assume also that when shear stresses at the support B are taken into account, the moment of resistance of the section is reduced to M_R which is less than the required moment M_c, as shown. At X, the point of maximum bending moment in the span, the shear force is zero and there will be no reduction in the moment of resistance, which therefore remains at M_p. If a new reactant line AQ is drawn to correspond with the available moment M_p at X, it can be seen that the required plastic moment at B could be reduced further if required, and the section would be adequate provided that

$$M_R \geq M_c - (M_p - M_c)/0.4142 \qquad (2.68)$$

It is not sufficient, however, merely to show that the section is adequate, because the reduced moment capacity at B will also affect the design of the adjacent internal span. It is therefore necessary to calculate a new value for the plastic moment M_B at B after redistribution. For this it is assumed that the ratio of the redistributed moment to the provided moment is the same at X as at B.

i.e. $(M_c + 0.4142m)/M_p = (M_c - m)/M_R$ \qquad (2.69)

from which m, the reduction to be applied at B, may be calculated.

Equations (2.68) and (2.69) have been developed to show the reasoning applied when a reduction in the plastic moment capacity of a section is involved. In an actual design the procedure would depend on the circumstances. An example is given in the next section.

2.10.13 **Settlement of supports**   In an elastic analysis of continuous and single span statically indeterminate beams the effect of translations and rotations at supports due to differential settlement is difficult to assess, mainly because the settlement displacements are either indeterminate or only capable of approximate estimation. At collapse however, the only effect of settlement is to change the amount of rotation taking place at some of the plastic hinges. It has no appreciable effect on the results of the analysis for the strength of the beam, although its effect on deflections and the possible consequent damage to cladding and internal partitions may be important.

2.11   DESIGN STUDIES

Example 2.11   Elastic design of a beam carrying a concrete floor
The floor of an office building consists of 125 mm pre-cast concrete planks, with a mass of 205 kg/m^2, topped with a 40 mm concrete screed

and 20 mm wood blocks. Lightweight partitions supported by the floor are equivalent to a superficial load of 1.0 kN/m^2. The floor is to rest on the top flanges of simply supported beams spanning 8.0 m, and at a pitch of 3.75 m. There is a suspended ceiling of mass 40 kg/m^2. The imposed load specified by CP 3 Ch. V, Pt 1 for offices is 2.5 kN/m^2. The beam will be designed in grade 43 steel. Assume a density of 2400 kg/m^3 for the concrete screed, and 900 kg/m^3 for the wood blocks.

LOADS

	kg/m^2	kN/m^2
Dead load (structural)		
Self-weight (assumed)	20	
125 mm pre-cast planks	205	
	---	
	225 x 9.81/10^3 =	2.21
Dead load (non-structural)		
40 mm concrete screed 2400 x 0.04 =	96	
20 mm wood blocks 900 x 0.02 =	18	
Suspended ceiling	40	
	---	
	154 x 9.81/10^3 =	1.51
Imposed loads		
CP3		2.50
Lightweight partitions		1.00

Total non-structural		5.01

Total load on span = (2.21 + 5.01) x 8 x 3.75 = 217 kN

Total non-structural load per metre = 5.01 x 3.75 = 18.8 kN/m

BENDING

M = WL/8 = 217 x 8/8 = 217 kN.m

Assuming full lateral support from slab, allowable bending stress = p_{bt} from Table 2 (Appendix) = 165 N/mm^2

Z_{ex} required = M/p_{bt}
$$= 217 \times 10^6/165 \times 10^3 = 1315 \text{ cm}^3$$

Try 457 x 191 x 74 U.B.

The following data relating to the section are obtained from the section tables (Appendix).

$D = 457.2$ mm, $t = 9.1$ mm, $T = 14.5$ mm, $r = 10.2$ mm
$d = 407.9$ mm, $I = 33388$ cm^4, $Z_{ex} = 1461$ cm^3

The section tables give two other beams of the same weight that could be considered. It is usual to select the stiffer beam in such cases, except when a shallower beam is required for architectural reasons. In this case the broader flange would give a better bearing for the reinforced concrete planks.

LATERAL RESTRAINT

The lateral restraint is provided by friction between the beam flange and the concrete slab. Assuming a coefficient of friction of 0.3, total frictional force $H = 0.3 \times 217 = \underline{65.1\ kN}$. This must not be less than 2.5 percent of the maximum force in the compression flange, i.e.

$H \geq 0.025\ M_{max}/(D - T)$ ... (Equation 2.33)

$M_{max} = 217$ kN.m, $D = 457.2$ mm, $T = 14.5$ mm

$0.025 \times 217 \times 10^3/(457.2 - 14.5) = \underline{12.3\ kN}$

The lateral restraint is clearly adequate. The coefficient of friction assumed is reasonable for concrete and unpainted steel, but in this case adequate support would have been provided by a much lower coefficient.

DEFLECTIONS

The deflection from non-structural loads must not exceed span/360.

Max deflection = $5wL^4/(384\ EI)$... Fig. 2.21(b)

   $w = 18.8$ kN/m, $L = 8$ m, $E = 210$ kN/mm^2

   $I = 33388$ cm^4

Transforming all quantities to kN-mm units,

deflection = $5 \times 0.0188 \times (8000)^4/(384 \times 210 \times 33388 \times 10^4) = \underline{14.3\ mm}$

Span/360 = 8000/360 = $\underline{22.2\ mm}$. The deflection is therefore acceptable.

<u>DESIGN OF WEB</u>   (See Section 2.6)

For this beam the load is uniformly distributed, so the maximum shear force is at the ends, and the only concentrated loads are the support reactions.

Max. shear force = end reaction = W/2 = 217/2 = <u>108.5 kN</u>

Shear resistance is given by Equation (2.37), i.e.

$P_q = p_q'Dt/1000$ (kN)

p_q' (Table 11, Appendix) = 100 N/mm^2 for grade 43 steel

D = 457.2 mm,   t = 9.1 mm, hence

P_q = 100 x 457.2 x 9.1/1000 = <u>416 kN</u>

Shear resistance is adequate.

Bearing resistance components can be obtained from the Safe Load Tables in the BCSA-Constrado Handbook, or from Equations (2.38), i.e.

Beam component C_1 = (T + r)cot 30° x tp_b/1000 (kN)

T = 14.5 mm,   r = 10.2 mm, cot 30° = 1.732,   t = 9.1 mm,

p_b (Table 9 Appendix) = 190 N/mm^2 for grade 43 steel

C_1 = (14.5 + 10.2) x 1.732 x 9.1 x 190/1000 = <u>74.0 kN</u>

As this is less than the applied reaction, obtain C_2, the stiff bearing component, i.e.

$C_2 = tp_b/1000$ (kN/mm)

= 9.1 x 190/1000 = 1.73 kN/mm

Now use Equation (2.41) to find the total length of stiff bearing required. There are no flange plates and the length of stiff bearing is not specified, so $P_b = C_1$.

$L_b = (R - C_1)/C_2$ (mm)

= (108.5 - 74.0)/1.73 = <u>19.9 mm</u>

This is quite a small stiff bearing and could easily be provided by an unstiffened angle cleat at the beam-to-stanchion connection.

Web buckling components can be obtained from the Safe Load Tables or from Equation (2.40)

Beam component $C_1 = Dtp_c/2000$ (kN)

$D = 457.2$ mm, $t = 9.1$ mm

p_c is obtained from Table 17, (Appendix) using $l/r = 1.732d/t$;
$d = 407.9$ mm from the section tables (Appendix).

$l/r = 1.732 \times 407.9/9.1 = 77.6$

From Table 17(a), $p_c = 106$ N/mm^2

Hence $C_1 = 457.2 \times 9.1 \times 106/2000 = \underline{221\ kN}$

The beam component exceeds the applied force (108.5 kN) and any stiff
bearing would therefore be adequate.

In this case the design of the web is governed by the bearing
resistance.

Example 2.12 Elastic design with concentrated loads   Part of the
support for a conveyor consists of a pair of identical beams as shown
in Fig. 2.39. Each beam is connected to a stanchion and at end A by a
welded end plate and is supported on a cross-beam at D, where the
connection is made by bolting through the flanges. Loads are applied:

    a) through a secondary beam at B, which also provides lateral
       restraint;
    b) through a short column at C, which is connected to the upper
       flange of the beam;
    c) by a connecting beam joining the free ends of the cantilever
       at E, but not prevented from moving laterally.

The beams are to be fabricated in grade 50 steel.

    If the simple design approach is adopted, any reactive moment at
A is ignored and the beam is designed as being simply supported at A
and D.

Reactions and shears

Taking moments about A, and then about D
$R_D = (2 \times 150 + 5 \times 300 + 10 \times 100)/8 = 350$ kN
$R_A = (6 \times 150 + 3 \times 300 - 2 \times 100)/8 = 200$ kN
                                            ---
                                            550 kN

Check:

Total load on beam = 150 + 100 + 300 = 550 kN

Hence the shear force diagram can be drawn as in Fig. 2.39.

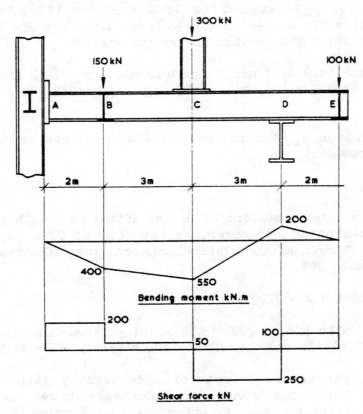

Fig. 2.39. Beam with concentrated loads.

Bending moments

$$
\begin{array}{ll}
 & \text{kNm} \\
M_B = 200 \times 2 = & 400 \\
M_C = 200 \times 5 - 150 \times 3 = & 550 \\
M_D = -100 \times 2 = & -200 \\
\end{array}
$$

Check M_B and M_C by taking moments of the forces to the right of the points. Hence the bending moment diagram, which is drawn on the tension side of the beam in Fig. 2.39.

Choice of section

A guide to the minimum depth of the section can be obtained from the span/depth ratios in Table 2.3 (k = 12 for grade 50 steel), hence

$$D = 8000/12 = 667 \text{ mm}$$

An estimate of the required elastic modulus can be obtained from the maximum bending moment. To allow for lateral buckling, assume initially that the allowable bending stress is about 3/4 of the maximum, i.e. p_{bc}(assumed) = 0.75 x 230 = 172 N/mm^2. Hence Z_{ex}(required) = 550 x 10^3/172 = 3198 cm^3. Try 686 x 254 x 125 UB (Z_{ex} = 3481), for which the section properties are:

D = 677.9 mm,   t = 11.7 mm,   T = 16.2 mm,   r = 15.2 mm,
I_x(gross) = 118003 cm^4, r_y = 5.24 cm, Z_{ex} = 3481 cm^3,
D/T = 41.9

The true value of p_{bc} for the trial section can now be obtained from Table 3b (Appendix).

Span ABCD

As there is a lateral support at B, the effective length is the length of the part-span BD. However, as the load at C is applied to the compression flange, without lateral support, the effective length must be increased by 20%, i.e.

l = 6000 x 1.2 = 7200 mm

Slenderness ratio l/r_y = 7200/(5.24 x 10) = 137.4. The parameters for entry into Table 3b are therefore l/r_y = 137.4, D/T = 41.9.

Since these parameters do not coincide exactly with the rows and columns of the table some designers save time, and obtain a conservative value for p_{bc}, by using the lowest value of p_{bc}, i.e for l/r_y = 140 and D/T = 50, giving p_{bc} = 142 N/mm^2. More accurately, the procedure is to interpolate linearly, first between rows and then between columns, or vice versa, which gives p_{bc} = 146 N/mm^2.

Moment of resistance = Z_{ex} x p_{bc}
                     = 3481 x 146/10^3 = 508 kNm

Since this is less than the maximum bending moment, the section is not suitable. Try a 686 x 254 x 140 UB, for which the section properties are:

D = 683.5 mm,   t = 12.4 mm,   T = 19.0 mm,   r = 15.2 mm,
I_x = 136276 cm^4, r_y = 5.38 cm, Z_{ex} = 3988 cm^3,   D/T = 36.0

Whence l/r_y = 7200/53.8 = 133.8, and from Table 3b, p_{bc} = 151 N/mm^2. Moment of resistance = 3988 x 151/10^3 = 602 kN.m.

This appears to be adequate for span ABCD, but the effect of the self-weight will need to be checked later. First examine the cantilever span DE.

Cantilever DE

By inspection it appears that the section is more than adequate for the cantilever; nevertheless for the purpose of demonstration the complete design procedure will be carried out.

The effective length factor is determined by comparing the support restraints and end conditions with Fig. 2.17, and then entering Table 2.2.

Support restraint:
Continuous and partially restrained against torsion (type C,P).

End condition:
Fully restrained against torsion by the connecting beam (type T).

Entering Table 2.2, the correct support restraint but with a free end (line 5) gives a factor of 2.0. However, as the end condition corresponds with line 2, a further factor of 0.75/0.85 must be applied, thus

$$l = 2000 \times 2.0 \times 0.75/0.85 = 3529 \text{ mm}$$

As this is less than the effective length of span ABCD and the maximum bending moment is also less, it is clear that the section is more than adequate for the cantilever as far as bending stress is concerned.

Self-weight

It is now necessary to check that the moment of resistance of the section is still adequate when the bending moments due to self-weight are added.

Mass of beam = 140 kg/m = $140 \times 9.81/10^3$ = 1.4 kN/m

Consider point D:
Total extra load on beam = 1.4×10 = 14 kN
Total load on cantilever = 1.4×2 = 2.8 kN
Taking moments about A, $R_D = 14 \times 5/8$ = 8.8 kN
Modified reaction R_D = 350 + 8.8 = 359 kN
Maximum shear force (in CD) = 250 + 8.8 - 2.8 = 256 kN

Consider point A:
Additional reaction R_A = 14 - 8.8 = 5.2 kN
Modified reaction R_A = 200 + 5.2 = 205 kN

Consider point C:
Increase in bending moment = $5.2 \times 5 - 1.4 \times 5 \times 2.5$ = 8.5 kNm
Modified M_C = 550 + 8.5 = 559 kNm

This is less than the moment of resistance (602 kNm) and is therefore acceptable.

Design of web

The critical points are the supports A and D, and the load point C.

Allowable average shear stress from BS 449, Table 11 (Appendix)

$$p_q' = 140 \text{ N/mm}^2$$

Shear capacity = $p_q'Dt$

$$= 140 \times 683.5 \times 12.4/10^3 = 1187 \text{ kN}$$

None of the shear forces exceed this value.

Web bearing:
As the method has been demonstrated in Example 2.11, the full working will not be given. Using Equation (2.38), or the Safe Load Tables from the BCSA-Constrado Handbook:

C_1 (end bearing) = 191 kN

The reaction at A exceeds this, but if a welded end plate is used, as illustrated, the shear force is distributed over the whole web section, and the calculation of stiff bearing is not relevant.

C_1 (continuous over bearing) = 191 x 2 = 382 kN

The reaction at D, and the applied load at C, are both less than this value, so no stiff bearing need be specified.

Web buckling:
From Equation (2.40), C_1 (end bearing) = 496 kN
Therefore web buckling is not critical.

Deflections

The deflection of span ABCD is not likely to be critical because the span/depth ratio is less than that of Table 2.5. Strictly the factors in the table are not applicable to concentrated loads, but the cantilever moment at D has the effect of reducing the deflections. Nevertheless the deflections due to non-structural loads would need to be calculated to prove compliance with BS 449. Similar calculations have been demonstrated in Example 2.8. For this example the reader is invited to check that the deflections due to the loads given are:

Mid-span in ABCD   11.3 mm
End of cantilever DE 6.5 mm (upwards)

The upwards deflection in DE is caused by the anti-clockwise rotation of the beam at D.

The maximum allowable deflections (span/360) are: span ABCD 22 mm, cantilever DE 5.6 mm.   The cantilever deflection is therefore slightly high.

<u>Example 2.13   Plastic design of two span beam</u>   The beam shown in Fig. 2.40(a) will be designed with a uniform section, using grade 50 steel.

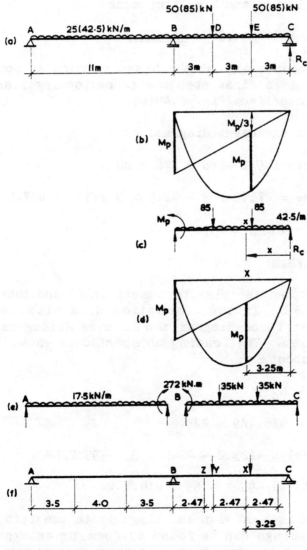

Fig. 2.40. Plastic design of 2 span beam.

LOADS

Using a load factor of 1.7 the ultimate loads shown in brackets are obtained.

PLASTIC MOMENT REQUIRED

To determine the required plastic moment of resistance for the section it is necessary to find the plastic moment at collapse in each span, and take the larger of the two.

Span AB behaves as a propped cantilever, for which $M_p = \gamma wL^2/11.66$

where $\gamma w = 42.5$ kN/m and L = 11 m, hence

$$M_p = 42.5 \times 11^2/11.66 = \underline{441 \text{ kN.m}}$$

For span BC one would expect the hinge to form at point E under the point load, and as a first step the corresponding plastic moment can easily be calculated from Fig. 2.40(b).

For the free bending moment diagram:

Reaction $R_c = 42.5 \times 9/2 + 85 = 276.3$ kN.m

Bending moment $M_E = 276.3 \times 3 - 42.5 \times 3 \times 1.5 = 637.5$ kN

For a hinge at E $4M_p/3 = 637.5$

i.e. $M_p = \underline{478.1 \text{ kN.m}}$

This is greater than the plastic moment in AB and therefore the hinge occurs in span BC. If the assumption of a plastic hinge at E is correct the shear force diagram should pass through zero at E. This can now be checked. The loading on span BC is shown in Fig. 2.40(c). Taking moments about B

$$R_C = 276.3 - M_p/9 \tag{1}$$

$$= 276.3 - 478.1/9 = 223.2$$

Shear at E (right) = -223.2 + 42.5 × 3 = -95.7 kN

Shear at E (left) = -95.7 + 85 = -10.7 kN

As there is no change of sign the hinge is to the left of E. The true position of the hinge can be found by forming an expression for the bending moment at the point of zero shear and equating it to M_p, as follows.

From Fig. 2.40(c), the shear force at X is

$$Q_x = -R_c + 42.5x + 85 = 0$$

Substituting for R_c from (1) and solving for M_p

$$M_p = 382.5 \ (4.5 - x) \tag{2}$$

The bending moment at X is M_p, i.e.

$$M_p = R_c x - 42.5x^2/2 - 85(x - 3)$$

Substitution for R_c from (1) and M_p from (2), and simplification, results in the quadratic equation

$$x^2 + 18x - 69 = 0$$

from which x = <u>3.247 m</u>

Back-substitution into (2) gives M_p = <u>479.1 kN.m</u>

The error arising from the assumption of a hinge at E is therefore very small.

The bending moment diagram at collapse is shown in Fig. 2.40(d). The bending moments in span AB are less than M_p and the span will not collapse.

<u>CHOICE OF SECTION</u>

It is now necessary to select a section with a moment of resistance equal to 479.1 kN.m or greater.

Assuming grade 50 steel, the yield stress (Table 2.6) is 355 N/mm^2 for material not exceeding 16 mm in thickness, so the plastic modulus required is

$$Z_p = 479.1 \times 10^3/355 = 1350 \ cm^3$$

Try 457 x 191 x 67 U.B. (Z_p = 1471). This beam has the same weight as the 457 x 152 x 67 U.B. (Z_p = 1441) and would appear to be a better choice. The section dimensions and properties are:

D = 453.6 mm, B = 189.9 mm, t = 8.5 mm, T = 12.7 mm, r = 10.2 mm, I_x = 29401 cm^4, r_y = 4.12 cm, D/T = 35.7, Z_{ex} = 1296 cm^3.

Checks:

Thickness of material < 16 mm.   Flange width/thickness (B/T) ratio = 189.9/12.7 = 14.95; this is less than the limiting value from Table 2.5 (15.25), so the section is acceptable for plastic design.

SHEAR

Average shear stress $f_{qp} = Q/d_w t$
The maximum shear force is at B in span BC

$$Q_B = 276.3 + M_p/9 = 276.3 + 479.1/9 = 330 \text{ kN}$$

$$d_w = D - 2T = 453.6 - 2 \times 12.7 = 428.2 \text{ mm}$$

$$f_{qp} = 330 \times 10^3/(428.2 \times 8.5) = \underline{90.7 \text{ N/mm}^2}$$

The overall depth/flange width (D/B) ratio = 453.6/189.9 = 2.39.   It is therefore not necessary to consider a reduction of the plastic moment of resistance unless f_{qp} exceeds $Y_s/3$ (Section 2.10.7), i.e. 355/3 = $\underline{118.3 \text{ N/mm}^2}$.   Reduction due to shear is therefore negligible.

DEFLECTIONS

For the sake of this example it will be assumed that the non-structural loads are 70 percent of the loads given, i.e. a uniformly distributed load of 17.5 kN/m and point loads of 35 kN.   The moments at the supports will be determined by the 'moment distribution' method of analysis.

The formulae for fixed end moments are:

for the u.d. load: FEM = $wL^2/12$
for a single point load: FEM (left) = $-Wb^2a/L^2$

where, for the left hand load: a = 3 m, b = 6 m
       for the right hand load: a = 6 m, b = 3 m

Both beams are symmetrically loaded

Span AB:
       FEM = $17.5 \times 11^2/12 = 176 \text{ kN.m}$

Span BC:
       FEM = $17.5 \times 9^2/12 + 35 (6^2 \times 3 + 3^2 \times 6)/9^2 = 188 \text{ kN.m}$

Distribution factors are:
       Span AB 9/20; span BC 11/20, giving the following table:

	A	9/20 B	11/20 C	D
FEM	−176	+176	−188	+188
Release ends	+176			−188
Carry over		+88	−94	
Net FEM	0	+264	−282	0
Balance		+8	+10	
	0	+272	−272	0

For each span the loadings are as shown in Fig. 2.40(e). The deflections are obtained by the formulae from Fig. 2.21, thus

$$I = 29401 \text{ cm}^4$$

$$EI = 210 \times 29401 \times 10^4 = 61742 \times 10^6 \text{ kN.mm}^2$$

Span AB:

$$wL = 17.5 \times 11 = 192.5 \text{ kN, } M_2 = 272 \text{ kN.m}$$

$$(\text{centre}) = 5 \times 192.5 \times (11000)^3/(384EI)$$

$$- 272 \times 10^3 \times (11000)^2/(16EI)$$

$$= 54.0 - 33.3 = \underline{20.7 \text{ mm}}$$

$$\text{Span}/360 = 11000/360 = \underline{30.6 \text{ mm}}$$

Span BC:

$$wL = 17.5 \times 9 = 157.5 \text{ kN, } M_1 = -272 \text{ kN.m, } W = 35 \text{ kN}$$

$$(\text{centre}) = 5 \times 157.5 \times (9000)^3/(384EI)$$

$$- 272 \times 10^3 \times (9000)^2/(16EI)$$

$$+ 2(1 - 4 \times (1/3)^3) \times 35 \times (9000)^3/(48EI)$$

$$= 24.2 - 22.3 + 14.7 = \underline{16.6 \text{ mm}}$$

$$\text{Span}/360 = 9000/360 = \underline{25.0 \text{ mm}}$$

The deflections are not therefore excessive.

DESIGN OF WEB

The design is completed by designing the web elastically using the full working load, thus.

Centre support:
The support moment at 70 percent of full load = 272 kNm (from moment distribution table). By simple proportion the moment at full load = 272/0.7 = 389 kNm at full load.

Shear at B in span AB = 25 x 11/2 + 389/11 = 173 kN
Shear at B in span BC = 25 x 9/2 + 50 + 389/9 = <u>206 kN</u>

Reaction at B = 173 + 206 = <u>379 kN</u>

Using the same procedure as in the previous example the following results are obtained:

Shear capacity of web = 540 kN, which exceeds the maximum shear at B and is therefore acceptable.

For the bearing and buckling values at B, the beam is continuous over the bearing and the values of C_1 will be twice those obtained from Equations (2.38) and (2.40), i.e.

C_1 = 2 x 87.7 = 175.4 kN
C_2 = 2.21 kN

The minimum length of stiff bearing required for web bearing = (379 - 175.4)/2.21 = 92 mm. This is well within what would normally be provided and is acceptable.

Web buckling

C_1 = 2 x 236 = 472 kN > 379 kN therefore satisfactory.

End supports

Reaction at A = 25 x 11/2 - 389/11 = 102 kN

Reaction at C = 25 x 9/2 + 50 - 389/9 = 119 kN

Design for 119 kN in bearing at each end

Minimum length of stiff bearing = (119 - 87.7)/2.21 = 14 mm which can easily be provided with normal connections.

Concentrated load points

The concentrated loads are less than 175.4 kN, so stiff bearing does not need to be designed.

LATERAL SUPPORTS

There will need to be lateral supports at B and at 3.25 m from C where the plastic hinge occurs in the span.   In the region of the hinges $1/r_y$(max) = 60 (Section 2.10.5) and hence $l = 60 \times 4.12/100 = 2.47$ m. Referring to Fig. 2.41(f) there would need to be lateral supports at points B, X, Y, and Z in span BC.   In the elastic parts of the span the spacing of the supports would be greater, and clearly need not be considered in this case.   Supports at points B and X are essential. Points Y and Z could be moved to give a more even distribution; provided the spacing of 2.47 m is not exceeded.   In span AB there is no plastic hinge, but some plastic deformation could occur.   Assume that the point of zero shear occurs at distance x from A.

Reaction $R_A = 42.5 \times 11/2 - 479.1/11 = 190.2$ kN.m

For zero shear x = 190.2/42.5 = 4.48 m

Maximum bending moment:

M(max) = $190.2 \times 4.48 - 42.5 \times (4.48)^2/2 = \underline{425.6 \text{ kN.m}}$

Max. stress = $425.6 \times 10^3/1296 = 328$ N/mm^2, i.e. elastic.

Dividing by load factor, 328/1.7 = $\underline{193 \text{ N/mm}^2}$

D/T ratio for the beam = 35.7, and from BS 449, Table 3(b) (Appendix), the corresponding value of $1/r_y$ is approximately 105.   Hence the maximum spacing of lateral supports = $105r_y = 105 \times 4.12/100 = 4.326$ m.   For a span of 11 m a practical arrangement, assuming lateral supports at A and B, would be two intermediate supports as shown in Fig. 2.40(f).   These could be provided by bracing if the supports naturally occurring in the design were insufficient.

Having regard to the economy of the design it is an interesting excercise to find the section that would be required for the elastic design - assuming maximum allowable bending stress, i.e. full lateral support.

Maximum bending moment (at B) = 389 kN.m

p_{bt} from Table 2 = 230 N/mm^2

Z_{ex} required = $389 \times 10^3/230 = 1691$ cm^3

The lightest suitable section is a 533 x 210 x 82 UB (Z_e = 1798 cm^3)

The plastic design therefore gives a saving in mass of steel of (82 – 67) x 20 = 300 kg per beam.

Example 2.14   Plastic design of three span beam

Using the plastic method, design the three span beam in Fig. 2.41(a), using grade 50 steel:

    (i) with a uniform section
    (ii) with reinforced end spans.

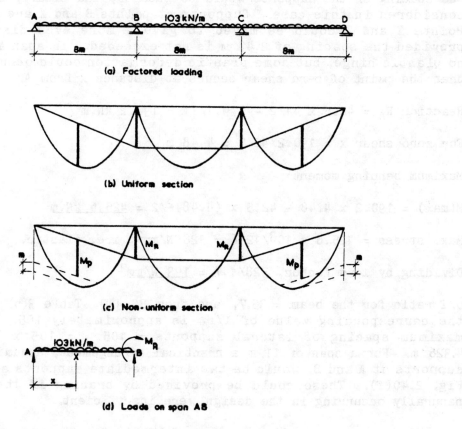

(a) Factored loading

(b) Uniform section

(c) Non-uniform section

(d) Loads on span AB

Fig. 2·41. Plastic design of 3 span beam.

UNIFORM SECTION

With a uniform section hinges form in the end spans and the centre span remains elastic, as indicated by the bending moment diagram in Fig. 2.41(b).

M_p(required) = $\gamma wL^2/11.66$

i.e. M_C = 103 x $8^2/11.66$ = <u>565 kNm</u>

$Z_p = M_p/Y_s$ = 565 x $10^3/355$ = 1592 cm^3

Try a 457 x 191 x 74 UB (Z_p = 1657)

i.e. M_p(provided) = 1657 x $355/10^3$ = <u>588 kNm</u>

The other relevant section properties are:
D = 457.2 mm,   B = 190.5 mm,   t = 9.1 mm,   T = 14.5 mm.

Check: B/T = 190.5/14.5 = 13.1
This is less than the limiting value (15.25) from Table 2.5, so the section is acceptable for plastic design.

Maximum shear force (to left of B):

Q = $\gamma wL/2 + M_p/L$, where $\gamma wL/2$ = 103 x 8/2 = <u>412 kN</u>

  = 412 + 565/8 = 483 kN

d_W = D - 2T = 457.2 - 2 x 14.5 = 428.2 mm

f_{qp} = $Q/(d_W t)$ = 483 x $10^3/(428.2$ x 9.1) = 124 N/mm^2

Since $f_{qp} > Y_s/3$ reduction of M_p due to shear should be taken into account (Section 2.10.7).

Reduction = $Z_{pw}(Y_s - Y_e)$

Z_{pw} = $td_W^2/4$ = 9.1 x $(428.2)^2/4$ = 417000 mm^3

$Y_e^2 = Y_s^2 - 3(f_{qp})^2$

    = $(355)^2 - 3(124)^2$ = 79900

Hence Y_e = 283 N/mm^2

Reduction = 417000$(355 - 283)/10^6$ = 30 kNm

i.e. M_R = 588 - 30 = <u>558 kNm</u>

This is less than the calculated value of plastic moment (M_c = 565kNm). However if Equation (2.68) is satisfied the section is adequate after re-distribution of moments, i.e. if

    $M_R \geq M_c - (M_p - M_c)/0.4142$

i.e. $M_R \geq 565 - (588 - 565)/0.4142$

i.e. if $558 \geq 509$, which is true, so the section is adequate.

Since the plastic moment at B has been reduced, it is in general necessary to check that the bending moment in the centre of span BC does not exceed the plastic moment of resistance of the section, i.e. that the free bending moment in BC does not exceed $M_p + M_B$, where M_B is the plastic moment at B, adjusted in accordance with Equation (2.69). In this case, with uniform spans and loading, the section is clearly adequate.

NON-UNIFORM SECTION

An alternative design is to choose a basic section capable of providing the required plastic moment for the middle span, and reinforce it with welded flange plates in the end spans.

Centre span

$$M_p(\text{required}) = \gamma w L^2/16$$

$$= 103 \times 8^2/16 = \underline{412 \text{ kNm}}$$

$$Z_p = 412 \times 10^3/355 = 1161 \text{ cm}^3$$

Try a 457 x 152 x 60 UB (Z_p = 1284 cm^2)

The other relevant section properties are
D = 454.7 mm,   B = 152.9 mm,   t = 8.0 mm,   T = 13.3 mm,
Z_{ex} = 1120 cm^3

Check: B/T = 152.9/13.3 = 11.5
This is less than 15.25 and is therefore acceptable.

Support moments

Flange plates will be designed so that plastic hinges form in the end spans. At the internal supports the plastic moment will be the plastic moment of resistance of the basic section, reduced to take acount of shear stress, if necessary.

$$M_p(\text{support}) = Y_s Z_p = 355 \times 1284/10^3 = \underline{456 \text{ kNm}}$$

Maximum shear force (at B in AB)

$$= \gamma w L/2 + M_p/8 = 412 + 456/8 = 469 \text{ kN}$$

$d_w = 454.7 - 2 \times 13.3 = 428.1$ mm

Average web shear $= 469/(428.1 \times 8.0) = 137$ N/mm^2

This is greater than $Y_s/3$, so shear must be considered.

$Z_{pw} = 8.0 \times (428.1)^2/4 = 366500$ mm^3

$Y_e^2 = 355^2 - 3(137)^2 = 69720$

$Y_e = 264$ N/mm^2

Reduction in $M_p = 366500(355 - 264)/10^6 = \underline{33 \text{ kNm}}$

$M_R = 456 - 33 = 423$ kNm

As this is greater than the M_p required for the centre span (412 kNm), the basic section is still adequate.

End spans

The bending moment diagram for the beam is as shown in Fig. 2.41(c). The maximum bending moment in the end span is greater than the plastic moment of resistance M_p of the basic section and it is necessary to provide reinforcement to give an additional moment m, as shown by the dotted lines. First the maximum bending moment in the end span will be determined, using the fact that maximum bending moment coincides with zero shear force. Consider Fig. 2.41(d) which shows the loading on span AB.

Shear force at x:

$R_A = \gamma wL/2 - M_R/L$

$\quad = 412 - 423/8 = 359$ kN

$Q_x = 359 - 103x = 0$ for maximum moment

i.e. $x = 359/103 = 3.487$ m

$M_x = R_A x - \gamma wx^2/2$

Putting $x = 3.487$

$M_x(\text{max}) = 359 \times 3.487 - 103 \times (3.487)^2/2 = \underline{626 \text{ kNm}}$

Since the shear force is zero, the plastic moment provided by the basic section will not be reduced at X, so the flange plates must provide m = 626 - 456 = <u>170 kNm</u>

The plastic modulus provided by two flange plates is equal to the area of one plate multiplied by the distance between their centroids. Computation is saved by using the depth D of the beam, which gives a conservative value for the plate area A_p. Hence:

$$m = A_p D Y_s$$

from which

$$A_p = 170 \times 10^6/(454.7 \times 355) = 1053 \text{ mm}^2$$

Assuming the plate has the same width as the beam, (it would most likely be wider to allow for welding) the flange plate thickness is given by

$$t_p = A_p/B = 1053/152.9 = 6.89 \text{ mm}$$

Use 8 mm plates as 7 mm plates are not available.

<u>Curtailment of flange plates</u>

If yielding in the basic section is to be avoided altogether the plates will need to be curtailed at the points where the bending moment reduces to the yield moment, i.e. when $M_x = M_y$

$$M_y = Y_s Z_{ex} = 355 \times 1120/10^3 = 397 \text{ kNm}$$

$$M_x = R_A x - \gamma w x^2/2 = M_y$$

Substituting in the values of R_A and γw, the following quadratic equation results:

$$x^2 - 6.97x + 7.71 = 0$$

from which x = 1.38 m and 5.60 m.

These are the theoretical curtailment points.   It would be necessary to extend the flange plate further, to allow the weld to transfer the full force in the plate, as described in Chapter 7.

The design is completed by checking the web for bearing and buckling at working load, and checking the deflections due to the non-structural loads.

REFERENCES

Astill, A.W. et al. (November 1980) - Web buckling of steel I beams, CIRIA Tech. Note 102.

BCSA and Constrado, (1980) - Structural steelwork handbook, properties and safe load tables, The British Constructional Steelwork Association Limited, and Constructional Steel Research and Development Organisation.

Constrado, (1972) -The Steel Designers' Manual, Fourth Edition, Crosby Lockwood Staples.

Horne, M.R., (1971) - Plastic Theory of Structures, Nelson.

Megson, T.H.G., (1980) - Strength of Materials for Civil Engineers, Nelson.

Morris, L.J. and Randall, A.L., (1979) - Plastic Design, Constrado.

Ryder, G.H., (1969) - Strength of Materials, Cleaver Hume, S.I. Edition.

Zbirohowski-Koscia, K., (1967) - Thin Walled Beams, Crosby Lockwood and Sons Limited, London.

3   Axially Loaded Members

3.1   GENERAL PRINCIPLES

3.1.1   <u>Introduction</u>   Axially loaded members are referred to by several different names depending on their position and function in a structure. Tension members are generally called 'ties', or 'tie beams' if they also carry lateral loads. 'Strut', 'stanchion' and 'column' are the terms most commonly applied to compression members. Strut is a general term, but stanchion and column are reserved for upright members, in particular those carrying the vertical loads from the roof and floors of a building. Compression members carrying lateral loads are described as 'beam-columns' or 'laterally loaded struts'.

The limiting stress at ultimate load is the yield stress Y_s, as defined in Chapter 1, which is virtually the same in tension and compression. The behaviour of members, however, is significantly different in tension and compression, and is the main factor influencing the specification of allowable stresses at working load. Also taken into account are the small, unavoidable departures from pure axial loading resulting from fabrication and erection tolerances, and the random lack of straightness arising from mill tolerances for rolled sections. For convenience allowable stresses are expressed as average stresses. This enables them to be compared directly with the average stresses obtained by dividing the axial forces at working load by the cross sectional areas of members, and avoids the problem of dealing with non-uniform stress distribution.

3.1.2   <u>Effect of holes - net cross section</u>   When a tension member is loaded progressively to failure, plastic deformation occurs first at sections which have been reduced by holes drilled for fasteners, or provided for some other purpose. At these sections further loading causes strain hardening, and stresses increase beyond the yield point. Then, as the load continues to increase, either the stress in the bulk of the member reaches the yield point and a ductile failure occurs, or the member fails suddenly at a reduced section, before the rest of the member has yielded. These two alternative methods of failure have been considered in proposals for a limit state code of practice. For design to BS449 tensile stresses are calculated on a reduced or 'net' section taken through a transverse line of holes, ignoring the effects of strain hardening.

For a section taken through the diameter of a single hole, the net area is determined by deducting the sectional area of the hole from the gross sectional area.   For black bolts in clearance holes the following areas in mm^2 should be deducted.

For bolts up to and including 24 mm diameter:

$$\text{Deducted area} = t(d + 2) \tag{3.1}$$

For bolts greater than 24 mm diameter:

$$\text{Deducted area} = t(d + 3) \tag{3.2}$$

where t = the thickness of the holed section or plate in mm.
　　　 d = the nominal diameter of the bolt in mm.

The same formulae can be used for rivets, although their use has largely been discontinued.   For countersunk holes the deducted area should include the section through the countersunk portion.   When there are several rows of holes, the area to be deducted is:

a)   the sum of the sectional areas of the maximum number of holes in any line at right angles to the direction of stress, or

b)   if the holes are staggered, the greater of that given by (a) above or the sum of the sectional areas of all the holes in any zig-zag line extending progressively across the member or part of the member, less an area equal to $s^2 t/(4g)$ for each gauge space in the chain of holes,

where s = the staggered pitch; i.e. the distance, measured parallel to the direction of stress in the member, between centres of holes in consecutive lines.
　　　 t = the thickness of the holed material.
　　　 g = the gauge; i.e. the distance, measured at right angles to the direction of stress, between centres of holes in consecutive lines.

Compression members usually fail by elastic buckling at average stresses below the yield stress, in which case the presence of holes is not likely to have much effect on the load carrying capacity of the member.   It is possible in short stocky members for yielding to occur at a reduced section before the onset of buckling, but if fasteners are present in the holes the amount of plastic deformation that can occur, before the bolts themselves are loaded in bearing, is limited to the clearance in the holes.   In the case of rivets the holes are completely filled during erection by the expansion of the hot rivets into the holes. BS 449 specifies that, in general, compressive stresses should be calculated on the gross sectional area, thus ignoring completely the effect of holes.   However, in a future limit state code it is likely that some limit of the ultimate load capacity will be required, to prevent excessive plastic deformation at reduced sections.

Example 3.1   For the plate shown in Fig. 3.1, calculate the net cross sectional area.   The plate is 20 mm thick and contains four lines of staggered holes drilled for 24 mm diameter black bolts.

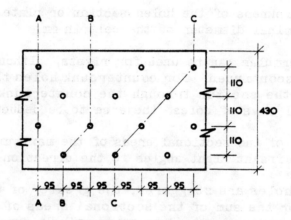

Fig. 3.1.   Plate with staggered holes.

In this case t = 20 mm,   s = 95 mm, and g = 110 mm
Diameter of holes = 24 + 2 = 26 mm
Gross cross sectional area = 20 x 430 = 8600 mm^2

As the bolts are staggered, any line such as AA, at right angles to the direction of stress, contains only two holes.   There are two feasible zig-zag lines, namely BB which contains three holes with one gauge space, and line CC which contains four bolts with three gauge spaces.   The area to be deducted is the greatest of:

Line AA: 2 x 20 x 26 = 1040 mm^2
Line BB: 3 x 20 x 26 - 1 x 95^2 x 20/(4 x 110) = 1150 mm^2
Line BC: 4 x 20 x 26 - 3 x 95^2 x 20/(4 x 110) = 849 mm^2

The net area is therefore given by line BB, i.e.

8600 - 1150 = 7450 mm^2

3.2 TENSION MEMBERS

3.2.1 <u>Allowable stress</u> Under axial tension a member tends to remain straight, or, if there is a small initial curvature, to straighten out as the load is increased. Small, unavoidable eccentricities of loading have little effect on the tensile capacity of a member at ultimate load. For each grade of steel there is a basic allowable stress which is set out in BS 449, Table 19 (Appendix). The reduced stresses correspond with the lower yield stress for thicker sections. The factor of safety against yielding varies slightly between steel grades, but has an average value of about 1.63.

All standard rolled sections, other than Universal columns, have thicknesses less than 40 mm and so qualify for the higher stresses. Some of the heavier Universal column sections have flange thicknesses greater than 40 mm, and the two heaviest sections have flanges thicker than 65 mm.

3.2.2 <u>Eccentric connections</u> When a member is connected to one side of a gusset plate, bending moments are introduced in addition to the direct axial force. In a tension member these moments produce lateral deflections which reduce the eccentricity of loading near the middle of the member. Thus under increasing load the bending stresses become concentrated more towards the ends. In angles connected by one leg the principal sectional axes are inclined to the plane containing the bending moment. Secondary deflections therefore occur normal to the plane of bending and, because of the restraints provided by the gusset plates, twisting also takes place.

For angles and tees experiments conducted by Nelson, for the British Constructional Steelwork Association in 1953, demonstrated that the above effects could be compensated for in design by reducing the net cross sectional area of the member. BS 449, Cl. 42 gives two empirical rules for calculation of the reduced areas. These can be expressed as follows:

1. For a single angle connected by one leg only:

$$\text{net area} = a_1[1 + 3a_2/(3a_1 + a_2)] \tag{3.3}$$

2. For a tee connected by its table, or a pair of angles connected back-to-back and acting as a compound tee

$$\text{net area} = a_1[1 + 5a_2/(5a_1 + a_2)] \tag{3.4}$$

where a_1 is the net sectional area of the connected leg
      a_2 is the sectional area of the unconnected leg.

In order to avoid overlapping areas when calculating a_1 and a_2, half the thickness of the angle, or tee, is subtracted from the leg length. Thus for the angle shown in Fig. 3.2, connected by its longer leg, the areas are given by:

$$a_1 = t(A - t/2) \text{ less sectional area of holes}$$
$$a_2 = t(B - t/2) \tag{3.5}$$

For other types of eccentric connection the bending stresses must be calculated and combined with the axial tensile stress, as described in Section 3.4.3. In many instances the true value of the eccentricity cannot be determined with certainty and in such cases it is common practice to take the perpendicular distance between the centroidal axes of the connected members.

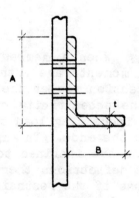

Fig. 3.2.   Eccentrically connected angle tie.

3.3   COMPRESSION MEMBERS

3.3.1   Allowable stress   An ideal pin-ended strut becomes elastically unstable and buckles at the Euler critical stress, as defined by the well known formula:

$$C_O = \pi^2 E/(l/r)^2 \tag{3.6}$$

where C_O = the Euler critical stress
E = Young's modulus
l = the length
r = the radius of gyration.

The term l/r is the slenderness ratio of the strut. The ideal conditions assumed by Euler, namely homogeneity and istotropy of the material, pure axial loading, and perfect straightness, cannot be relied upon to occur in practice, so real struts tend to buckle at stresses below those predicted by Euler. Small unavoidable eccentricities of loading and lack of initial straightness can be simulated mathematically by assuming an initial curvature which produces a small central deflection v, empirically defined, as shown by the dotted line in Fig. 3.3. When an axial load P is applied the member buckles and the deflection increases to V, as shown.

Fig. 3.3. Strut with initial curvature.

The maximum compressive stress, p, in the extreme fibres on the concave side of the strut, is then given by:

$$p = P/A + Pv/Z_e \qquad (3.7)$$

where A = the cross sectional area
Z_e = the elastic section modulus

The critical buckling load is reached when $p = Y_s$, i.e. when yielding commences in the extreme fibres of the strut. By adopting a simple sinusoidal function for the initial curvature, it can be shown that:

$$p_{cr} = \beta - (\beta^2 - Y_s C_o)^{1/2} \qquad (3.8)$$

where p_{cr} = the critical average compressive stress, obtained by dividing the critical buckling load by the gross cross sectional area of the strut
β = $[Y_s - C_o(\eta + 1)]/2$
Y_s = the yield stress as defined in Chapter 1
η = an empirical function representing the initial curvature
C_o = the Euler critical stress obtained from Equation (3.6), using E = 210 000 N/mm^2

Equation (3.8) is known as the Perry-Robertson formula and is the formula given in Appendix B of BS 449. The value of the function η originally obtained experimentally by Perry (1925) is given by

$$\eta = 0.003(l/r) \qquad (3.9)$$

However, the value currently adopted by BS 449 is based on experimental work by Duthiel in France and was suggested by Godfrey (1962) to give more economical designs. It is given by

$$\eta = 0.3(1/100r)^2 \qquad\qquad (3.10)$$

where l = the effective length of the strut. For a strut with both
ends pinned, l is equal to the actual length. For other
end restraints it is defined in Section 3.3.4
      r = the radius of gyration

As the lowest value of the central deflection v in Equation (3.7) is
limited by the initial curvature, bending moments are generated as
soon as an axial load is applied; and buckling starts immediately.
There is, therefore, no condition of elastic instability as defined by
Euler; and the average compressive stress can never reach the Euler
critical stress for a strut of finite length. Nevertheless the
failure of a compression member by buckling is sudden when compared
with the ductile failure of a tension member.

The allowable average compressive stress p_c at working load is
obtained by dividing the critical stress from Equation (3.8) by a load
factor of 1.7, thus:

$$p_c = p_{cr}/1.7 \qquad\qquad (3.11)$$

For a particular grade of steel, the allowable stress is a
function of the slenderness ratio only, and is tabulated against
slenderness ratio in BS 449, Tables 17 (a), (b) and (c) - (Appendix).
Values of p_c for slenderness ratios between zero and 30 are obtained
by linear interpolation, using values of 155 N/mm^2, 215 N/mm^2, and 265
N/mm^2 for steel grades 43, 50 and 55 respectively when l/r = 0. This
modification gives slightly greater economy than Perry-Roberston at
low slenderness ratios, and for a strut with a slenderness ratio of
zero makes the allowable stress in compression equal to the allowable
stress in tension. Special provisions are made for thicker sections,
either by placing an upper limit on the stress obtained from Table 17,
or, for grade 50 steel, by using Equation (3.8) with the lower value
of Y_s.

The relationship between allowable stress p_c, from Table 17, and
slenderness ratio is shown graphically in Fig. 3.4 which also shows
the Euler critical stress divided by 1.7, for comparison. The figure
shows that at lower values of slenderness ratio, where the allowable
stress is influenced considerably by the yield stress, there would be
a considerable weight advantage from using the higher grades of steel.
This advantage is progressively reduced as the slenderness ratio
increases, because of the increasing influence on p_c of Young's
modulus, which is a constant for all grades of steel.

Current European practice is to provide three non-dimensional
curves for determining the critical stress. Each curve represents a
group of sections and takes into account the effects of strain

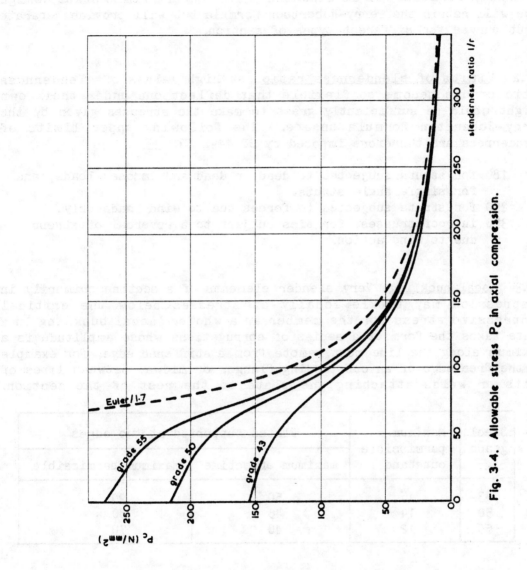

Fig. 3-4. Allowable stress Pc in axial compression.

hardening, section geometry, the axis of buckling, and the pattern of residual stresses generally associated with the sections in the group. These factors are not explicitly taken into account by BS 449, which only specifies a single general expression for η in the Perry-Roberston formula.   It is expected that a future limit state design code will retain the Perry-Robertson formula but will provide separate strut curves for different types of section.

3.3.2 **Limits of slenderness ratio**   At high values of slenderness ratio struts become so flexible that deflections under their own weight would be sufficiently great to make the stresses given by the Perry-Roberston formula unsafe.   The following upper limits of slenderness are therefore imposed by BS 449, Cl. 33.

> 180 for struts subjected to dead or dead and imposed loads, and for single angle struts.
> 250 for struts subjected to forces due to wind loads only.
> 350 in roof trusses, for ties subject to a reversal of stress due to wind action.

3.3.3 **Local buckling**   Very slender elements of a section primarily in compression may buckle locally at stresses below the critical compressive stress for the member as a whole.   Local buckling in a plate takes the form of a series of corrugations whose amplitude is a maximum along the line most remote from a stiffened edge, for example at the free edge of an outstanding flange, or midway between lines of bolts or welds attaching the plate to the rest of the section.

Steel grade	Maximum permissible outstand	Plates supported at two edges	
		maximum effective	maximum permissible
43	16	50	90
50	14	45	80
55	12	40	70

Table 3.1   Slenderness limits for elements of strut sections

Buckled parts of the section are relieved of some of the load that they would normally support, with the result that the compressive stress on other parts of the section is increased, thus reducing the ultimate load capacity of the member.   It is preferable to design sections in which local buckling does not occur, by ensuring that the slenderness ratios of plates, defined as the ratio of width to thickness, do not exceed the limits prescribed by BS 449, Cl. 32(a), and given in columns 2 and 3 of Table 3.1 above. For plates supported at two edges, greater slenderness ratios than those prescribed in column 3 are permissible, up to the maximum values given in column 4, but the section then becomes less effective.

Consider, for example, the section shown in Fig. 3.5, assuming steel grade 50. In welded construction the critical dimensions are measured from the faces of the connected plates, as shown.

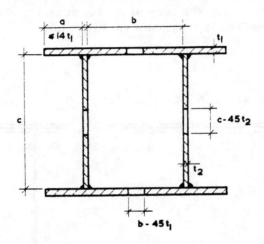

Fig. 3.5.   Effective section in grade 50 steel.

Based on the outstands, the minimum permissible thickness of the flange plates is obtained from column 2 of Table 3.1, i.e. $t_1 \geq a/14$. The maximum clear width of the flange plate between the webs, without providing additional stiffeners, is obtained from column 4 i.e. $b_{max} = 80t_1$. Similarly $c_{max} = 80t_2$, or alternatively, $t_2(min) = c/80$.

Column 3 of the table gives the maximum slenderness for a fully effective element. Hence if b exceeds $45t_1$ it is assumed that the central portion of the flange plate is affected by local buckling and is no longer effective in resisting compression. The effective part of the flange is shown hatched. The width of the buckled zone is therefore assumed to be $b - 45t_1$. A similar conclusion can be drawn for the web plate if c exceeds $45t_2$, as shown in the figure.

For calculation of the sectional area and radius of gyration, only the effective (hatched) parts of the section are taken into account. The strut is then designed in the usual way, using these properties. Other sectional properties are assumed to be unaffected.

If it should be necessary to calculate bending stresses on an effective section, the method is described in BS 449, Cl. 27(a) for

the flanges of plate girders. In a compression member the maximum
bending stress in the compression flange is initially calculated, as
for a beam, using the whole cross section. This stress is then
increased in proportion to the ratio of gross sectional area/effective
sectional area, of the compression flange of the beam, where the
effective flange area is calculated from the effective width, as
described above. The compressive stresses due to axial force and
bending are then combined as described in Section 3.4.3.

   In bolted construction plate widths are measured between adjacent
lines of fasteners; outstands are measured from the outermost
fastener. Some examples are given in Fig. 3.6.

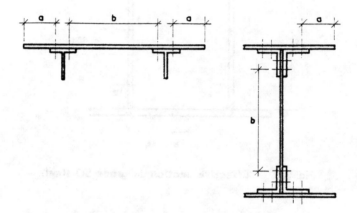

Fig. 3.6.   Critical dimensions in plate elements.

   The method described above for dealing with local buckling is
called the 'effective section method'. It tends to be conservative,
except at low values of slenderness ratio, and becomes increasingly so
as the slenderness ratio of the strut increases. It is probable that
when BS 449 is replaced by a new code of practice, designers will be
given the opportunity to use other available methods, should they
prove to be more economical. The results of applying alternative
methods are compared by Dwight (1978).

3.3.4 Effective length The Perry-Robertson formula is based on the
assumption that the ends of the strut are provided with frictionless
joints which are prevented from lateral movement, but which can rotate
freely about any axis of the section. It can be shown however, that a
variety of end conditions can be simulated by replacing the actual
length of the strut by an effective length. It is convenient to
consider four combinations of ideal end conditions:

a) Both ends rigidly built in.
b) One end rigidly built in, the other end pinned.
c) Both ends pinned.
d) One end built in, but the other end free to rotate and move
    laterally.

These conditions are shown diagrammatically in Fig. 3.7, where
the theoretical values of the effective length, denoted by l, are
given.   The effective length is the length of an analogous pin-ended
strut in which, for cases (a) and (b) the points of contraflexure in
the buckled shape take the place of the pinned ends.   Case (c) is the
ideal case for which the formula was derived, in which the effective
length is equal to the actual length.   In case (d) the built-in end
corresponds with the mid-point of the analogous strut.

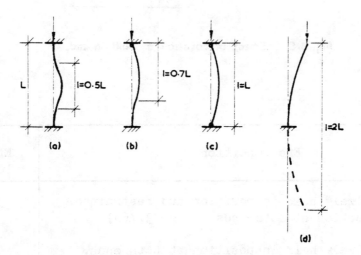

Fig. 3.7.   Theoretical effective lengths of struts.

In practice the full rigidity of the built-in condition is never
achieved and some rotation always occurs due to the flexibility of the
connection or the part of the structure to which the strut is
connected.   Only a very small amount of rotation is necessary to
release a built-in end and convert it to the pinned condition.   On the
other hand small translational movements are not so significant and
are more easily prevented by fairly light supporting members.

BS 449, Cl. 31(a) allows for some rotation and specifies
correspondingly longer effective lengths for the first two conditions
of Fig. 3.7, and includes a further case, less severe than case (d).
The effect of rotation at the built-in end in case (b) is illustrated
in Fig. 3.8.   The phrase 'restrained in direction' is used to describe
an end at which rotation is restricted by connection to relatively

stiff supporting members or a large foundation. An end which is prevented from lateral or transverse movement is described as being 'effectively held in position'.

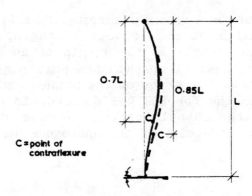

C = point of contraflexure

Fig. 3·8. Effect of rotation at built-in end.

End condition	Effective length factor
Effectively held in position and restrained in direction at both ends - Fig. 3.7(a).	0.7
Effectively held in position at both ends, but restrained in direction at one end only - Fig. 3.7(b).	0.85
Effectively held in position at both ends, but not restrained in direction - Fig. 3.7(c).	1.0
Effectively held in position and restrained in direction at one end, and at the other end partially restrained in direction but not held in position.	1.5
Effectively held in position and restrained in direction at one end, but not held in position or restrained in direction at the other end - Fig. 3.7(d).	2.0

Table 3.2 Effective length factors for real struts

Table 3.2 above gives the factors by which the actual length must be multiplied to produce the effective length, the actual length being defined as the distance from centre to centre of intersections with supporting members.

Supporting members, at the ends of a strut, usually lie in the direction of the rectangular axes x-x and y-y of the strut section. The degree of restraint provided by these members may differ, in which case a different value of effective length is obtained for each axis.

Judgement of the effectiveness of end restraints is not always easy and BS 449 gives guidance with beam-to-stanchion connections and with angle struts. These will be discussed later in this chapter. In general the design procedure can be summarised as follows.

1. Evaluate the degree of positional and directional restraint provided at both ends of the strut, considering each axis in turn; then select the appropriate effective length factors from Table 3.2.

2. Calculate both slenderness ratios, using the effective length and radius of gyration appropriate to each axis.

3. Obtain the allowable stress p_c from BS 449, Table 17 (Appendix) using the larger slenderness ratio.

3.3.5 _Angles as struts_ For an angle connected by one leg, the conditions at the ends, other than the direction of the axial force, are the same in compression as in tension. In tests on two mild steel lattice girders composed of angles, Mackey and Wilkinson (1953) used single and double angles as struts with maximum slenderness ratios (based on the radius of gyration about the minor principal axis) in the order of 130. Results of the tests showed that in single angle struts the bending stresses near the end connections were considerably greater when single bolted connections were used, as compared with connections having two bolts in line. The deflections of the trusses, excluding that due to bolt slip, were also increased when single bolts were used. It was also shown that the connections provided almost complete restraint against bending in the plane of the truss, but owing to the low resistance of the gusset plates to bending normal to their own planes, directional restraints in this direction were negligible. It was found that if buckling about the minor principal axis was to be adopted as the criterion for design, the effective length of single angle struts varied, depending on the relative lengths of the connected and outstanding legs. An effective length of 0.8 times the distance between intersections was recommended. In the case of struts consisting of pairs of angles back to back and forming the continuous compression boom, buckling at right angles to the plane of the truss occurred in single curvature between lateral supports, alternately concave and convex in adjacent panels. Effective lengths were therefore equal to the lengths between the intersections of lateral supports.

The specifications of BS 449, Cl. 30(c) for discontinuous angle struts in general are set out in Table 3.3 below, where p_c is the allowable stress in compression obtained from BS 449, Table 17 (Appendix), using the radius of gyration about the minor principal axis as the basis for calculation of slenderness ratio. For double angles the principal axes coincide with the rectangular axes.

Type of strut	Number of bolts in line in end connections	Effective length factor	Allowable stress
Single angle	2	0.85	p_c
Single angle (max. l/r = 180)	1	1.0	$0.8p_c$
Double angle back to back connected to both sides of gusset plate	2	0.7 to 0.85	p_c
Double angle back to back connected to one side only of gusset plate	1 or more in each angle	0.85	p_c

Table 3.3 Design data for discontinuous angle struts

For double angles back to back, the angles must be connected together at intervals as specified in BS 449, Cl. 37. For double angles connected to both sides of a gusset plate it is left to the designer to select a suitable effective length factor between the limits prescribed, having regard to the degree of restraint provided. The rules for bolted joints apply equally to rivets and to equivalent welded joints. Double angles are rarely used now that large tees cut from Universal beams and columns are available. The only advantage of using double angles is the avoidance of eccentricity by connecting to both sides of a gusset plate, but this does not compensate for the additional cost of providing intermediate connections between the component angles.

The design of braced frames and trusses is described in Chapter 6 However it is relevant to observe that the range of effective length factors is given in BS 449, Cl. 43a(ii) as between 0.7 and 1.0. When angles are used, the rules of Cl. 30(c), as depicted in Table 3.3

should also be observed. Continuous rafters are subjected to combined bending and axial stresses and are designed in accordance with Section 3.4.3.

3.3.6   Stanchions Stanchions are vertical members carrying the loads from the roof and floors of a building. In multistorey buildings the stanchions form a grid in plan as shown diagrammatically in Fig. 3.9.

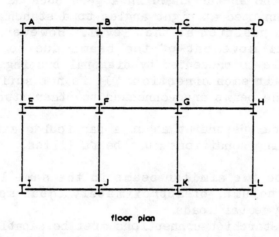

floor plan

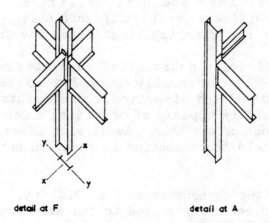

detail at F   detail at A

Fig. 3.9.   Typical beam - stanchion connections.

At each floor level the loads are transmitted to the stanchions through wall and floor beams. In most cases stanchions are continuous throughout their height. However in tall buildings they are spliced at intervals of two or more storeys to facilitate fabrication and erection, and to allow for changes in section. Stanchions are designed as separate elements between floors, bearing in mind the restraints provided by the beams at each floor and roof level and by the bases connecting the stanchions to their foundations.

Assessment of the effective length of a stanchion in a multistorey building is, with one or two exceptions, fairly straightforward provided that the general principles are understood. These will now be discussed.

When stanchions are arranged in a grid such as that that of Fig. 3.9, two beams connected at right angles to a stanchion are sufficient to restrain it in position at that level. However, it is important that longitudinal movement of the beams due to sidesway of the building as a whole is prevented by diagonal bracing or shear walls in at least one bay, in each direction. It is not sufficient just that the far ends of the beams are connected to other stanchions.

For directional restraint about a particular axis to be assumed, all of the following conditions must be fulfilled.

1.   There must be two similar beams in the same line, one on each side of the column, of approximately equal spans and carrying approximately equal loads.
2.   The beam-to-stanchion connections must be capable of transmitting moments; i.e. the connections should be made with top and bottom flange cleats or better.
3.   The beams must have adequate stiffness. Beams carrying substantial floor loads are usually sufficient, but light beams used as ties or for carrying light loads are not.

At foundation level a substantial gussetted base of the type shown in Fig. 4.29(b) is usually regarded as being capable of providing both positional and directional restraints, provided that the foundation itself is capable of providing a degree of restraint commensurate with that of the base. Small slab bases as shown in Fig. 4.20(a) effectively hold the stanchion in position but do not restrain it in direction.

Consider first the connection at F in Fig. 3.9. The detail shows floor beams in two directions, bolted to the stanchion through welded end plates, and of roughly equal spans and loading on either side of the stanchion. At this connection therefore the stanchion is effectively held in position and restrained in direction about both the x-x and y-y axes.

Now consider the corner stanchion at A. The light tie or wall beam between A and B is not sufficiently stiff to provide directional restraint about the y-y axis. The beam from A to E carries floor loads, is adequately stiff, and has a moment connection. However, because it is not balanced by another beam on the other side of the stanchion, there will be a rotation at its end which will be transmitted to the stanchion. The beam does not therefore provide directional restraint about the x-x axis. As there are two beams at right angles the stanchion is effectively held in position.

Applying similar reasoning to other connections, the reader can deduce, for example, that stanchion E is restrained in direction about the x-x axis. With regard to the y-y axis however, the same reasoning can be applied to beam EF as to beam AE above. Stanchion B is not restrained in direction about the x-x axis for the same reason, neither is it restrained in direction about the y-y axis, because, although beams AB and BC are equal in span and loading, they are light wall beams with insufficient stiffness.

Having assessed the restraints at each end of a stanchion, its effective length factor with respect to each axis can be obtained from Table 3.2. In the case of a single stiff beam at each end of the stanchion the factor 1.0 is obtained. However if the beams are identical or nearly so, both ends of the stanchion rotate in the same direction, producing double curvature and reducing the effective length. BS 449 allows a reduction from 1.0L to 0.85L in such cases. The results for the four stanchions just discussed, assuming them to be between identical floors, are set out in Table 3.4, where L is the actual length, taken as the distance from centre to centre of the beams intersecting at each end, in the direction considered. r_x and r_y are the radii of gyration about the x-x and y-y axes respectively.

Stanchion	End restraints			Slenderness ratios	
	Position	Direction x-x	Direction y-y	Axis x-x	Axis y-y
F	Yes	Yes	Yes	$0.7L/r_x$	$0.7L/r_y$
A	Yes	No	No	$0.85L/r_x$	$1.0L/r_y$
E	Yes	Yes	No	$0.7L/r_x$	$0.85L/r_y$
B	Yes	No	No	$0.85L/r_x$	$1.0L/r_y$

Table 3.4 Slenderness ratios of stanchions in Fig. 3.9

As an exercise the reader may deduce the effective length factors for the same stanchions between the first floor and the foundation, assuming:

(a) a substantial gussetted base giving positional and directional restraints.

(b) a small slab base giving positional restraints only.

The results are given in Table 3.5.

Stanchion	Substantial base		Small slab base	
	x-x	y-y	x-x	y-y
F	0.7	0.7	0.85	0.85
A	0.85	0.85	1.0	1.0
E	0.7	0.85	0.85	1.0
B	0.85	0.85	1.0	1.0

Table 3.5   Effective length factors for ground-floor stanchions of Fig. 3.9

In general it is necessary to calculate the slenderness ratio relative to both the x-x and y-y axes of the stanchion. However, if the effective length factors are in the range of 0.7 to 1.0, and the actual length is the same in both cases, l/r_y is greater than l/r_x for all standard UC and UB sections and it is then only necessary to calculate l/r_y.

BS 449, Appendix D gives illustrations showing the slenderness ratio to be used in design for a number of specific combinations of end connections that occur in multistorey structures. The same results can be obtained by application of the general principles just described.

A simple example of a single-storey stanchion is given in Fig. 3.10. In the plane of the diagram the caps of the stanchions are tied together by the roof truss, but are not otherwise held in position or restrained in direction. This corresponds approximately to the 4th condition in Table 3.2, so the effective length with respect to the x-x axis is 1.5L. For stability about the y-y axis, the stanchion is held in position at the top by the eaves girder, at height L_3 by the crane gantry girder, and at height L_1 by a continuous battened angle rail. These three members are connected through to a braced bay. The sheeting rails are not continuous, but the sheeting itself provides some resistance to buckling; and some directional restraint is provided by the eaves girder and the base. BS 449,

Appendix D currently allows $0.75L_1$, or $0.75L_2$, whichever is the greater, to be used as the effective length for the calculation of l/r_y. In the absence of the continuous battened rail, the effective length would be $0.75L_3$, where L_3 is the height to the gantry girder. The same effective length factor is allowed for single-storey stanchions supported laterally by brickwork. A number of additional examples are given in BS 449, Appendix D.

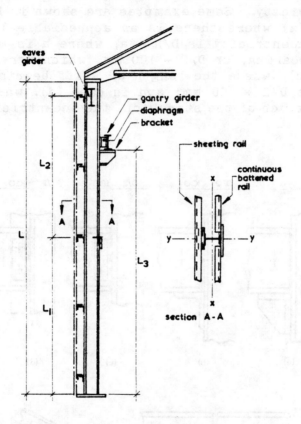

Fig. 3.10.   Single-storey side stanchion
with crane gantry girder.

At the caps of stanchions in single-storey buildings the deflections due to lateral forces should not exceed 1/325 of the height of the stanchion unless it can be shown that greater deflections will not impair the strength and efficiency of the structure, or result in damage to finishes.

3.3.7 <u>Eccentricity for stanchions and solid columns</u> The bending moments transmitted through the connections to a stanchion or column can only be calculated accurately when the connections are designed to transmit moments without slip or plastic deformation. In such cases the bending moments obtained from an accurate analysis would be used in design.

In the simple design method beams are designed as simply supported, but bending moments are induced in the stanchion because of the eccentricity of the connections. BS 449, Cl. 34(a) specifies that the beam reactions should be assumed to occur at 100 mm from the face of the stanchion, or at the centre of the bearing, whichever gives the greater eccentricity. Some examples are shown in Fig. 3.11 (a) to (d). In case (a) where there is an appreciable length of stiff bearing, the eccentricity is $D/2 + a$, where a is measured to the centre of the bearing, or $D/2 + 100$ mm, whichever is greater. In cases (b) and (c), where the length of stiff bearing is small the eccentricity is $D/2 + 100$ mm; and in case (d), where the beam is connected to the web of the stanchion, the eccentricity is $t/2 + 100$ mm.

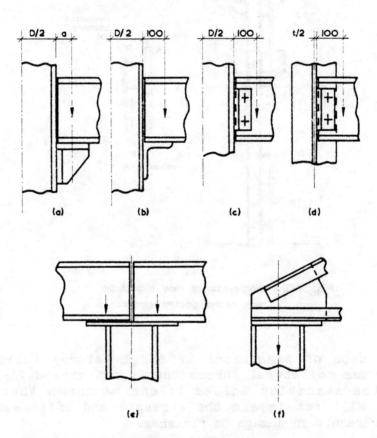

Fig. 3.11.   Eccentricities for stanchions.

Exceptions are simple capping beams and roof truss bearings without moment connections, as shown in Fig. 3.11 (e) and (f). For these cases the loads on the stanchion are assumed to act along the lines indicated.

Simple rules are also prescribed in Cl. 34(b) for the distribution of bending moments between the upper and lower stanchions at a junction. Provided that the two stanchions are continuous or effectively jointed, the bending moment is divided equally between the stanchions, unless the ratio of their stiffnesses exceeds 1.5, when the moment is distributed in proportion to the stiffnesses of the stanchions. For this purpose stiffness is defined as the second moment of area divided by the actual length. Any carry-over of moment to the far end of the stanchions is ignored. The method is illustrated in the following example.

Example 3.2 The first three storeys of a corner stanchion are shown diagrammatically in Fig. 3.12. The beam reactions in kN are indicated on each beam. Determine the bending moments in stanchions (1) and (2).

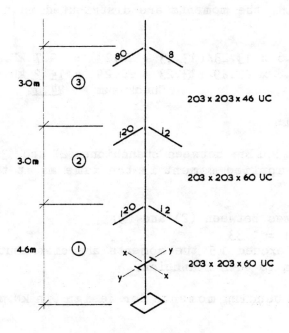

Fig. 3.12. Multistorey corner stanchion.

Section properties

203 x 203 x 46 UC:
D = 203.2 mm,   t = 7.3 mm, I_x = 4564 cm^4, I_y = 1539 cm^4
203 x 203 x 60 UC:
D = 209.6 mm,   t = 9.3 mm, I_x = 6088 cm^4, I_y = 2041 cm^4

Bending moments about the x-x axis

Stiffness (in cm units):

Stanchion (1): I_x/L = 6088/460 = 13.23
Stanchion (2): I_x/L = 6088/300 = 20.29
Stanchion (3): I_x/L = 4564/300 = 15.21

First floor level:

Applied moment = beam reaction x $(100 + D/2)/10^3$ kN.m
            = $120(100 + 209.6/2)/10^3$
            = 24.6 kN.m

Ratio of stiffness between stanchions (2) and (1)
     = 20.29/13.23 = 1.53
As this exceeds 1.5, the moments are distributed in proportion to the stiffness, thus:

Stanchion (1)   24.6 x 13.23/(13.23 + 20.29) =   9.7 kN.m
Stanchion (2)   24.6 x 20.29/(13.23 + 20.29) = <u>14.9</u> kN.m
                              Check sum   <u>24.6</u>

Second floor level:

Assuming that the splice between stanchions (2) and (3) lies above the floor level, the applied moment is the same as at the first floor, i.e. 24.6 kN.m.

Ratio of stiffnesses between (2) and (3)
     = 20.29/15.21 = 1.33
As this does not exceed 1.5 the moments are distributed equally, i.e. 24.6/2 = 12.3 kN.m to each stanchion.

Summarising, the bending moments for design (in kN.m) are tabulated below

Stanchion	Bottom	Top
1	0.0	9.7
2	14.9	12.3

The moment at the base is assumed to be zero because the carry-over is ignored.

Bending moments about the y-y axis

In this case the beams are connected to the web of the stanchion and

Applied moment = beam reaction x $(100 + t/2)/10^3$ kN.m

For example, at first floor level

Applied moment = $12(100 +9.3/2)/10^3$ = 1.3 kN.m

Distribution of moments is carried out as before, using the stiffness I_y/L in this case.

The complete design of a stanchion is given in Example 3.3.

3.4 DESIGN PROCEDURE

The objective is to select a section which, while satisfying the design criteria, is the lightest possible for economy. An iterative, trial and error, procedure is necessary except in the case of tension members under pure axial load, when an explicit solution is possible.

3.4.1 Axially loaded tension members
The design criterion for tension members can be expressed as

$$f_t/p_t \leq 1.0 \qquad (3.12)$$

where f_t = the average tensile stress at working load calculated on the net cross sectional area, or, for eccentrically connected angles and tees, the reduced area from Equations (3.3) and (3.4)

p_t = the allowable stress in axial tension appropriate to the type and thickness of section and grade of steel, from BS 449, Table 19 (Appendix).

As the allowable tensile stress does not depend on the slenderness ratio of a member, the required cross section can be obtained without iteration, thus

$$A = F_t/p_t \qquad (3.13)$$

where A = the minimum required cross sectional area
F_t = the applied axial tensile force.

3.4.2   <u>Axially loaded compression members</u>   The design criterion for compression members can be expressed as

$$f_c/p_c \leq 1.0 \qquad\qquad (3.14)$$

where f_c = the average axial compressive stress at working load calculated on the gross cross sectional area.

p_c = the allowable stress in axial compression appropriate to the slenderness ratio of the member and the grade of steel, from BS 449, Table 17 (Appendix).

The required section cannot be obtained directly because f_c and p_c depend on different section properties.   The following iterative procedure can be adopted.

1.   Select a section by intuition or experience.
2.   Calculate the slenderness ratio of the member.
3.   Obtain p_c from BS 449, Table 17 (Appendix).
4.   Calculate $f_c = F_c/A$, where F_c is the applied axial compressive force.
5.   Evaluate Equation (3.14).
6.   Return to stage 1 if the result is greater than or much less than unity.

The design is most economic when the result of stage 5 is as near to 1.0 as possible.

3.4.3   <u>Members subject to combined stresses</u>   The analysis of members subjected to combinations of axial force and bending moments is generally very complex and non linear, and except for some simple sectional shapes, cannot be resolved into design formulae unless simplifying assumptions justified by experimental evidence are made. Simple linear interaction formulae based on allowable stresses are provided by BS 449, Cl. 14(a) and (b).   The formulae allow bending and axial stresses to be calculated separately.

For combined bending and tension:

$$f_t/p_t + f_{bt}/p_{bt} \leq 1.0 \qquad\qquad (3.15)$$

where f_t and p_t are as defined above for axially loaded members.

f_{bt} = the resultant maximum tensile stress due to bending about both principal axes.

p_{bt} = the appropriate allowable tensile stress in bending, from BS 449, Table 2 (Appendix).

For combined bending and compression:

$$f_c/p_c + f_{bc}/p_{bc} \leq 1.0 \tag{3.16}$$

where f_c and p_c are as defined above for axially loaded members.

    f_{bc} = the resultant maximum compressive stress due to bending about both rectangular axes.

    p_{bc} = the appropriate compressive stress for members subject to bending, from BS 449, Table 3 (Appendix).

As axial and bending stresses can vary independently along the length of a member, it may be necessary to examine several sections in order to find the combination of stresses giving the greatest (most unfavourable) result. The design procedure is similar to that described above for axially loaded compression members and is demonstrated in the following example.

<u>Example 3.3 Two-storey corner stanchion</u> The two-storey corner stanchion in Fig. 3.13 is to be designed with a uniform section throughout, in grade 43 steel. The beams have been previously designed as simply supported, using the simple approach. Their reactions are therefore known for the load case under consideration, and are shown in kN.

The beams framing into the web of the stanchion are light beams connected by web cleats only. The two beams carrying the floor and and roof loads are bolted to the stanchion flanges through welded end plates. The base of the stanchion is substantial and capable of providing both positional and directional restraints.

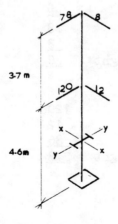

Fig. 3.13.   2-storey corner stanchion.

Fire protection for steelwork frequently takes the form of lightweight casing, or intumescent paint. However for this example it will be assumed that the stanchion is protected against fire and mechanical damage by a reinforced concrete casing. Two designs will be carried out: first considering the casing as additional dead load, and secondly making use of the provisions of BS 449, Cl. 30(b) and Cl. 21 which permit the strength of the casing to be taken into account.

It is convenient to start the design with the bottom stanchion, which is usually the more heavily loaded.

Effective lengths

The stanchion is similar to stanchion A in Fig. 3.12 for which the effective length factors in the first two storeys are as follows

Storey	Axis	
	x-x	y-y
1	0.85	0.85
2	0.85	1.0

Clearly the greatest slenderness ratios will be obtained by considering the y-y axis, and the effective lengths are therefore:

Stanchion (1)	4600 x 0.85 = 3910 mm
Stanchion (2)	3700 x 1.0 = 3700 mm

Axial loads

First floor   kN
    Floor beam   78
    Wall beam   8
    Total, excluding self-weight   86

Ground floor
    Floor beam   120
    Wall beam   12
    From upper storey   86
    Total, excluding self-weight   218

Trial section

Stanchion (1)

Max. permissible l/r (Section 3.3.2) = 180
Hence minimum r_y = 3910/180 = 21.7 mm.

All the Universal sections satisfy this criterion. Universal Columns have a higher r_y/r_x ratio than Universal beams, which makes them more economical for loads which are predominantly axial. Selection of a trial section is a matter of experience but a minimum section can be obtained from the Safe Load Tables of the BCSA-Constrado Handbook, which give the maximum safe concentric loads for various effective lengths of each section. A heavier section must be chosen when bending moments are to be considered.

Try a 152 x 152 x 37 UC, for which the safe concentric load on a 4 m effective length is 357 kN. The relevant section properties are:

D = 161.8 mm, B = 154.4 mm, t = 8.1 mm, A = 47.4 cm^2,
r_y = 3.87 cm, Z_{ex} = 274.2 cm^3, Z_{ey} = 91.78 cm^3,
D/T = 14

The self-weight of the beams is known and is included in their reactions, but until a section has been selected, the self-weights of the stanchion and casing are unknown. The additional load can now be calculated.

Steel:
Mass/metre = 37 kg.
Self-weight of stanchion (2) = 37 x 3.7 x 9.81/10^3 = <u>1.3 kN</u>

Casing:
Assume minimum concrete cover = 50 mm

D = 161.8 + 100 = 261.8 (say 270 mm)
B = 154.4 + 100 = 254.4 (say 260 mm)

Net sectional area of concrete = (270 x 260) − 47.4 x 10^2 = 65460 mm^2
Assuming concrete density = 2400 kg/m^3
Weight of concrete = 3.7 x 65460 x 2400 x 9.81/10^9 = <u>5.7 kN</u>
Total self-weight = 1.3 + 5.7 = <u>7 kN</u>
Modified axial load on stanchion (1)
    F_c = 218 + 7 = <u>225 kN</u>

<u>Bending moments</u>

Axis x-x: M_x = beam reaction x (100 + D/2)/10^3 kN.m
          = 120 x (100 + 161.8/2)/10^3 = 21.71 kN.m

Axis y-y: M_y = beam reaction x (100 + t/2)/10^3 kN.m
          = 12 x (100 + 8.1/2) = 1.25 kN.m

For a uniform section the stiffnesses are in inverse proportion to the lengths of the stanchions so the ratio is 4.6/3.7 = 1.24. As this does not exceed 1.5 the applied moment can be divided equally between the stanchions, i.e.

```
x-x    21.71/2 = 10.85 kN.m
y-y     1.25/2 =  0.63 kN.m
```

Actual stresses

Axial: $f_c = F_c/A = (225/47.40) \times 10 = \underline{47.5 \ N/mm^2}$

Bending: $f_{bc} = M_x/Z_{ex} + M_y/Z_{ey}$

$$= (10.85/274.2 + 0.63/91.78) \times 10^3 = \underline{46.4 \ N/mm^2}$$

Allowable stresses

Axial:

$l/r_y = 3910/38.7 = 101$

From BS 449, Table 17 (Appendix), $p_c = \underline{78 \ N/mm^2}$

Bending:

Considering the stanchion as a beam, the base provides restraints against torsion and lateral bending. The beams at the first floor provide restraint against torsion, but only partial restraint against lateral bending. The effective length factor from Table 2.1 is therefore 0.85 giving a slenderness ratio l/r_y of 101. From BS 449, Table 3(a) (Appendix), $p_{bc} = \underline{165 \ N/mm^2}$.

Combined stresses

The axial and bending stresses must now be combined in the interaction formula - Equation (3.16)

$f_c/p_c + f_{bc}/p_{bc} \leq 1.0$

i.e. $47.5/78 + 46.4/165 = 0.89$

A 152 x 152 x 37 UC is therefore satisfactory and would be quite economical, as the result is close to unity.

As an exercise the reader is invited to repeat the design for the higher grades of steel and for the lighter sections in the group. The results are as follows.

Section	grade 43	grade 50	grade 55
152 x 152 x 37	0.89	0.72	0.66
152 x 152 x 30	> 1.0	0.92	0.84
152 x 152 x 23	> 1.0	> 1.0	> 1.0

From these results it can be seen that if grade 50 steel is specified, a lighter UC section can be used. There is no additional advantage to be gained from the use of grade 55 steel. The reader is invited to check that the same section is adequate for the upper stanchion. In this case the axial force is reduced, but the whole of the bending moment from the roof beam is applied to the stanchion and is greater than the bending moment in the lower stanchion.

<u>Cased strut</u> (Assuming grade 50 steel)

If the concrete casing complies with BS 449, Cl. 30(b) (Appendix) it can be assumed to carry some of the load and to increase the radius of gyration of the section. Try a 152 x 152 x 23 UC as the steel core. The relevant section properties are:

D = 152.4 mm, B = 152.4 mm, t = 6.1 mm, A = 29.8 cm^2,
r_x = 6.51 cm, r_y = 3.68 cm, Z_{ex} = 165.7 cm^3,
Z_{ey} = 52.95 cm^3, D/T = 22.3

Minimum casing = 152.4 + 100, say 260 mm square.

Steel area A_s = 2980 mm^2

Concrete area A_c = 260 x 260 - 2980 = 64620 mm^2

From Cl. 30(b), r_y = 0.2(B + 100)

$$= 0.2(152.4 + 100) = 50.48 \text{ mm.}$$

l/r_y = 3910/50.48 = <u>77.5</u>

l/r_x (based on steel core) = 3910/65.1 = 60.1

$l/r_y > l/r_x$ and is therefore used to obtain p_c.

From BS 449, Table 17(b) (Appendix), p_c = 134

From Cl. 30(b) the allowable stress in the concrete is taken as $p_c/(0.19$ x p_{bc} from BS 449, Table 2)

i.e. p_{cc} = 134/(0.19 x 230) = 3.07 N/mm^2

The total axial force that can be supported by the stanchion is therefore

$$P_c = p_c A_s + p_{cc} A_c$$

$$= (134 \text{ x } 2980 + 3.07 \text{ x } 64620)/10^3 = \underline{597 \text{ kN}}$$

However this value must not exceed twice the axial load capacity of the steel core.

Slenderness ratio l/r_y of core = 3910/36.8 = 106
From Table 17(b) p_c = 84 N/mm^2
$\qquad$ $2p_cA_s$ = 2 x 84 x 2980/10^3 = <u>501 kN</u>

This lower value is therefore the maximum axial load capacity of the cased stanchion.

The bending moment due to the beam reactions is based on an eccentricity of 100 mm from the face of the casing, i.e. 100 + 260/2 = 230 mm.

The bending moment in each stanchion (half the applied moment) is therefore:

$\qquad$ Axis x-x $\quad$ M_x = 120 x 230/2 = 13800 kN.mm
$\qquad$ Axis y-y $\quad$ M_y = $\;$ 12 x 230/2 = $\;\;$ 1380 kN.mm

Bending stress (based on steel core)

$\qquad$ $f_{bc} = M_x/Z_{ex} + M_y/Z_{ey}$

$\qquad\qquad$ = 13800/165.7 + 1380/52.95

$\qquad\qquad$ = <u>109 N/mm^2</u>

The allowable bending stress, from BS 449, Table 3(b) is based on l/r_y for the cased section and D/T for the steel core in accordance with Cl. 21, BS 449 for cased beams (Appendix). Entering Table 3(b) (Appendix) with l/r_y = 77.5 and D/T = 22.3, p_{bc} = 230 N/mm^2. This must not exceed 1.5 x p_{bc} for the steel core. Entering Table 3(b) with l/r_y = 106, p_{bc} = 201 N/mm^2 for the steel core. <u>230 N/mm^2</u> is therefore acceptable.

Equation (3.16) can be re-arranged by substituting the axial load and axial load capacity of the strut for f_c and p_c, thus

$\qquad$ $F_c/P_c + f_{bc}/p_{bc} \leq 1.0$

i.e. 225/510 + 109/230 = 0.92

Note: 225 kN for F_c is slightly conservative, because the self-weight of the stanchion is less than in the previous case. Re-calculation would not be justified however, unless the result of Equation (3.16) exceeded unity.

The cased section is therefore satisfactory. Application of the provisions of Cl. 30(b) of BS 449 for cased struts and Cl. 21 for cased beams thus enables a lighter steel section to be used than when the concrete casing is merely considered as additional dead load.

REFERENCES

Dwight, J.B. (1978) - Design of axially loaded columns including interactive buckling. The background to British Standards for structural steelwork. Imperial College, London, and Constrado.

Godfrey, G.B. (March 1962) - The allowable stress in axially loaded steel struts, Structural Engineer, pp. 97-112.

Mackay, M.E. and Wilkinson, N.W. (1953) - Report on experimental investigation of two mild steel lattice girders, BCSA publication No. 7.

Nelson, H.M. (1953) - Angles in tension, BCSA publication No. 7.

Robertson, A. (1925) - The strength of struts, Selected Engineering Paper No. 28, Inst. Civ. Engrs.

4 Structural Connections

4.1 INTRODUCTION

This chapter is concerned with the design of welded and bolted structural steel connections. British industry prefers to weld steel members in the controlled conditions of the workshop and to use bolted connections on site. Welding can be carried out successfully on site but it needs to be carefully supervised and is therefore expensive.

All structural connections are semi-rigid, i.e. there are always relative linear and rotational movements between the connected members. If all the relative movements are small the connection is described as rigid. If the rotational movement is large and the linear deformations are small then the connection is described as pinned. These distinctions between connections are important in structural design because they affect the choice of method of analysis of the complete structure, as explained in Chapter 5.

4.2 THE IDEAL STRUCTURAL CONNECTION

The types of structural connections in common use have been developed and modified in the past to suit the manufacturing and assembly processes. Many connections satisfy the ideal requirements which are as follows. A structural connection should be:

(1) Simple to manufacture and assemble.
(2) Standardised for situations where the dimensions and loads are similar thus avoiding a multiplicity of dimensions, plate thicknesses, weld sizes and bolts.
(3) Manufactured from materials and components that are readily available.
(4) Designed and detailed so that work is from the top of the joint not from below where the workman's arms will be above his head. There should also be sufficient room to locate a spanner, or space to weld if required.
(5) Designed so that welding is generally confined to the workshops to ensure a good quality and reduce costs.
(6) Detailed to allow sufficient clearance and adjustment to accommodate the lack of accuracy in site dimensions.
(7) Designed to withstand not only the normal working loads but also the erection forces.

(8) Designed to avoid the use of temporary supports to the structure during erection.
(9) Designed to develop the required load-deformation characteristics at the service load and at ultimate load.
(10) Detailed to resist corrosion and to be of reasonable appearance.
(11) Low in cost and cheap to maintain.

4.3 WELDING AND TYPES OF WELD

Welding is a method of connecting components by heating the materials to a suitable temperature so that fusion occurs. The most common method for heating steelwork is by means of an electric arc between a coated wire electrode and the materials being joined. The electrical circuit is shown in Fig. 4.1(a). During the process, which is illustrated in Fig. 4.1(b) the coated electrode is consumed, the wire becomes the filler material, and the coating is converted partly into a shielding gas, partly into slag, and some part is absorbed by the weld metal. This method, known as the manual metal arc welding process was the original method, and is still the most common for structural connections.

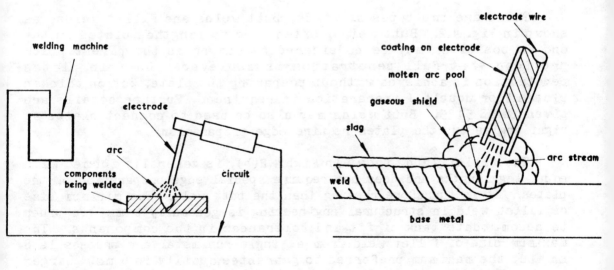

(a) Arc welding circuit   (b) Shielded arc welding

Fig. 4·1. Shielded metal arc welding.

For manual metal arc welding the electrode should comply with BS 639 where the specified minimum yield stress is 330 N/mm^2 for type E43 electrode, and 360 N/mm^2 for type E51 electrode. Information on other mechanical properties is given in BS 639. The main reason for the flux covering to the electrode in the manual metal arc welding process is to provide an inert gas which shields the molten metal from atmospheric contamination. In addition the flux forms a slag to protect the weld until it has cooled to room temperature, when the slag should be easily detachable. Other functions of the flux include, arc stabilisation, control of surface profile, control of weld metal composition, alloying, and deoxidisation. On the other hand it should be noted that the flux covering can be a source of hydrogen contamination from absorbed and chemically combined moisture. The absorbed moisture can be removed by drying. Further detailed information on composition and choice of electrode is available in Gourd.

The particular advantage of welding is that it produces a rigid connection. However, the method requires more skill and supervision than bolted connections. Most welded structural connections are effected using the manual metal arc process but long continuous welds, which occur for example in built-up girders, are laid down with automatic welding equipment. The automatic processes achieve the exclusion of the atmospheric pollution by gas shielding, flux core, or submerged arc. Further details are given in Gourd.

There are two types of welds, butt welds and fillet welds, as shown in Fig. 4.2. Butt welds, often used to lengthen plates in the end on position, may be considered as strong as the parent plate provided that full penetration is achieved. On thin plates penetration is achieved without preparing the plate, but on thicker plates V or double J preparation is required. Further details are given in BS 5135. Butt welds may also be used to connect plates at right angles but the plates require edge preparation.

The fillet weld, shown in Fig. 4.2(b), is generally formed with equal leg lengths and does not require special edge preparation of the plates. It is therefore cheaper than the butt weld. The minimum size of fillet weld in structural engineering is generally 6 mm, in order to accommodate lack of fit and tolerances in the components. The maximum size of fillet weld from a single run metal arc process is 8 mm, but the maximum preferred to guarantee quality is 6 mm. Larger sizes of fillet weld are used where required but they have to be formed by multiple weld runs.

The use of intermittent butt and fillet welds is permitted by BS 449 but they are not favoured in structural engineering because they introduce stress discontinuities, act as stress raisers, may induce fatigue cracks, may act as corrosion pockets, and are difficult to produce with an automatic welding machine. The detailed requirements covering length and spacing of intermittent fillet welds are given in Cl. 54 BS 449.

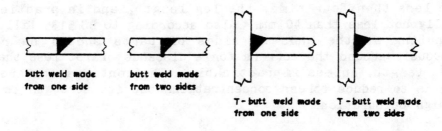

butt weld made from one side

butt weld made from two sides

T-butt weld made from one side

T-butt weld made from two sides

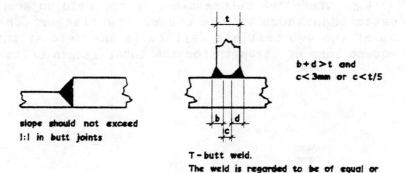

slope should not exceed 1:1 in butt joints

$b+d>t$ and $c<3mm$ or $c<t/5$

T-butt weld.
The weld is regarded to be of equal or higher strength than the vertical member

(a) Examples of butt welds

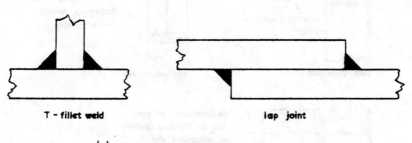

T - fillet weld

lap joint

(b) Examples of fillet welds

Fig. 4·2. Types of welds in structural connections.

4.4 EFFECTIVE LENGTH OF A WELD

The strengths of the 'start' and 'stop' sections of a weld are unreliable and it is common practice to reduce the actual length of the weld to a hypothetical 'effective length' for design purposes. The effective length is the actual length minus twice the leg length. According to BS 5135 the effective length to transmit loading should not be less than four times the leg length, and in practice it is generally not less than 40 mm. Also according to BS 5135 fillet welds terminating at the ends or sides of parts should be returned continuously around the corners for a distance of not less than twice the leg length, unless impracticable. The continuation around the corner is to reduce stress concentrations but its strength is ignored in design calculations.

The effective length may be reduced further if a welded component distorts under load. This can occur in situations similar to that shown in Fig. 4.3, where the deformations in the weld adjacent to the web are greater than those at the end of the flange. The larger deformations at the web initiate failure in the weld at this point with a consequent loss of strength for the total length of the weld.

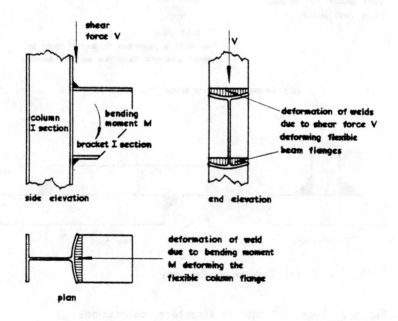

Fig. 4·3. Reduction in strength of welds associated
with flange deformations.

This effect was investigated experimentally first by Elzen and later by Rolloos, and the International Institute of Welding design rules recommend that the effective length b_{we} of a weld associated with a steel section, of web thickness t and flange thickness T, is

$$b_{we} = 2t + CT \qquad (4.1)$$

where C is given in Table 4.1.

Factor C I section	Factor C Box section	Position	Type of steel
7	5	tension flange	F_e 360 (Y_s = 235 N/mm^2)
10	7	compression flange	F_e 360 (Y_s = 235 N/mm^2)
5	4	tension flange	F_e 510 (Y_s = 355 N/mm^2)
7	6	compression flange	F_e 510 (Y_s = 355 N/mm^2)

Table 4.1   Values of factor C associated with effective weld lengths

4.5   FAILURE CRITERIA FOR FILLET WELDS

The real external forces acting on a fillet weld are probably similar to those shown in Fig. 4.4(a) according to Clarke (1971). Experiments by Biggs et al on right angled fillet welds of equal leg length loaded to failure show that the fracture plane angle varies between 10° and 80° depending on the combination of external forces. The actual distribution of stress on the failure plane is uncertain but a theoretical distribution by Kato and Morita shows peak stresses at the root of the weld which reduce towards the face of the weld. This distribution appears to be confirmed by experimental observations of cracks initiating at the root. The situation is further complicated by residual stresses and such variables as the type of electrode, type of steel, the ratio of the size of the weld to the plate thickness, the quality of the weld, and whether the loading is static or dynamic. If stresses on the failure plane are assumed to be uniform then the relationship between the average shear stress and tensile stress on the failure plane has been shown by Biggs et al to approximate to an ellipse. An ellipse of failure stresses combined with a variable fracture angle can be used theoretically to predict the magnitude of the external forces but the method is too complicated for general design purposes.

For design purposes a complex system of external forces acting on a fillet weld is reduced to forces acting in three perpendicular directions on a unit length of weld as shown in Fig. 4.4(b). The stresses related to each force are calculated using the throat thickness a_w; and the value of the shear stress p_w, when $F_x = F_y = 0$, may be considered as the basic design stress. The throat thickness a_w $\simeq 0.7$(leg length), for equal leg lengths and 90° between the fusion faces.

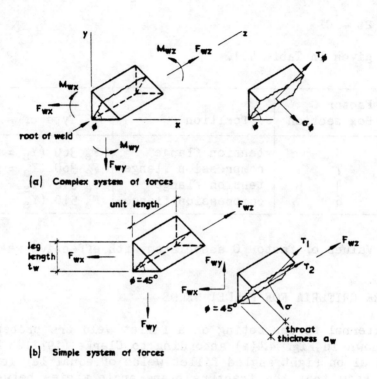

(a) Complex system of forces

(b) Simple system of forces

Fig. 4·4. Forces acting on a fillet weld.

For design purposes the forces F_{wx}, F_{wy} and F_{wz} are assumed to be related by the vector addition stress formula recommended in BS 5135

$$(F_{wx}/a_w)^2 + (F_{wy}/a_w)^2 + (F_{wz}/a_w)^2 = p_w^2 \qquad (4.2)$$

This relationship is approximately correct when compared with experimental results of Biggs et al.

Equation (4.2) may be rearranged as

$$F_{wx}^2 + F_{wy}^2 + F_{wz}^2 = (a_w p_w)^2 \qquad (4.3)$$

Equation 4.3 is more convenient for design purposes because the magnitude of the applied external forces per unit length, F_{wx}, F_{wy} and F_{wz}, can be determined independently of the weld size, and related to the design shear resistance of the weld $a_w p_w$.

The strength of fillet welds has been shown by Ligtenberg to be

related to the strength of the parent material. The v.
elastic allowable design stresses given in Cl. 53 BS 449 ar.
related to the grade of steel as shown in Table 4.2.

Grade of steel	Allowable stress N/mm^2
43	115
50	160
55	195

Table 4.2   Allowable shear stresses for welds

The correct type and strength of electrode must be used for each grade of steel as specified in BS 639.

The values of design stress may appear to be conservative but they have been reduced to allow for the inaccuracies in the use of simple theory, lack of knowledge of fatigue and brittle fracture, and the different relative strengths of side and end fillet welds. It is convenient for design purposes to prepare a table of design values of the strength of fillet welds per unit length as the leg length varies. The values given in Table 4.3 are calculated from

P_W = 0.7 x (leg length) x (allowable shear stress).

Equal leg lengths mm	Force per unit length		
	grade 43 p_W = 115N/mm^2 kN/mm	grade 50 p_W = 160N/mm^2 kN/mm	grade 55 p_W = 195N/mm^2 kN/mm
6	0.483	0.672	0.819
8	0.644	0.896	1.092
10	0.805	1.120	1.365
12	0.966	1.344	1.638
15	1.208	1.680	2.048
18	1.449	2.016	2.459
20	1.610	2.240	2.730
22	1.771	2.464	3.003
25	2.013	2.800	3.413

Table 4.3   Allowable forces per unit length of fillet weld (BS 449)

It should be noted that although the strength of the weld is calculated using the throat thickness the weld is specified by the leg length.

An alternative method of design recommended by the European Commission XV also assumes a fixed 45° critical plane but calculates the normal and shear stresses on this plane and combines them using a more accurate failure criterion. This method is more laborious and introduces the possibility of further errors when resolving forces onto the 45° plane. The relationship between the forces F_{wx}, F_{wy} and F_{wz} for this method can be expressed in the same form as Equation (4.3) and is shown in Holmes and Martin. The size of weld using this method is slightly less than using the vector addition method.

4.6 LOAD-DEFORMATION RELATIONSHIPS FOR FILLET WELDS

The strength of the weld in a connection is of primary importance but the load-deformation characteristics of the weld must also be considered. The magnitude of the deformation at the maximum load varies from approximately 0.6 mm to 1.4 mm depending on the orientation of the weld in relatiom to the applied load as shown by Clarke (1970) in Fig. 4.5. The maximum deformation is for a side fillet weld parallel to the applied load and the minimum deformatiom is for an end fillet weld perpendicular to the applied load. The allowable design stress is based on the weaker side fillet weld and as shown in Fig. 4.5 at this stress the disparity in deformations is small.

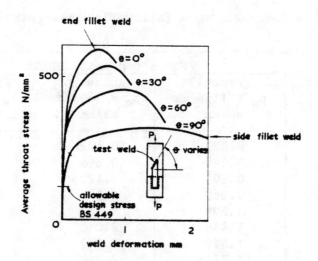

Fig. 4·5. Load - deformation relationships for an 8mm fillet weld by Clarke A.

The load-deformation relationship is linear in the elastic range of behaviour, and for elastic allowable stress methods of analysis theoretical linear load-deformation relationships are assumed.

4.7 FACTORS AFFECTING THE STRENGTH OF WELDED CONNECTIONS

The structural designer should be aware of the following conditions which can affect the strength of welded connections:

(a) The weld may not be formed correctly and there may be cavities and slag inclusions. These can be detected using non-destructive testing as described in Section 1.25.

(b) The use of an incorrect welding electrode. Electrodes are related to the parent steel grade as specified in BS 639.

(c) There may be excessive clearance, or lack of fit, between members which effectively reduces the size, and consequently the strength, of the weld.

(d) Stress concentrations, associated with welded connections, combined with oscillating loads producing fatigue, may reduce the strength of welded connections. This is described in more detail in Section 1.20.

(e) Residual stresses, introduced from differential heating during welding, may reduce the strength of the weld. This is described in more detail in Section 1.22.

(f) Hydrogen cracks, as shown in the connection of two plates in Fig. 4.6, are associated with welding and occur when the cooling rate is too rapid. Excessive hardening occurs in the heat affected zone which cracks under the action of residual stresses if sufficient hydrogen is present in the weld. This defect can be avoided by controlling the cooling and the hydrogen input to the weld as described in BS 5135.

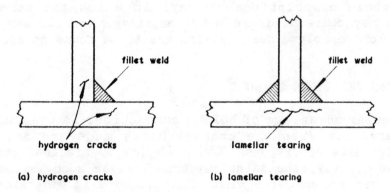

(a) hydrogen cracks   (b) lamellar tearing

Fig. 4·6. Faults associated with welding.

(g) Lamellar tearing cracks occur when welding a plate connection of the type shown in Fig. 4.6(b). Further examples are given in BS 5135. The cracks are produced by a combination of low ductility in the plate in the transverse direction and high joint restraint in the weld, which

induces tensile forces adjacent to the connection. The low
ductility in the plate is produced by inclusions of non-
metallic substances formed in the steel making process.
When the ingot is rolled to make steel these inclusions form
as plates parallel to the direction of rolling. Only a
small percentage of plates are susceptible to lamellar
tearing, and where it occurs, joint details can be changed,
as recommended in BS 5135, to reduce the chances of it
affecting the strength of the connection. Research into
lamellar tearing has been carried out by Farrar and Dolby.

(h)   Brittle fracture can occur with a welded connection as
described in Section 1.23.

(i)   Corrosion can occur, which reduces the size of components
and causes pitting, which may initiate fatigue cracks.

(j)   Insufficient 'penetration' of the parent metal leads to a
reduction in strength of the weld. In simple terms the
welder uses a voltage and arc length which produces a stable
arc and a satisfactory weld profile. The current then
becomes a main factor in controlling penetration. Another
important factor in depth of penetration is edge
preparation. Plates of 6 mm and square edges can be butt
welded from one side but the edges of thicker material must
be bevelled to provide access for the arc.

(k)   Lack of sidewall fusion also produces a weak connection, and
this occurs if there is a poor bond between the parent metal
and the weld metal. Good bonding can only occur when the
surface of the parent metal has been melted before the weld
metal is allowed to flow into the joint.

The above descriptions of ways in which the strength of a
connection may be reduced are not exhaustive, and further sources of
error and more complete descriptions are to be found in Gourd.

4.8   BOLTING AND TYPES OF BOLT

The particular advantage of bolted connections is that they require
less supervision when the connection is made and are therefore
suitable for site conditions. Other advantages are that the joint can
be fastened quickly, the joint supports load as soon as the bolts are
tightened, and accommodates minor discrepancies in dimensions.

Disadvantages of bolted connections are that for large forces the
space required for the joint is extensive and the connection is not as
rigid as a welded connection, even when friction-grip bolts are used.

Steel bolts are identified by the gross diameter and the
preferred sizes of steel bolts in general use are 16, 20, 22, 24, 27,
30, 33 and 36 mm diameter. The most common size in connections is the
20 mm diameter. Bolts are also identified by their strength, i.e.

grade, and also by special characteristics, e.g. friction-grip.   The types of steel bolts in general use are:

(1)   International Standards Organisation (ISO) metric black bolts, grade 4.6, to BS 4190.   This group also includes foundation bolts.   Details are given in Appendix.
(2)   ISO precision bolts, strength grade 8.8, to BS 3692.
(3)   High strength friction grip bolts to BS 4395, of two types:
      (a)   General grade strength 8.8.
      (b)   Higher grade strength 10.9.

The bolt grade figures when multiplied together refer to the minimum guaranteed yield strength, e.g. for grade 4.6 the 4 = $f_u/10$ and the 6 = $(Y_s/f_u) \times 10$.   The yield strength is thus $4 \times 6 = 24$ kgf/mm^2 (approx 235 N/mm^2).

The grade 4.6 bolt is low in cost, can be placed with the use of only simple tools, and requires little supervision during the erection stage.   At fracture the bolt has the relatively large percentage elongation of 25%, a property which is preferred in plastic collapse conditions and which is a useful attribute if a bolt has to be forced into position on site.

Where forces are large, or where space is restricted, or where costs of erection can be reduced by using fewer bolts, then the higher strength grade 8.8 bolts are used.   The percentage elongation at fracture is only 12%, but is still acceptable for design purposes.

Where a more rigid connection is required, e.g. in plastic methods of design, high strength friction-grip (HSFG) bolts are used. The strength of HSFG bolts is equal to, or greater than that of grade 8.8, but there is an increased cost for the additional site supervision which is necessary to ensure that the bolts are axially pre-stressed or pre-loaded in tension to the design values.   The object of pre-stressing a bolt is to ensure that when a connection is subject to shear forces the friction between the 'faying' surfaces prevents slip.   The prestress also reduces rotation of the connection when subject to a bending moment.   For further explanation see later design examples.

HSFG bolts are axially stressed or pre-loaded by tightening the nut using a spanner with a large lever arm.   Methods of tightening are specified in BS 4604.   If the spanner is in the form of a 'torque wrench' it can be calibrated in relation to the required axial force in the bolt.   A simpler method of measuring the force in the bolt, but less accurate according to laboratory investigation by Bahia and Martin, is to use load indicating washers.   In this method special washers under the head of the nut reduce in thickness to a specified value as the axial force in the bolt increases.   A further alternative method of ensuring that the bolt is stressed is to specify 'turns of the nut'. American investigations, reported by Fisher and Struik, showed that in general the clamping force produced by calibrated wrench and turn of the nut methods on site, exceeded the specified value.

Close tolerance turned bolts are used only where accurate alignment of parts is needed. The shank of the bolt is at least 2 mm greater in diameter than the threaded portion and the hole is 0.15 mm greater than the shank diameter. This small tolerance necessitates the use of special methods to ensure that all the holes align correctly.

Foundation bolts or holding down bolts are used for connecting steel members to concrete pads or to concrete foundations. The bolts are cast into the concrete before erection and thus require careful setting out. Most of these bolts are grade 4.6 but the higher strengths may be used if required.

Rivets were used extensively both in the fabrication shop and on site and had the advantage that as they were driven hot they expanded to fill the hole and resulted in a rigid connection. However due to labour costs they have now been almost entirely superseded by welding and bolting.

4.9 WASHERS

In British practice most bolts have steel washers under the head and the nut, to distribute the bolt force and to prevent the nut or head from damaging the component or member. Washers are not essential in all cases according to the European Convention for Structural Steelwork (ECSS). Washers for grade 4.6 and 8.8 bolts are specified in BS 4320 and hardened washers for HSFG bolts are specified in BS 4395 Pt 1.

4.10 BOLT HOLES (Cl. 59 BS 449)

Bolt holes are usually drilled, but may be punched full size or under size and reamed. Holes should never be formed by gas cutting.

Bolt holes may be punched full size, according to BS 449, Cl. 59, in light roof members, light framing, connecting angles, plates other than splices, for strength grade below grade 8.8, provided that punching does not unduly distort the material. Bolt holes are made larger in diameter than the bolt diameter to accommodate inaccuracies in manufacture and to facilitate assembly. The clearance is 2 mm for bolts not exceeding 24 mm diameter and 3 mm for bolts exceeding 24 mm diameter. An exception to this is a close tolerance turned bolt where the clearance is 0.15 mm.

Bolt holes reduce the gross cross sectional area of a member to the net cross sectional area and therefore the net value is used in calculations where members, or parts of members, are in tension. (See Section 3.1.2). Bolt holes also produce stress concentrations but it is argued that these are offset by the fact that at ultimate load the highly stressed net cross section will work harden before fracture and yield will by then have occurred at adjacent cross sections. (See also Section 2.10.6). The gross cross sectional area is used however for calculations where members, or parts of members, are in compression, because at ultimate load the bolt hole deforms and the shank of the bolt resists part of the load in bearing. (See also Section 3.1.2).

4.11   SPACING OF BOLTS (Cl. 51 BS 449)

The distance between the centre lines of bolts in the direction of stress is the pitch. The minimum pitch is specified to prevent excessive reduction in cross sectional area, to ensure that there is sufficient space to tighten the bolts, and to prevent overlapping of washers. The maximum pitch is specified to prevent buckling of plates in compression between bolts, to ensure that bolts act together as a group to resist forces, and to prevent corrosion.

The following table of pitch values is based on Cl. 51 BS 449.

Situation	Pitch
minimum pitch	2.5 x nom. bolt dia.
maximum pitch (corrosive condition)	32t or 300 mm
maximum pitch (tensile stress)	16t or 200 mm
maximum pitch (compressive stress)	12t or 200 mm
maximum pitch (two bolts in line adjacent and parallel to an edge of an outside plate, in compression or tension members)	(100 mm + 4t) or 200 mm

note: t is the smaller thickness of the connected material

Table 4.4   Spacing of bolts

There are also recommended positions, spacing and diameter of holes in standard sections as shown in a Table taken from the Structural Steel Handbook (Appendix). The distances are based on allowing sufficient clearance adjacent to the web, and edge distances as described in Section 4.13.

4.12 STRENGTH PROPERTIES OF SINGLE BOLTS

Bolted connections generally consist of two or more bolts; and individual bolts in connections may be subject to any combination of axial, shear or bearing forces which may produce failure. It is important therefore to consider single bolt strengths in more detail.

4.12.1 Tensile Properties
A bolt when subject to a tensile force fails on the net area at the root of the thread. The net area, which is approximately 80% of the gross area, is related to the pitch of the threads and the minor diameter as given in BS 3643. The net areas of bolts in common use are given in Tables taken from the Structural Steel Handbook (Appendix).

Failure is not the only design criterion. It is also necessary to limit deformations at working load and ultimate load. The load-deformation relationship for a grade 4.6 bolt in tension is linear-elastic followed by a non-linear plastic stage, work hardening, and finally fracture as shown in Fig. 1.4 in Chapter 1. In design calculations at ultimate load the maximum tensile stress in a bolt should not exceed the minimum guaranteed yield stress. This value prevents the deformation that occurs in the plastic stage, and work hardening provides an additional safety factor against fracture. The minimum guaranteed tensile yield stress for a grade 4.6 bolt is 235 N/mm^2, as specified in BS 3692.

The allowable design tensile stress in a bolt at working load is set so that stresses in service do not exceed the elastic limit in order to prevent permanent deformation. The allowable tensile design stresses specified in BS 449 are summarised in Table 4.5.

	Tension	Shear	Bearing			
			on bolt	on grade 43 steel	on grade 50 steel	on grade 55 steel
Grade 4.6 bolts	120	80	250	250	250	250
Grade 8.8 bolts	280	187	585	250	350	400

Table 4.5 Allowable stresses in N/mm^2 for bolts

The allowable tensile stress must also have an adequate safety factor in relation to the minimum guaranteed tensile yield stress, and for grade 4.6 bolts the safety factor = 235/120 = 1.96.

The minimum guaranteed tensile yield stress for grade 8.8 bolts is the lesser value of 0.7(tensile strength) = 0.7 x 785 = 549.5 N/mm^2 or 628 N/mm^2 which is the stress at 0.2% strain as specified in BS 3692.

The allowable tensile stress for grade 8.8 bolts, according to BS 449, is related to the value for grade 4.6 bolts in proportion to the yield stress, i.e. (549.5/235) x 120 = 280 N/mm^2. This ensures that the safety factor, which is 235/120 = 1.96 for grade 4.6 bolts, is the same value for grade 8.8 bolts. The allowable design tensile stresses are given in Table 4.5 and values of the allowable design tensile forces on grade 4.6 and 8.8 bolts are given in Tables 4.7 and 4.8.

To develop the frictional shear resistance of a high strength friction bolt, the bolt is preloaded in tension to approximately 65% of the specified minimum guaranteed yield strength. Values of the preload, as specified in BS 4395, are given in Table 4.9.   These values of preload are chosen to be less than the elastic limit so that there is no permanent deformation.   Some HSFG bolts in connections are subject to tensile forces at service load and the maximum allowable tensile force is (preload)/(load factor).   The load factor in tension is 1.7 as specified in BS 4604 Pt 1, Cl. 3.1.2, but where fatigue conditions exist the load factor is 2.0.   Values of the allowable tensile forces in HSFG bolts general grade are given in Table 4.9.

4.12.2 _Shear properties_ The bolts in some multi-bolt groups are subject to shear forces, and the load-shear deformation relationships are therefore important. Experimental relationships for various bolts, as given by Bahia and Martin, are shown in Fig. 4.7. A linear relationship is followed by a non-linear relation to fracture. The non-linearity is dependent on the shear deformation of the bolt and the elongation of the hole in the plate, which depends on the thickness of the plate. If a bolt is in single shear, further deformation occurs because of bending of the bolt. It is important for the designer to notice from Fig. 4.7 that the strength of a bolt in single shear is independent of the thickness of the plate.

At failure the average shear stress on a bolt is related to the tensile stress. From the Huber-Von-Mises-Hencky shear distortion strain energy theory the ratio of shear to tensile stress k_b, is equal to 0.57, as described in Section 1.16. Experimental values of k_b however vary and are higher. Chesson et al give an experimental value of $k_b = 0.62$, and Bahia and Martin give a value of $k_b = 0.68$ for bolts on the threaded portion. A value of 0.7 is recommended by ECCS and a value of 0.7071 by BS 5400. The value adopted for use in BS 449 is 0.67, as shown by the ratio of shear stress to tensile stress in Table 4.5.

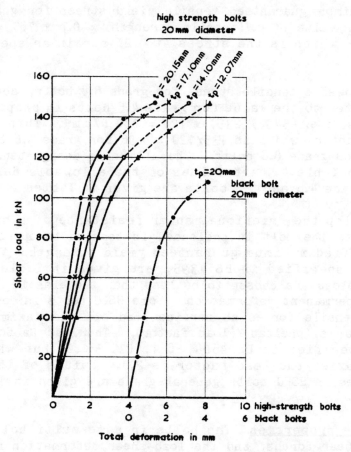

Fig. 4·7. Relationship between shear load and
deformation for single bolt tests.
(Bahia and Martin)

The shear resistance of a bolt depends on the position of the shear plane. If the bolt fails across the threads then the net cross sectional area of the bolt is used. If it fails across the shank then the gross cross sectional area is used. Generally in structural engineering most of the threaded portion lies outside the connected parts and the bolt is therefore assumed to fail across the gross cross sectional area. Values of the allowable design shear strength of grade 4.6 and grade 8.8 bolts based on the gross cross sectional area are given in Tables 4.7 and 4.8 respectively.

The shear strength of a 'non slip' HSFG connection is related to the coefficient of friction between the faying surfaces. In most design situations the surface condition of the steel may be described as weathered and clear of loose mill scale, loose rust, oil, dirt or paint. In this condition BS 4604 Pt 1 Cl. 3.1.1 specifies a value of $\mu_s = 0.45$ provided the surfaces fit together. Values of the coefficient of friction, or slip factors, for other surface conditions are given in Table 4.6 taken from BS 5400 Pt 3. If surfaces in contact are coated, or otherwise treated including the use of a machined surface, the slip factor should be determined by tests as described in BS 4604 (Appendix A).

Surface conditions	Slip factor
Weathered, clear of all mill scale and loose rust	0.45
Blasted with shot or grit and loose rust removed	0.50
Sprayed with aluminium	0.50
Sprayed with zinc	0.40
Treated with zinc silicate paint	0.35
Treated with etch primer	0.25

Table 4.6   Slip factors for friction-grip fasteners (BS 5400)

For HSFG connections subject to shear forces in the plane of the friction forces the allowable shear resistance according to BS 4604 Pt 1 Cl. 3.1.1 is

$$P_V = (\text{slip factor}) \times (\text{no. of eff. interfaces}) \times (\text{proof load})/(\text{load factor}) \qquad (4.4)$$

The load factor is 1.4 without wind and 1.2 with wind, provided the connections are adequate when wind forces are neglected. No additional factor is required to take account of fatigue conditions. Values of the allowable design shear resistance of general grade HSFG bolts for $\mu_s = 0.45$ are given in Table 4.9.

4.12.3   <u>Tension-shear properties</u>   In some connections, e.g. a bracket on a column, the bolts are subject to a combination of tensile and shear forces. The experimental relationship obtained by Chesson et al between the shear and tensile stresses at failure is given in Fig. 4.8. A curve that empirically fits these results with a reasonable degree of accuracy is an ellipse

$$(f_{bt}/f_{btu})^2 + (f_{bv}/0.62f_{btu})^2 = 1 \qquad (4.5)$$

The elliptical relationship is recommended in ECSS and BS 5400 Pt 3.

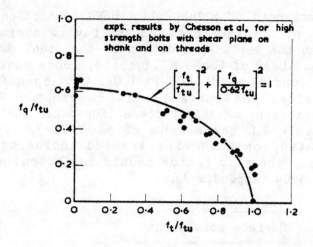

expt. results by Chesson et al, for high
strength bolts with shear plane on
shank and on threads

$$\left[\frac{f_t}{f_{tu}}\right]^2 + \left[\frac{f_q}{0.62 f_{tu}}\right]^2 = 1$$

Fig. 4·8. **Relationship between shear and tensile
stresses for bolts. (Based on the net
cross-sectional area of the bolt.)**

The interaction formula between the allowable tensile and shear
stresses in a bolt specified by BS 449 Cl. 48d is tri-linear

$$f_{bt}/p_{bt} + f_{bq}/p_{bq} \leqslant 1.4 \tag{4.6}$$

$$f_{bt}/p_{bt} \leqslant 1 \tag{4.7}$$

$$f_{bq}/p_{bq} \leqslant 1 \tag{4.8}$$

Equations (4.6) to (4.8) may be conveniently expressed in terms of
bolt forces for design purposes.

$$F_{bt}/P_{bt} + F_{bv}/P_{bq} \leqslant 1.4 \tag{4.6a}$$

$$F_{bt}/P_{bt} \leqslant 1 \tag{4.7a}$$

$$F_{bq}/P_{bq} \leqslant 1 \tag{4.8a}$$

The allowable tensile strength of a bolt, P_{bt}, is calculated
using the net cross sectional area of the bolt at the root of the
thread. The allowable shear strength of a bolt, P_{bq}, is calculated
using the gross cross sectional area. Values of P_{bt} and P_{bq} for
common bolt sizes are given in Table 4.7 for grade 4.6 bolts and
Table 4.8 for grade 8.8 bolts. Equations (4.6a) to (4.8a) may also be
expressed in graphical form for each type and size of bolt.

The application of an external tensile force F_{bt} to a HSFG bolt reduces the preload force P_{bs} and consequently reduces the frictional resistance F_{bq}. This is expressed algebraically in BS 4604 Pt 1 Cl. 3.1.3 as an allowable shear load incorporating a load factor of 1.4 for shear and 1.7 for tension

$$F_{bq} \leqslant (\mu_s/1.4)(P_{bs} - 1.7 F_{bt})$$

rearranging

$$F_{bq}/(\mu_s P_{bs}/1.4) + F_{bt}/(P_{bs}/1.7) \leqslant 1$$

which can be expressed as an interaction formula similar to Equation (4.6a), thus

$$F_{bq}/P_{bq} + F_{bt}/P_{bt} \leqslant 1 \tag{4.9}$$

where P_{bq} is the allowable shear force and P_{bt} is the allowable tensile force on a bolt, which are given in Table 4.9 for bolt sizes in common use.

4.12.4   Bearing properties   A bolt subjected to a shear force develops enclosed bearing stresses between the bolt and the plate as shown in Fig. 4.9(b). The real distribution of bearing stresses is non-uniform, as shown in Fig. 4.9(c) for a bolt in double shear, but this is too complicated for design purposes, and the stress distribution is therefore assumed to be uniform as shown in Fig. 4.9(d). The experimental relationship found by Bahia and Martin, between average bearing stress and deformation for a bolt that failed in single shear, is shown in Fig. 4.10. The higher the bearing stresses the greater the elongation of the bolt hole. It should be noted that the bearing stresses at failure of the bolt in single shear are greater than the yield stress of the plate, which explains why allowable values given in Table 4.5 are relatively high.

The allowable bearing stresses according to BS 449 Cl. 50c for grade 4.6 and 8.8 bolts are given in Table 4.5. It is important for the designer to note that the critical bearing stress for grade 4.6 bolts is on the bolt because it is weaker than the plate, while for the grade 8.8 bolt the weaker material is always the plate. These bearing stresses are applicable when the end distance, e_b, as shown in Fig. 4.9, is twice the diameter of the bolt. In design situations bearing stresses are one control on the thickness of the plate. A thickness of ply required to resist the maximum allowed shear force, for maximum allowable bearing stress is given in Tables 4.7 and 4.8 calculated from the equation

$$P_{bb} = d_b t_p p_b \tag{4.10}$$

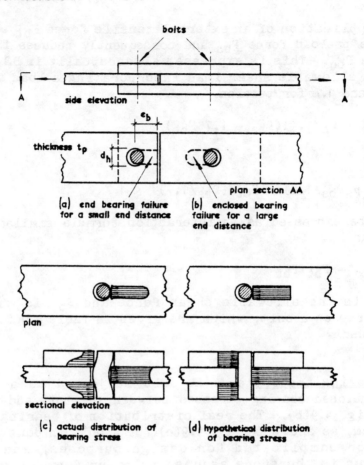

**Fig. 4·9. Bearing stresses
in bolts.**

Bearing stresses are not applicable to HSFG bolts at working load
because the connection is designed not to slip into bearing. BS 4604
Cl. 3.4 however states that no outer ply, and preferably no inner ply,
should be smaller in thickness than half the bolt diameter, or 10 mm,
whichever is less. This is to avoid crumpling, tearing, or bending of
plies.

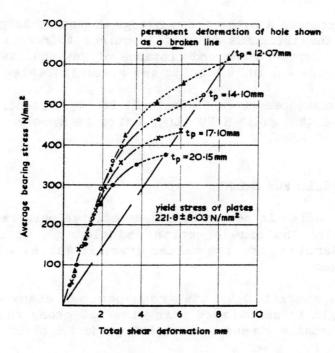

note: bolts failed on the threads in single shear at a shear
stress of 600 N/mm²

**Fig. 4·10. Relationship between bearing stress and
deformation for an M20 high-strength single bolt
(tensile test) by Bahia & Martin.**

4.13   EDGE AND END DISTANCES FOR BOLT HOLES

To develop the required strength for a connection the end and edge
distances of holes must not be too small. The edge distance is the
distance at right angles to the direction of stress from the centre
line of a hole to the outside edge of the plate. Minimum edge
distances specified in BS 449 and codes of practice in other countries
vary from 1.2 to 1.7 times the diameter of the hole. The values
prescribed by BS 449 are approximately 1.6 to 1.7 times the diameter
of the hole and are given in Table 21 (Appendix).

The end distance is from the centre line of the bolt hole to the
end of the plate in the direction of the load. This distance is
specified in design to prevent shear failure along the dotted lines
shown in Fig. 4.9(a). In common practice the end distance is
generally $2d_b$ for grade 4.6 and 8.8 bolts and this distance allows the
maximum allowable bearing stress p_b to be used, and consequently the
minimum thickness of plate. If the end distance e_b is reduced to less
than $2d_b$, then the maximum allowable bearing stress is reduced in
proportion to $(e_b/2d_b)p_b$, and consequently the plate is thicker. The

minimum end distances are the minimum edge distances as given in Table 21 (Appendix). The thickness of plate required to resist the maximum allowable shear force for an end distance of $2d_b$ and the minimum end distance for grade 4.6 and 8.8 bolts are given in Tables 4.7 and 4.8.

The edge distances given in Table 21 BS 449 (Appendix) also apply to HSFG bolts although BS 5400 Pt 3, which is more recent, gives a minimum value of $1.5d_b$.

4.14   DESIGN TABLES FOR BOLTS

The use of the tables in design avoids repetitive calculations for the determination of the size of bolts and plates. It is important however to understand how the values given in Tables 4.7, 4.8 and 4.9 have been derived.

A bolt is specified by its gross nominal diameter, but the tensile strength is calculated using the net cross sectional area. The allowable tensile strength of an M20 grade 4.6 black bolt is

$$P_{bt} = \text{(net tensile stress area)} \times \text{(allowable tensile stress)}$$
$$= A_b P_{bt} = 245 \times 120 \times 10^{-3} = 29.4 \text{ kN}$$

This value is shown in Table 4.7. The net tensile stress area of 245 mm^2 used in the calculations is given in a Table taken from the Structural Steel Handbook (Appendix).

The shear capacity of a bolt depends on whether the shear plane intercepts the threaded or the gross cross section. Generally in design the area on the gross diameter is assumed to resist the shear force. The allowable shear strength of an M20 grade 4.6 black bolt is

$$P_{bq} = \text{(gross cross section)} \times \text{(allowable shear stress)}$$
$$= (\quad/4)d_b^2 p_q = (\quad/4)(20)^2 \times 80 \times 10^{-3} = 25.13 \text{ kN}$$

This value is given in Table 4.7.

The minimum plate thickness required to resist the maximum allowable single shear strength of an M20 grade 4.6 bolt bearing on a grade 43 steel plate, depends on the bearing stress.

An end distance of twice the bolt diameter allows the use of the thinnest plate, and the maximum allowable bearing stress. The allowable single shear force

$$P_{bb} = \text{(bolt diameter)} \times \text{(plate thickness)} \times \text{(allowable bearing stress)}$$
$$P_{bb} = d_b t_p p_b$$
$$25.13 \times 10^3 = 20 \times t_p \times 250$$
$$t_p = 5.03 \text{ mm}$$

The thickness of plate given in Table 4.7 is 5.0 mm. If the end distance is reduced to the minimum of 30 mm as specified in Table 21, BS 449 (Appendix) then the bearing stress is decreased and the plate thickness is increased.

$$P_{bb} = d_b t_p (e_b/2d_b) p_b$$
$$25.13 \times 10^3 = 20 \times t_p \times (30/(2 \times 20))\ 250$$
$$t_p = 6.70\ mm$$

This value is also given in Table 4.7.

If a bolt is in double shear then there is an inner ply and two outer plies as shown in Fig. 4.9. Each thinner outer ply must be at least the thickness previously calculated to develop the single shear strength of the bolt. The inner ply thickness must be twice each outer ply thickness.

The allowable shear strength of a general grade M20 HSFG bolt for a single ply is

$$P_{bq} = (\text{coefficient of friction}) \times (\text{proof load})/(\text{load factor})$$
$$= \mu_s P_{bs}/\gamma = 0.45 \times 144/1.4 = 46.29\ kN.$$

The value given in Table 4.9 is 46.3 kN. The proof load of 144 kN is given in a Table taken from the Structural Steelwork Handbook (Appendix) and the value of the load factor $\gamma = 1.4$ is for loading conditions without wind. The shear resistance for conditions with wind is $P_{bq} = 0.45 \times 144/1.2 = 54.0\ kN.$

The allowable tensile strength of a general grade M20 HSFG bolt is

$$P_{bt} = (\text{proof load})/(\text{load factor})$$
$$= P_s/\gamma = 144/1.7 = 84.71\ kN$$

The value given in Table 4.9 is 84.7 kN.

Bolt size	Allowable tensile strength	Allowable single shear strength	Min. thickness of steel plies (mm) to develop shear capacity in bearing		
			End dist. $2d_b$ & min *	grade 43 steel p_b = 250 N/mm^2	
				outer	inner
	kN	kN	mm	mm	mm
M12	10.1	9.1	24 22	3.0 3.3	6.0 6.6
M16	18.8	16.1	32 26	4.0 5.0	8.0 10.0
M20	29.4	25.1	40 30	5.0 6.7	10.0 13.4
M24	42.4	36.2	48 36	6.0 8.0	12.0 16.0
M30	67.3	56.5	60 50	7.5 9.1	15.0 18.2
M36	98.0	81.4	72 62	9.1 10.5	18.2 21.0
M42	134.4	110.8	84 72	10.6 12.3	21.2 24.6
M48	176.4	144.8	96 82	12.1 14.1	24.2 28.1

Note: $P_{bt} = A_{net} \times 120$, $P_{bq} = A_{gross} \times 80$, $P_{bb} = d_b t_p p_b$ Newtons
  * min. end dist. from Table 21 BS 449, or $1.6d_h$

Table 4.7   Allowable design strengths for grade 4.6 bolts.

Bolt size	Allow. tens. str.	Allow. sing. shear str.	Minimum thickness of steel plies in mm to develop shear capacity in bearing						
			End dist. $2d_b$ & min *	grade 43 p_b=250N/mm^2		grade 50 p_b=350N/mm^2		grade 55 p_b=400N/mm^2	
				outer	inner	outer	inner	outer	inner
	kN	kN	mm	mm	mm	mm	mm	mm	mm
M12	23.7	21.2	24	7.1	14.2	5.0	10.0	4.4	8.8
			22	7.7	15.4	5.5	11.0	4.8	9.6
M16	44.1	37.6	32	9.4	18.8	6.7	13.4	5.9	11.8
			26	11.6	23.2	8.3	16.6	7.2	14.4
M20	68.7	58.8	40	11.8	23.6	8.4	16.8	7.3	14.6
			30	15.7	31.4	11.2	22.4	9.7	19.4
M24	99.1	84.6	48	14.1	28.2	10.1	20.2	8.8	17.6
			36	18.8	37.6	13.4	26.8	11.8	23.6
M30	157.4	132.2	60	17.6	35.2	12.6	25.2	11.0	22.0
			50	21.2	42.4	15.1	30.2	13.2	26.4
M36	229.3	190.4	72	21.2	42.4	15.1	30.2	13.2	26.4
			62	24.6	49.2	17.6	35.2	15.4	30.8
M42	314.3	259.2	84	24.7	49.4	17.6	35.2	15.4	30.8
			72	28.8	57.6	20.6	41.2	18.0	36.0
M48	412.5	339.5	96	28.2	56.4	20.2	40.4	17.6	35.2
			82	33.0	66.0	23.6	47.2	20.6	41.2

Note: P_{bt} = A_{net} x 280.6, P_{bq} = A_{gross} x 187.1, P_{bb} = $d_b t_p p_b$ Newtons
 * min. end dist. from Table 21 BS 449, or 1.6d_h

Table 4.8   Allowable design strengths for grade 8.8 bolts.

Bolt size	Proof load kN	Allowable tensile strength (static load) $\gamma = 1.7$ kN	Allowable tensile strength (impact load) $\gamma = 2.0$ kN	Allowable shear strength (without wind) $\gamma = 1.4$ kN	Allowable shear strength (with wind) $\gamma = 1.2$ kN
M12	49.4	29.1	24.7	15.9	18.5
M16	92.1	54.2	46.1	29.6	34.5
M20	144	84.7	72.0	46.3	54.0
M22	177	104.1	88.5	56.9	66.4
M24	207	121.8	103.5	66.5	77.6
M27	234	137.6	117.0	75.2	87.8
M30	286	168.2	143.0	91.9	107.3
M36	418	245.9	209.0	134.4	156.8

Note: $P_{bt} = (\text{proof load})/\gamma$; $P_{bq} = 0.45(\text{proof load})/\gamma$

Table 4.9 Allowable design strengths for general grade HSFG bolts.

4.15 STEEL PACKINGS (Cl. 48d BS 449)

To ensure that a bolt is effective, there should be one thread projecting beyond the nut, and one thread between the nut and the unthreaded shank, after tightening. Where connections do not fit, or where different sizes of components are used, then steel packings are inserted. The effect of the packing on a bolt is to introduce additional bending moments which reduce the strength of the bolt. No reduction need be made for up to 6 mm thick packings, or friction grip fasteners in non-slip connections, but for other thicknesses the shear capacity is reduced by 1.25% for each additional millimetre thickness.

4.16 BOLTED CONNECTIONS WITH PRYING FORCES

A prying force is an axial force that is induced in a bolt in addition to the external applied force and has been investigated by Douty and McGuire. The simplest type of connection where this occurs is a tee stub joint as shown in Fig. 4.11, although this connection does not occur often in practice. The prying force occurs with all types of bolts, with and without preload. If the tee stub joint is subjected to an external tensile force F_e then this induces a tensile force F_{bt} in the bolt and prying force Q_b. From a consideration of the equilibrium of forces the tensile force on the bolt

$$F_{bt} = F_e + Q_b \tag{4.11}$$

The magnitude of Q_b is greatest for a thin plate and may be zero for a very thick plate. Prying forces are also present in beam-to-column connections where an end plate is welded to a beam and bolted to the column, as shown in Fig. 4.11(b).

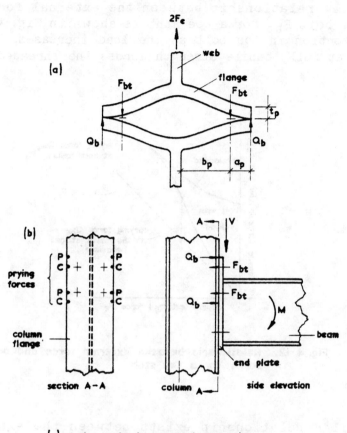

(a) prying action in a tee stub connection
(b) prying forces in a beam-to-column connection

Fig. 4·11. Prying forces.

The prying force develops because flange plates are in contact in the tee stub connections, or because the end plate is in contact with the column flange in the beam-to-column connections. The contact areas in the tee stubs are clearly seen to be at the ends of the flange plates. In the beam-to-column connections the contact areas are not so well defined, but limited experimental evidence suggests that they are in pairs and may be any combination of those labelled C and P as shown in Fig. 4.11. If the end plate and column flange are very thick, i.e. greater than 25 mm, then there is often no contact area when the connection is subjected to a bending moment. If the column flange is very thin, 6 mm, and the end plate is 15 mm thick, then there appear to be eight contact areas as shown by experiments by

Packer and Morris. In experiments by Bahia et al there appeared to be
four contact areas in the elastic stage of behaviour labelled C in
Fig. 4.11, and two contact areas at the bottom of the end plate
labelled P at fracture of the bolts.

A typical relationship between the external force F_e and the
force in the bolt F_{bt} for a tee stub is shown in Fig. 4.12. Despite
bending distortion in the bolt as the load increases, failure of the
bolt occurs at full tensile strength across the threaded portion.

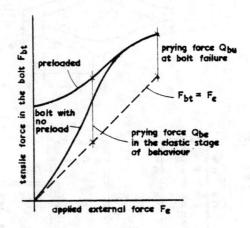

Fig. 4·12.   Relationship between external force and bolt
force for a tee stub.

A similar relationship exists between the external applied
bending moment and the moment of resistance of the forces in the bolts
for a beam-to-column connection. The tensile forces in the bolts
adjacent to the tension flange of the beam are approximately equal,
and the bolts adjacent to the compression flange only resist a small
part of the bending moment. The prying force in the elastic stage of
behaviour may be greater than at ultimate load and depends on the
number of contact areas.

4.17   ELASTIC THEORY FOR PRYING FORCES

In situations as shown in Fig. 4.11 the bolts must be designed to
resist the external forces plus the prying forces, and it is therefore
necessary to develop a theory to calculate the magnitude of the prying
force. For an allowable stress method of design the prying force Q_{be}
is related to the external force F_e, assuming linear elastic behaviour
of the components.

The theoretical model in Fig. 4.13 shows an end plate of thickness t_p and of cantilever length $(a_p + b_p)$. The extremity of the plate is in contact with a column flange, for example, and this introduces the prying force Q_{be}. The external force F_e is assumed to be balanced by the prying force Q_{be} and an axial force F_{bt} in the bolt. The axial force in the bolt produces an extension of the bolt of δ_b.

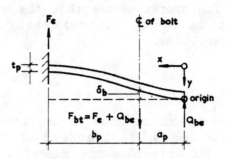

Fig. 4·13. Forces acting in the elastic stage for prying force theory.

Applying Macaulay's method for the deflection of a beam with the origin at 0 and the deflection positive downwards.

$$EI(d^2y/dx^2) = - Q_{be}x + (F_e + Q_{be}) [x - a_p] \qquad (4.12)$$

Integrating twice and applying the following boundary conditions; $dy/d_x = 0$ when $x = a_p + b_p$, $y = 0$ when $x = 0$, and $y = -\delta_b$ when $x = a_p$ the following solution is obtained

$$Q_{be} = \frac{F_e - 2EI\delta_b/a_pb_p^2}{2(a_p/b_p) + (2/3)(a_p/b_p)^2} \qquad (4.13)$$

The extension of the bolt originally preloaded with a force F_{bs} is

$$\delta_b = (F_{bt} - F_{bs})g_p/A_bE_b = (F_e + Q_{be}-F_{bs})g_p/A_bE_b \qquad (4.14)$$

Combining Equations (4.13) and (4.14) and rearranging

$$Q_{be} = \frac{F_e(1-k_e) + k_eF_{bs}}{2(a_p/b_p) + (2/3)(a_p/b_p)^2 + k_e} \qquad (4.15)$$

where $k_e = E_pw_pt_p^3g_p/(6a_pb_p^2A_bE_b)$ \qquad (4.16)

Equation (4.15) is applicable in the linear elastic range of behaviour and relates the prying force Q_{be} to the external applied force F_e. A more detailed development of the theory is given in Holmes and Martin.

In design situations the width of plate per bolt w_p in the tee stub theoretical model is clearly defined and is also easily determined in the case of an end plate extension in beam-to-column connection. Prying forces, however, also develop in relation to deformations in a column flange similar to the tee stub model based on the elastic theory. Timoshenko shows that the value of an effective width w_c equivalent to w_p is approximately $(b_c + c_c/2)$ or $2b_c$ whichever is the lesser value.

4.18   GUSSET PLATES

Gusset plates are used in connections to provide sufficient space for bolts or to distribute loads, in brackets or foundation bases, as shown in Fig. 4.14(a). In this case the gusset plate is triangular in shape and the loading is applied to one edge as shown in Fig. 4.14(b). BS 449 gives no information on how to deal with the design of gusset plates and the thickness of the plate is generally decided on past experience. The mode of failure is by elastic-plastic buckling, and empirical solutions have been produced by Salmon et al.

An alternative approximate theoretical method of design for non-slender gusset plates which has been substantiated by experiments is given by Martin. The basic structural shape of a gusset plate is shown in Fig. 4.14(b), and for theoretical purposes the plate is assumed to act as a series of fixed ended struts parallel to the free edge. The distribution of direct stress across the width W_g is shown on an element of the gusset plate in Fig. 4.14(c). The stress varies from yield stress for a short strut at the inside right angle, to a stress at the free edge depending on the slenderness ratio. The curve connecting these two stresses is complicated but it can be replaced by a straight line as shown in Fig. 4.14(d), provided that the slenderness ratio of the free edge $l_g/r_g < 185$.

Taking moments of forces about a theoretical hinge at 0 for conditions at buckling and ignoring the moment of resistance of the base plate as shown by Robinson and Martin

$$P_{gu}s_g = \int_0^W w_g f_g t_g \ dw_g \qquad (4.17)$$

For each strip the buckling stress of f_g is linearly related to the slenderness ratio (l_g/r_g). The effective length $l_g = w_g$ when $L_g = H_g$, and from experiments by Martin this was found to be approximately correct when $L_g \neq H_g$.

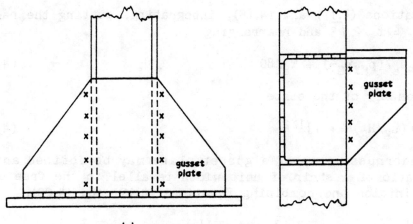

(a) examples of gusset plates

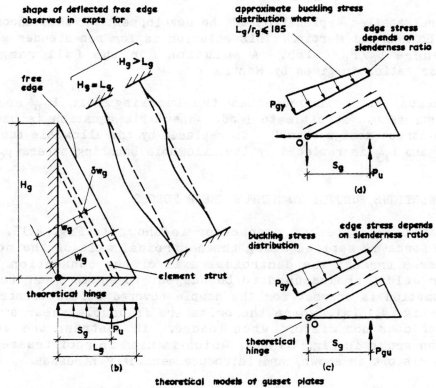

theoretical models of gusset plates

Fig. 4·14. Gusset plates.

The buckling stress for each strip is

$$f_g = p_{gy}[1 - (w_g/(185r_g))] \text{ provided that } l_g/r_g < 185 \qquad (4.18)$$

Combining Equations (4.17) and (4.18), integrating, putting the radius of gyration $r_g = t_g/2\sqrt{3}$ and rearranging

$$t_g = 2P_{gu}s_g/(p_{gy}W_g^2) + W_g/80 \qquad (4.19)$$

where from geometry of the plate

$$W_g = L_g/[(L_g/H_g)^2 + 1]^{1/2} \qquad (4.20)$$

The slenderness ratio of a gusset plate may be defined as the slenderness ratio of a strip of unit width parallel to the free edge. From this definition and combining Equations (4.19) and (4.20),

$$\frac{l_g}{r_g} = \frac{W_g}{(t_g/2\sqrt{3})} = \frac{2\sqrt{3}}{[(L_g/H_g)^2 + 1]^{1/2}} \left(\frac{L_g}{t_g}\right) \qquad (4.21)$$

A more detailed explanation of the development of this theory is given in Holmes and Martin. This solution is for non-slender gusset plates where $l_g/r_g < 185$. A solution for the full range of slenderness ratios is given by Martin.

Equation (4.19) incorporates the buckling load P_{gu} and the buckling stress p_{gy} at ultimate load. When this equation is used in allowable stress design then P_{gu} is replaced by the allowable buckling load P_{ge}, and p_{gy} is replaced by the allowable buckling stress p_c.

4.19   CONNECTIONS SUBJECT TO SIMPLE SHEAR FORCES

One the simplest types of connection is shown in Fig. 4.15. The external force is applied along the centroidal axis of the member which passes through the centroidal axis of the connection. The bolts, or welds, are assumed to be subject to equal shear forces. This assumption is correct for the double covered bolted connections shown in Fig. 4.15(a), where the bolts are in double shear and the connection does not distort when loaded. In contrast the welded connection shown in Fig. 4.15(b), which is used in roof trusses, is likely to distort as shown, and introduce secondary stresses.

When the type of connection shown in Fig. 4.15(a) is long in the direction of the force the bolts at the end resist a greater force than the elements at the centre. This is because the inner ply, extends further than the outer plies, i.e. they are not compatible. BS 449 does not give advice for this situation but theoretical work is given in Fisher and Struik, and an empirical formula for bolts is given in BS 5400 Pt 3, expressed as a factor which reduces the full strength of the connection.

$$\text{shear strength factor} = 1 - (L_j - 15d_b)/(200d_b) \qquad (4.22)$$

where L_j is the joint length in metres on one side of the connection. The joint length is the length over which the load is transferred to the splice plate as shown in Fig. 4.15(a). The shear strength factor should not be less than 0.75, and L_j should be greater than $15d_b$ when applying Equation (4.22).

For the welded joint shown in Fig. 4.15(b) the overlap distance l, should preferably be equal to or greater than b, according to BS 5400 Pt 3 and ECSS, and not less than four times the thickness of the thicker part connected, according to BS 5135. If only side welds are used for this connection then there should be end returns of not less than twice the weld size. Side welds however are susceptible to corrosion in exposed conditions.

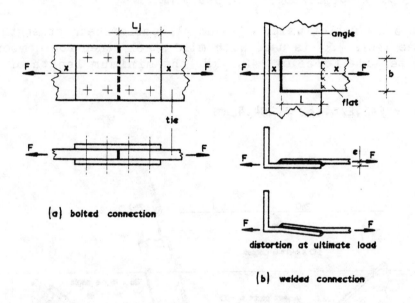

(a) bolted connection

(b) welded connection

Fig. 4·15. Examples of connections subject to simple shear forces.

Example 4.1 Connection subject to a simple shear force Design the connection for joint 4 in the roof truss design in Chapter 6. The size of angles and tees are shown in Fig. 4.16 and have been determined from the forces obtained from the frame analysis in Chapter 6.

The length and breadth of the gusset plate is obtained from a scale drawing where overlaps of angle and tee sections are at least equal to the width of the connected leg as explained for the welded connection shown in Fig. 4.15(b). The thickness of the gusset plate must not be less than 6 mm to allow for corrosion. As a rule of thumb the thickness of the gusset plate should also be at least equal to the thickness of the angle and tees at the connection. A rectangular gusset plate has been chosen because it is low in cost, simple to mark out, and simple to cut. Alternative complicated shapes may be aesthetically more acceptable and more economical in material, but are higher in fabrication cost.

To avoid distortions in the gusset plate according to BS 5400 Pt 3, the maximum unsupported length b_g as shown in Fig. 4.16 is

$$b_g/t_g \leqslant 50[355/Y_s]^{1/2}$$

For this example if t_g = 10 mm

$$b_g \leqslant 10 \times 50[355/250]^{1/2} = 595.8 \text{ mm}$$

If a 6 mm fillet weld, with an allowable shear strength of 0.483 kN/mm (see Table 4.3) is used with mild steel, then for member 24 with an axial tensile force of 142.2 kN, the minimum length of the weld required is

$$T/F_w = 142.2/0.483 = 294.4 \text{ mm}$$

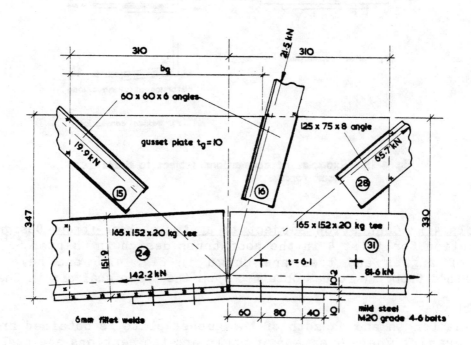

Fig. 4.16. Detail of connection no. 4 of roof truss.

Two side welds each of length 150 mm with end returns are therefore satisfactory. The length of weld for members 15, 16 and 28 are determined using the same method. This type of calculation ignores the small eccentricity of the force in relation to the centroid of the weld group. If the eccentricity is taken into account as shown in Section 4.20 then the increase in weld size is small.

The connection for member 31 is site bolted and the arrangement of bolts is shown in Fig. 4.16. The four M20 bolts in single shear resist a shear force of P_{bq} = 4 x 25.1 = 100.4 > 81.6 kN therefore satisfactory. The single shear resistance of an M20 grade 4.6 bolt is obtained from Table 4.7. Also from Table 4.7 the thickness of ply required to resist this shear force for an end distance of 40 mm, is 5.0 mm. The thickness of the gusset plate and web of the 165 x 152 x 20 kg tee sections are adequate to resist bearing stresses.

4.20 CONNECTIONS SUBJECT TO ECCENTRIC SHEAR FORCES

Connections, such as those shown in Fig. 4.17, are subject to eccentric shear forces where the bolts, or welds, are subject to eccentric shear forces due to the moment $M = V_e$ and direct shear forces due to the shear force V.

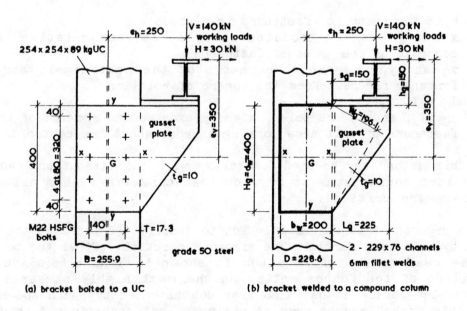

(a) bracket bolted to a UC   (b) bracket welded to a compound column

Fig. 4.17. Column bracket subject to an eccentric shear force.

Connections subject to eccentric shear forces rotate about an instantaneous centre of rotation, as shown in Fig. 4.18(a). The position of the instantaneous centre of rotation varies with the magnitude of the external forces, e.g. when M = Ve = V = 0 the centre is at infinity on the y-y axis through the centroid of the bolt group and when H = 0 and V acts at infinity then the centre is at the centroid of the group of fasteners. In the linear-elastic stage of behaviour it is assumed that the shear force acting on a fastener is proportional to the relative movement (slip) of the plates, which is proportional to its distance from the centre of rotation, provided that the connected materials behave as rigid bodies. One fastener, furthest from the centre of rotation, will therefore be subject to a greater shear force than the others.

The force on the most highly loaded fastener may be obtained by a simple theory in common use that assumes that rotation occurs about the centroid of the fastener group. For convenience the forces acting on a fastener are determined in two perpendicular directions and then combined vectorially, as shown in Fig. 4.18(b). Algebraically the resultant force on a fastener furthest from the centroid of the group may be expressed as:

$$F_R = [F_y{}^2 + F_x{}^2]^{1/2}$$

$$= [(V/n + Mx_G/I_G)^2 + (H/n + My_G/I_G)^2]^{1/2} \qquad (4.23)$$

where n is the number of fasteners in a group
   x and y are the coordinates of a fastener relative to the centroid of the group of fasteners
   x_G and y_G are the coordinates of the most highly stressed fastener furthest from the centre of rotation
   $I_G = I_x + I_y$
   I_G, I_x, and I_y are polar, x and y axes second moments of area of fasteners of unit area forming a group about the centroid.

This method may be used for bolts, with and without preload, and for welded connections at service load conditions when allowable stresses are critical.

A rigorous theoretical solution to justify the previous theory is given by Bahia and Martin. The rigorous solution defines the position of the real centre of rotation as shown in Fig. 4.18(a) but the magnitude of the forces acting on the most highly stressed bolt remains the same. It has also been demonstrated by Bahia and Martin that this method may be used at ultimate load conditions although it is slightly conservative.

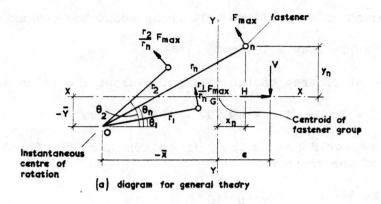

(a) diagram for general theory

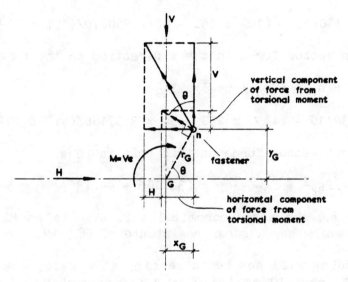

(b) vector diagram for simple vector addition theory
for a typical fastener

**Fig. 4·18.. Diagrams for the theory for a connection
subject to an eccentric shear force.**

Example 4.2 **Column bracket subject to eccentric shear forces**
Determine the thickness of the plate, size of fasteners required to
connect the bracket to the column at working load as shown in Fig.
4.17. Use grade 50 steel and design as a rigid connection.

Example 4.2.1 **Bolted connection** The first design uses HSFG bolts
spaced as shown in Fig. 4.17(a). For bolts of unit cross sectional
area the properties of the bolt group are:

The second moment of area of the bolt group about the x-x axis

$$I_{bx} = 4(80^2 + 160^2) = 128 \times 10^3 \text{ mm}^4$$

The second moment of area of the bolt group about the y-y axis

$$I_{by} = 10 \times 70^2 = 49 \times 10^3 \text{ mm}^4$$

The second moment of area of the bolt group about the polar axis

$$I_{bG} = I_{bx} + I_{by} = (128 + 49)10^3 = 177 \times 10^3 \text{ mm}^4$$

The maximum vector force in the y direction on a bolt furthest from the centroid of the group is

$$F_{by} = V/n_b + (Ve_h + He_v)x_n/I_{bG}$$

$$= 140/10 + (140 \times 250 + 30 \times 350)70/(177 \times 10^3) = 32 \text{ kN}$$

The maximum vector force in the x direction on the same bolt is

$$F_{bx} = H/n_b + (Ve_h + He_v)y_n/I_{bG}$$

$$= 30/10 + (1440 \times 250 + 30 \times 350)160/(177 \times 10^3) = 44.13 \text{ kN}$$

The resultant vector force on this bolt is thus

$$F_{br} = [F_{by}^2 + F_{bx}^2]^{1/2} = [32.00^2 + 44.13^2]^{1/2} = 54.50 \text{ kN}$$

From Table 4.9 the most economical bolt size is an M22 general grade HSFG bolt which has a shear resistance of 56.9 kN.

The HSFG bolts will not be in bearing at working load but the plate must be at least 10 mm thick to avoid crumpling as specified in BS 4604.

Example 4.2.2 Welded Connection An alternative welded connection is necessary where it is not possible to bolt to the column, e.g. the compound channel column shown in Fig. 4.17(b).

The design properties of the weld group composed of welds of unit size are:

The total length of weld

$$L_w = 2(d_w + b_w) = 2(400 + 200) = 1200 \text{ mm}$$

The second moment of area of the weld group about the x-x axis

$$I_{wx} = 2[d_w^3/12 + b_w(d_w/2)^2] = 2[400^3/12 + 200(400/2)^2]$$

$$= 26.67 \times 10^6 \text{ mm}^4$$

The second moment of area of the weld group about the y-y axis

$$I_{wy} = 2[b_w^3/12 + d_w(b_w/2)^2] = 2[200^3/12 + 400(200/2)^2]$$

$$= 9.33 \times 10^6 mm^4$$

The second moment of area of the weld group about the polar axis

$$I_{wG} = I_{wx} + I_{wy} = (26.67 + 9.33)10^6 = 36 \times 10^6 mm^4$$

The maximum applied vector force per unit length in the y direction, on a weld element furthest from the centroid of the weld group is

$$F_{wy} = V/L_w + (Ve_h + He_v)x_n/I_{wG}$$

$$F_{wy} = 140/1200 + (140 \times 250 + 30 \times 350)100/(36 \times 10^6) = 0.243 \text{ kN/mm}$$

The maximum applied vector force per unit length on the same weld element in the x direction is

$$F_{wx} = H/L_w + (Ve_h + He_v)y_n/I_{wG}$$

$$= 30/1200 + (140 \times 250 + 30 \times 350)200/(36 \times 10^6) = 0.278 \text{ kN/mm}$$

The resultant vector force per unit length on this weld element is

$$F_{wR} = [F_{wy}^2 + F_{wx}^2]^{1/2} = [0.243^2 + 0.278^2]^{1/2} = 0.369 \text{ kN/mm}$$

The weld size can now be obtained from Table 4.3. A 6 mm fillet weld for grade 50 steel resists a shear force of 0.672 kN/mm.

The thickness of the triangular part of the gusset plate shown in Fig. 4.17 may be determined from the theory in Section 4.18.

From Equation (4.20) the width of the gusset plate perpendicular to the free edge

$$W_g = L_g/[(L_g/H_g)^2 + 1]^{1/2}$$

$$= 225/[(225/400)^2 + 1]^{1/2} = 196.1 \text{ mm}$$

The thickness of the gusset plate is obtained from Equation (4.19) provided the buckling moment $P_g s_g$ is replaced by $Vs_g + Hh_g$ the appropriate value of the buckling moment in this case.

$$t_g = 2(Vs_g + Hh_g)/(p_g W_g^2) + W_g/80$$

$$= 2(140 \times 150 + 30 \times 150) \times 10^3/(215 \times 196.1^2) + 196.1/80$$

= 8.62 mm

A 10 mm thick grade 50 steel plate is suitable.

Check the slenderness ratio of the gusset plate from Equation (4.21).

$$l_g/r_g = 2 \sqrt{3} W_g/t_g = 2 \sqrt{3} \times 196.1/10$$

= 67.93 < 185 the slenderness ratio limit for the application of this theory.

4.21   CONNECTIONS WITH END BEARING

Some connections involve members butting together as shown in Fig. 4.19.   The diagram is general and represents beam-to-column, bracket-to-column, beam-to-beam, and column-to-base.   When a shear force, bending moment and axial force are applied to the beam, it rotates about the axis O located at a stiff bearing near the outer edge of the beam.   The bending moment M and axial force H are resisted by the forces R and R_O.   Force R is the resultant force of the fasteners, e.g. bolts or welds and force R_O is the reaction at the stiff bearing.

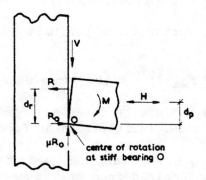

Fig. 4.19. Connections with end bearing.

Rotation takes place about the bearing at O because the structural stiffness of the connection changes at the interface of the column and the beam.   It should be emphasised, however, that the bearing must be stiff, i.e. it should not distort, especially in the line of action R_O.   If the bearing is not stiff it should be reinforced, otherwise rotation will occur about an axis closer to the mid-depth of the beam, and the strength of the connection will be reduced.

The beam is in contact with the support at O and a friction force μR_O is developed.   If the frictional force is large there will be negligible movement in the direction of V and the connection will rotate about O, the frictional force will resist the entire shear force , and the shear forces on the fasteners will be zero. Alternatively if the frictional force is small, rotation and slip occur simultaneously, and the shear force V is resisted partly by the friction and partly by the fasteners. The welds, or bolts, therefore do not resist all of the forces.

If no slip occurs then

$$V - \mu R_o < 0 \tag{4.24}$$

And, taking moments of forces about R

$$M \pm H(d_r - d_p) - R_o d_r = 0 \tag{4.25}$$

Combining Equations (4.24) and (4.25)

$$\mu[M \pm H(d_r - d_p)]/(V d_r) > 1 \tag{4.26}$$

Most connections with end bearing do not slip and the left hand side of Equation (4.26) is greater than unity, but one exception is a bracket fastened to a column where the eccentricity of the load is small.

Traditionally, the frictional force associated with end bearing connections has been ignored because BS 5135 states that it should not be assumed that parts joined are in contact under the joint. Recent research by Bahia et al has shown however that friction does exist, and the latest ECSS recommendations allow friction to be taken into account where contact pressure is known to exist. Irrespective of whether slip occurs the tensile forces in the bolts are determined by taking moments of forces about the axis of rotation O-O. If the fasteners are the same size and have the same deformation characteristics, then in the elastic stage of behaviour rigid-body rotation is assumed and forces on the fasteners are assumed to be proportional to the distance from the axis of rotation. The fastener furthest from the centre of rotation therefore resists the largest tensile force the magnitude of which may be determined by taking moments of forces about the axis O-O, i.e.

$$F_{t(max)} = [M \pm H(d_p)]y_{max}/I_o + Q_{be} \tag{4.27}$$

where $I_o = \Sigma(y)^2$ is the second moment of area of unit area fasteners about axis O-O, and Q_{be} is a prying force for a bolt if it exists.

If slip occurs, as determined by Equation (4.26) and fasteners are of the same size, the shear force is assumed to be distributed uniformly to the fasteners. The shear force on a fastener is then

$$F_q = (V - \mu_s R_o)/n = \{V - \mu_s[M/d_r \pm H(1 - d_p/d_r)]\}/n \qquad (4.28)$$

The distance d_r to the position of the resultant force in the fasteners must be determined for each situation by taking moments of the areas of the fasteners about the axis O-O.

If the fasteners are welds then F_t and F_q are combined vectorially to determine the resultant force per unit length and the size of the weld is obtained from Table 4.3. If the fasteners are grade 4.6 or 8.8 bolts the values of F_t and F_q are combined using Equations (4.6a) to (4.8a).

When HSFG bolts are used then at working load the shear force is resisted by the friction developed by the preload in the bolt and no slip occurs. The frictional resistance of a HSFG bolt is reduced however by a tensile force as described in Section 4.12.3.

4.22   PINNED CONNECTIONS

When applying the simple design method prescribed by BS 449 it is assumed that frictionless pins exist at the connections. Practical pinned connections, examples of which are shown in Figs. 4.20 and 4.21, are in fact semi-rigid but their moment of resistance is low. When using pinned joints it is therefore important to realise that the stiffness of the structure with respect to horizontal forces is low and bracing or stiffening shear walls should be introduced where necessary.

4.22.1 <u>Pinned beam-to-column connection</u>   A typical pinned beam-to-column connection is shown in Fig. 4.20. The shear force is resisted by the bottom bolts connecting the bottom cleat to the column flange. During erection this bottom cleat is often initially bolted or welded to the column and acts as a marker for the crane driver. The beam is then placed in position on the cleat and bolted through the bottom flange. The top cleat is assumed to resist no vertical load but it does provide torsional resistance which is important for lateral stability. The resistance of the web to shear, bearing and buckling, must be checked, as described in Chapters 2 and 7.

<u>Example 4.3</u>   <u>Pinned beam-to-column connection</u>   Determine the size of angle cleats and bolts, or welds, for the cleated connection shown in Fig. 4.20. Use grade 50 steel and grade 8.8 bolts.

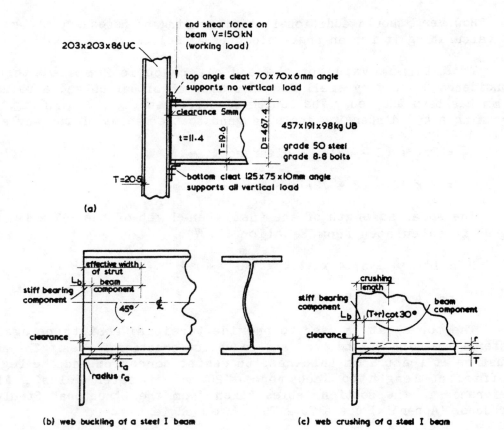

Fig. 4.20. Pinned beam-to-column connection.

The shear force of 150 kN is resisted by the four bolts in the vertical leg of the bottom angle cleat. From Table 4.8, four M20 grade 8.8 bolts in single shear resist a shear force of

$$P_q = 4 \times 58.8 = 235.2 \; > \; 150 \; kN \quad \text{therefore satisfactory.}$$

From Table 4.8 the minimum bearing thickness of the angle and the column flange for an M20 bolt and grade 50 steel is 8.4 mm. A 125 x 75 x 10 mm angle, grade 50 steel, will accommodate the bolts after checking the bolt spacings given in a Table which is taken from the Structural Steelwork Handbook (Appendix).

A check is now made on the web buckling and web bearing strengths of the beam. Using Equations (2.38) and (2.40), or the factors given in the Safe Load Tables of the BCSA-Constrado Handbook, it can be shown that in this case the resistance of the beam itself, without the aid of additional stiff bearing, is adequate for the shear force of 150 kN. For web buckling the beam component C_1 = 442 kN. For web bearing the beam component C_1 = 153 kN.

However, should additional stiff bearing be necessary the method of calculating it for an angle cleat is as follows.

The minimum value of the end clearance is 2 mm but this is considered to be very small for this type of situation and a value of 5 mm has been adopted. The length of stiff bearing shown in Fig. 4.20 assuming a 45° dispersion angle tangent to the radius of the angle.

$$L_b = 2t_a + (2 - \sqrt{2})r_a - \text{clearance} \qquad\qquad (4.29)$$

$$= 2 \times 10 + (2 - \sqrt{2})11 - 5 = 21.44 \text{ mm.}$$

The shear strength of the unstiffened web of the 457 x 191 x 98 kg UB is calculated from Equation (2.37).

$$Q = Dtp_q' = 467.4 \times 11.4 \times 140 \times 10^{-3}$$

$$= 746.0 \ > \ 150 \text{ kN.}$$

The top cleat is used to provide torsional resistance against lateral buckling of the beam and to resist erection forces. The angle must be at least 6 mm thickness to resist corrosion and the legs of sufficient length to accommodate 20 mm diameter bolts. After reference to the section tables taken from the Structural Steelwork Handbook (Appendix) use a 70 x 70 x 6 mm angle.

As an alternative, the four bolts in the vertical leg of the 125 x 75 x 10 mm angle could be replaced by 6 mm fillet welds along the two vertical edges of the cleat with end returns. The shear resistance of the 6 mm fillet welds from Table 4.3 is

$$P_W = 2l_wF_W = 2 \times 125 \times 0.672 = 168 \ > \ 150 \text{ kN}$$

4.22.2 **Pinned-beam-to-beam connection** Transverse secondary beams in simple floor systems are often connected to the main longitudinal beams as shown in Fig. 4.21. This type of simple web connection using angle cleats allows rotation at the ends of the secondary beams which are then designed as though simply supported at the ends. Grade 4.6 bolts are generally used because they allow rotation of the connection in the bolt holes, but grade 8.8 bolts can be used if space is limited. The torsional resistance of the main beam is low compared with a rigid support, and if there is a transverse beam on one side only this helps to justify the theoretical assumption of a pinned end.

The end shear force from the transverse secondary beam A in Fig. 4.21 is transferred to the main longitudinal beam B, via the bolts labelled 'b' and is assumed to act along the interface between the cleat and the web. These bolts are therefore designed for single shear and bearing on the web of beam B and on the web cleats.

It follows that the reaction from beam B is applied eccentrically to the bolts labelled 'a' as shown in Fig. 4.21. These bolts are in double shear and bearing on the web of beam A and the angle cleats. The maximum resultant force on a bolt is determined using the vector addition method described in Section 4.20.

An alternative mode of failure that can occur with this type of connection is associated with holes in the web of the transverse beam A. The web may fail in shear along a line connecting the bolt holes and a check on shear resistance along the bolt line is necessary. Also an excessive clearance should not be specified at the ends of the transverse beam A, otherwise there may be an end bearing failure of the bolts if large rotations occur.

Steel fabricators and erectors often prefer a welded end plate connection as an alternative to the two web cleats. This type of connection is more rigid and will introduce an end moment to beam A if there are transverse beams on both sides of the main beam B. If there is a transverse beam on one side only the end moment on beam A will be small because generally the torsional stiffness of the main beam B is low.

Example 4.4   Pinned beam-to-beam connection   Determine the size of bolts and angles required to connect the secondary beam to the main beam as shown in Fig. 4.21. The shear force at the end of the secondary beam is 100 kN at working load. Use grade 43 steel. The details are taken from the BCSA Structural Steel Handbook.

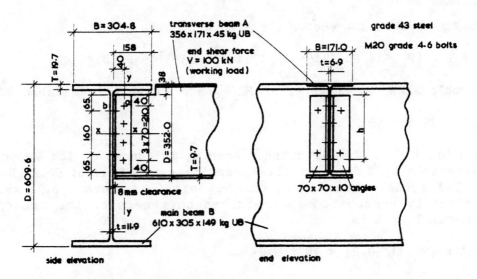

Fig. 4·21. Pinned beam-to-beam connection.

Assuming that the bolts in the web of the main beam are subjected to single shear forces then the number of M20 grade 4.6 bolts required is

$$n_b = V/P_{bv} = 100/25.1 = 3.98. \quad \text{Use 4 M20 grade 4.6 bolts.}$$

The value of P_{bv} = 25.1 kN is obtained from Table 4.7. Also from Table 4.7 the web bearing thickness of the main beam and the bearing thickness of the angles must be greater than 5 mm for an end distance of 40 mm. The thickness of the web of the main beam is 11.9 mm which is satisfactory. Use 70 x 70 x 10 mm angles as cleats.

The bolts connecting the cleats to the transverse beam A are in double shear and assumed to be subject to an eccentric shear force. The second moment of area of the single line of bolts about the x-x and y-y axes shown in Fig. 4.21 are

$$I_{bx} = 2(35^2 + 105^2) = 24.5 \times 10^3 \text{ mm}^2$$

$$I_{by} = 0$$

$$I_{bz} = I_{bx} + I_{by} = 24.5 \times 10^3 \text{ mm}^2$$

The force on a bolt in the x direction is

$$F_{bx} = (Ve)y_n/I_{bz} = (100 \times 40)105/24.5 \times 10^3 = 17.14 \text{ kN}$$

The force on a bolt in the y direction is

$$F_{by} = V/n_b = 100/4 = 25 \text{ kN}$$

The maximum resultant vector force on a bolt is

$$F_{bR} = [F_{bx}^2 + F_{by}^2]^{1/2} = [17.14^2 + 25^2]^{1/2} = 30.31 \text{ kN}$$

The double shear strength of an M20 grade 4.6 bolt from Table 4.7 is

$$2 \times 25.1 = 50.2 \text{ kN.}$$

From Table 4.7 the required inner bearing thickness of the web of the transverse beam A for a double shear force on a bolt of 50.2 kN is 10 mm. The actual shear force on the bolt is however 30.31 kN, and therefore by proportion the required thickness of the web of the transverse beam A is

$$t = (30.31/50.2) \times 10 = 6.04 \text{ mm}$$

The actual thickness of the web of the beam is 6.9 mm.

The shear resistance of the web of the transverse beam A along a vertical line through the holes is

$$Q = (h - 3.5d_h)tp_q'$$

$$= (250 - 3.5 \times 22) \times 6.9 \times 100 \times 10^{-3} = 119.4 \text{ kN}$$

This is greater than the applied shear force of 100 kN.

4.22.3 Pinned column-to-foundation connections (Cl. 38b BS 449) A column may be connected to a reinforced concrete foundation as shown in Fig. 4.22. The function of the two foundation bolts is to resist shear forces and the resistance to applied bending moments is small when rotating about the axis y-y. The connection is therefore assumed to act as a pin. The length and breadth of the base plate is related to the allowable bearing pressure on the bedding material between the steel base plate and the concrete. If the bearing stresses are low the base plate is made slightly larger than the column dimension, and the plate thickness is then chosen to be approximately equal to the thickness of the column flanges. If the bearing stresses are high the base plate dimensions are greater than the column dimensions, then the thickness of the base plate is related to the cantilever length A shown in Fig. 4.22. The actual relationship is complex but by making simplifying assumptions the formula given in BS 449 Cl. 38b to determine the thickness of the base plate can be derived as follows.

Consider an element of the steel base plate adjacent to the edge of the column flange as shown in Fig. 4.22(a). This element is subject to two mutually perpendicular bending stresses f_x and f_y associated with the cantilever length A and B. If A is the greater dimension and the maximum strain in the x direction is limited to p_{bct}/E, then

$$f_x/E - \nu f_y/E < p_{bct}/E \tag{4.30}$$

From the simple theory of elastic bending for a unit width of steel base plate

$$f_x = M/Z = (wA^2/2)/(t_p^2/6) \tag{4.31}$$

and $$f_y = M/Z = (wB^2/2)/(t_p^2/6) \tag{4.32}$$

Combining Equations (4.30), (4.31) and (4.32) and rearranging,

$$t > [(3w/p_{bct})(A^2 - \nu B^2)]^{1/2} \tag{4.33}$$

In BS 449 Poisson's ratio $\nu = 1/4$ and $p_{bct} = 185 \text{ N/mm}^2$. In practice the allowable bearing pressure w is generally assumed to be $0.25f_{cu}$.

The size of the fillet weld connecting a machined column end to the base plate is the minimum recommended, i.e. 6 mm. If the end of the column is not machined the welds must be designed to resist all of the axial force.

Example 4.5   Pinned column-to-foundation connection Determine the dimensions of a steel slab base to transmit an axial load of 1000 kN at working load to a concrete foundation as shown in Fig. 4.22. The column size is a 254 x 254 x 107 UC and the allowable bearing pressure w on the bedding material is $0.25f_{cu}$, i.e. w = 0.25 x 20 = 5 N/mm^2. The minimum area of base required is

$$A_p = N/w = 1000 \times 10^3/5 = 200 \times 10^3 \text{ mm}^2$$

If the base plate is square

$$D_p = (A_p)^{1/2} = (200 \times 10^3)^{1/2} = 447.2 \text{ mm, therefore use 450 mm.}$$

The maximum base plate extension

$$A = (D_p - B)/2 = (450 - 258.3)/2 = 95.85 \text{ mm}$$

The minimum base plate extension

$$B = (D_p - D)/2 = (450 - 266.7)/2 = 91.65 \text{ mm}$$

The base plate thickness from Equation (4.33)

$$t_p = [(3w/p_{bct})(A^2 - vB^2)]^{1/2}$$

$$t_p = [(3 \times 5/185)(95.85^2 - (91.65^2/4))]^{1/2} = 23.97; \text{ use 25 mm.}$$

If the end of the column is machined the load is assumed to be transferred directly to the base plate and a minimum size of fillet weld of 6 mm is used to connect the base plate to the column. Alternatively if no transfer of load by bearing is assumed then the force per unit length of weld is approximately

$$F_w = N/(4B + 2D) = 1000/(4 \times 258.3 + 2 \times 266.7) = 0.638 \text{ kN/mm}$$

From Table 4.3 an 8 mm fillet weld resists 0.644 kN/mm for grade 43 steel.

The base is subjected to a compressive force which is not transferred to the holding down bolts. The bolts are therefore only subject to erection forces and the diameter is often related to the base plate thickness. Use M24 grade 4.6 holding down bolts.

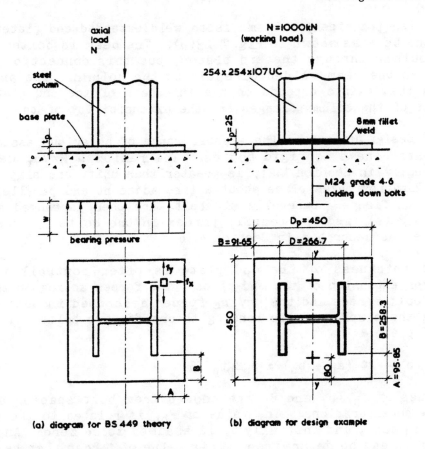

(a) diagram for BS 449 theory   (b) diagram for design example

Fig. 4·22. Pinned column-to-foundation connection.

4.23   RIGID CONNECTIONS

Rigid connections exhibit smaller linear and rotational displacements than pinned connections and are preferred when it is necessary to resist bending moments or to limit deflections of structures. They are also essential to resist fatigue and impact loading.

4.23.1   Rigid beam to column connections The most rigid beam-to-column connection is where the beam is welded directly to the column. The beam can be welded to the column on site, but this is expensive and difficult to control the quality of the welds. Alternatively stub cantilever beams can be welded to the column in the workshops and a suspended beam bolted between the ends of the cantilevers as shown in Fig. 4.23(a). The bolted connection is positioned at or close to a point of contraflexure. This method is common in America for multistorey buildings but is not used extensively in British practice. Stub brackets however are welded directly to columns as shown in Fig. 4.24.

British practice for beams is to weld extended end plates to the end of the beam as shown in Fig. 4.23(b). The beam is bolted on site to the column through the end plates, but this connection is less rigid than the beam welded directly to the column. The amount of rotation that exists depends on the thickness of the end plates, the thickness of the column flanges and the extensibility of the bolts.

The design method in the elastic range of behaviour assumes that components behave as rigid bodies. The ratio $\mu_s M / V d_r$ developed theoretically in Section 4.21, is greater than unity and slip does not occur. Rotation takes place about a line adjacent and parallel to the compression flange. Generally six bolts are used, arranged as shown in Fig. 4.23(b), and the tensile forces in the bolts and welds are determined as described in Section 4.16.

The thickness of the end plate is often controlled by the cantilever extension of the end plate. The forces acting on the plate are the bolt forces and the prying forces described in Section 4.16. Applying the simple elastic bending theory for two bolts and the full width of plate B_p

$$2F_{bt}b_p - 2Q_b(a_p + b_p) = p_b B_p t_p^2 / 6 \qquad (4.34)$$

The values of a_p, b_p and B_p are chosen from bolt spacing and edge distance considerations. The value of F_{bt} is related to the external bending moment, and initially Q_b is assumed to be zero. An initial value of t_p can be determined. This value of t_p can then be reduced when the value of the prying force Q_b has been determined from Equation (4.34). This iterative process is necessary because Q_b is a function of t_p.

The thickness of the end plate may also be controlled by the forces from the other two bolts adjacent to the tension flange of the beam. The bending stresses in the plate may be determined by considering an equivalent cantilever of width b_c based on a 45° dispersion angle as shown in Fig. 4.23(b). An equation similar to the one for the end plate in tension, given in Equation (4.34), can be formed and the thickness of the end plate determined. The equation also includes a prying force.

This method of determining the thickness of the end plate in the elastic range of behaviour may also be used for determining the thickness of the column flange required for a satisfactory connection. If there is a large axial stress in the column flange this will reduce the strength of the column flange in bending.

If a compression stiffener is not included, the web of the column may buckle due to the force from the compression flange of the beam. The method of design is described in Chapter 2. If the web is likely to buckle then stiffeners must be added as described in Chapter 7.

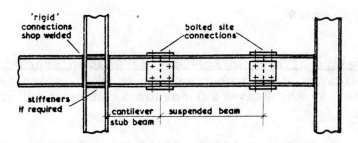

(a) American practice for multi-storey buildings.

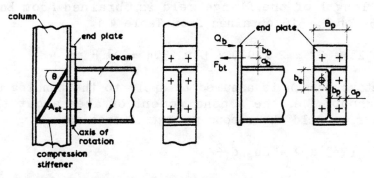

(b) British practice for low-rise structures

Fig. 4.23. Rigid beam-to-column connections.

A check must also be made for bearing stresses at the root of the web of the column as described in Chapter 2. If the bearing stresses are excessive then a stiffener improves the situation. The web bearing capacity is reduced by the direct stress in the web and flange of the column as described previously in Chapter 2.

The web of the column may be subjected to an excessive combination of direct stress and shear stress. The direct stress f_a is obtained from the axial load in the column. The shear stress f_q in the column web is obtained from the unbalanced flange forces in the beams.

$$f_q = (F_1 \sim F_2)/(d_c t_c) \simeq [(M_1/d_{f1}) \sim (M_2/d_{f2})]/(d_c t_c) \qquad (4.35)$$

These stresses are combined using the equivalent stress formula given in Cl. 14c BS 449 and explained in Section 1.16.

If this criterion is not satisfied then a diagonal tension, or compression, stiffener is added as shown in Fig. 4.23(b). If the diagonal stiffener shown in Fig. 4.23(b) is treated as a member of a pin-jointed frame then the shear force resisted by the stiffener is

$$V_{st} = A_{st}p_{st}\cos\theta \qquad\qquad (4.36)$$

Example 4.6   Design of a Rigid End-Bearing Column Bracket Determine the size of welds required to connect the bracket, which is cut from a Universal beam, to the Universal column as shown in Fig. 4.24.   Use grade 50 steel.

Example 4.6.1   Assuming rotation about axis O–O If there are no tension stiffeners in the column the strength of the flange welds is reduced because of the flexibility of the column flanges.   The effective length of the flange weld is obtained from Equation (4.1) with C = 5 for F_e 510 obtained from Table 4.1.

$$b_{we} = 2t + CT = 2 \times 9.9 + 5 \times 15.4 = 96.8 \text{ mm}$$

This effective length is assumed to apply to the bending properties of the weld group, i.e. the second moment of area about the axis O–O. For a unit size weld the second moment of area is

$$I_{woe} = (2/3)d_w^3 + 2b_{we}d_f^2$$

$$= (2/3)(834.9 - 2 \times 18.8)^3 + 2 \times 96.8 \times (834.9 - 18.8)^2$$

$$= (337.9 + 128.9)10^6 = 466.8 \times 10^6 \text{ mm}^4$$

If there is vertical slip at the hinge line O–O then the strength of the weld is affected by the flexibility of the beam flanges.   The effective length of the weld obtained from Equation (4.1) is

$$b_{we} = 2t + CT = 2 \times 14 + 5 \times 18.8 = 122 \text{ mm}$$

This effective length is assumed to apply to the shear properties of the weld, i.e. the length of the weld.   For a unit size weld

$$L_{we} = 4b_{we} + 2d_w = 4 \times 122 + 2 \times (834.9 - 2 \times 18.8)$$

$$= 2082.6 \text{ mm}$$

The distance from the axis O–O to the resultant force in the weld, d_{wr}, is determined from equating the moment of the forces in the weld group about the axis O–O shown in Fig. 4.24

moment of the whole = moment of the parts

$$(2b_{we} + 1/2 \times 2d_f)F_{wx}d_{wr} = F_{wx}I_{oe}/d_f$$

rearranging and putting $I_{oe} = (2/3)d_f^3 + 2b_{we}d_f^2$

$$d_{wr}/d_f = [(2/3) + 2b_{we}/d_f]/(1 + 2b_{wc}/d_f)$$

$$= \frac{2/3 + 2 \times 96.8/(834.9 - 18.8)}{1 + 2 \times 96.8/(834.9 - 18.8)} = 0.7306$$

The maximum force per unit length on a weld element in the x direction assuming rotation about the axis 0-0 is

$$F_{wx} = (Ve)d_f/I_{woe} = (270 \times 600)(834.9 - 18.8)/(466.8 \times 10^6)$$

$$= 0.2832 \text{ kN/mm}$$

If slip occurs there is a force on the weld in the y direction. From Equation (4.26)

$$(\mu_s/\gamma)e/d_{wr} = (0.45/1.4)600/[.7306(834.9 - 18.8)]$$

$= 0.3235 < 1$ slip occurs and therefore there is a force on the weld group in the y direction.

The average force per unit length on a weld element in the y direction from vertical slip is

$$F_{wy} = \frac{V}{L_{we}} - \frac{\mu_s R}{\gamma L_{we}} = \frac{V}{L_{we}} - \frac{\mu_s(Ve/d_{wr})}{\gamma L_{we}}$$

$$= \frac{270}{2082.6} - \frac{0.45\{270 \times 600/[0.7306(834.9 - 18.8)]\}}{1.4 \times 2082.6}$$

$$= 0.1296 - 0.0419 = 0.0877 \text{ kN/mm}$$

The resultant vector force per unit length on this weld element is

$$F_{wR} = [F_{wx}^2 + F_{wy}^2]^{1/2} = [0.2832^2 + 0.0877^2]^{1/2} = 0.296 \text{ kN/mm}$$

A 6mm fillet weld with grade 50 steel will resist a shear force of 0.672 kN/mm as shown in Table 4.3.

Check the thickness of the web of the UB section for buckling as a gusset plate. From Equation (4.20)

$$W_g = L_g/[(L_g/H_g)^2 + 1]^{1/2}$$

$$= 700/[(700/797.3)^2 + 1]^{1/2} = 526.0 \text{ mm}$$

The thickness of the web of the UB section acting as a gusset is obtained from Equation (4.19)

$$t_g = 2F_g s_g/(p_g W_g^2) + W_g/80$$

$$= 2 \times 270 \times 10^3 \times 600/(215 \times 526^2) + 526/80$$

$= 12.02 < 14$ mm and the thickness of the web of the UB section is therefore satisfactory.

Check the slenderness ratio of the gusset plate from Equation (4.21)

$$l_g/r_g = 2 \sqrt{3} W_g/t_g = 2 \sqrt{3} \times 526/14$$

$= 130 < 185$ the limit of the application of the theory, therefore acceptable.

The reaction R_o at the hinge may buckle or crush the web of the UC. The magnitude R_o is obtained from Equation (4.25)

$$R_o = Ve/d_{wr} = 270 \times 600/[0.7306(834.9 - 18.8)] = 271.7 \text{ kN}$$

Check the web buckling strength of the UC at the axis O-O. The value of C, as expressed in Equation (2.40) and given in tables in the Structural Steel Handbook is $C_1 = 582$ kN for grade 50 steel continuous over the bearing. This value is greater than $R_o = 271.7$ kN and therefore the web will not buckle.

Check the web bearing strength of the UC section at the axis O-O. The strength of the column component C_1 may be obtained from Equation (2.38) or from the BCSA Structural Steel Tables $C_1 = 272$ kN. This value is greater than $R_o = 271.7$ kN and therefore the allowable web bearing stresses are not exceeded.

The force R_o produces bearing and shear stresses in the web of the column which interact with the axial and bending stresses in the column. These stresses are combined as follows.

The bearing stress at the root radius of the column web

$$f_b = R_o/[(T_b + 2(T_c + r_c)\cot 30°)t]$$

$$= 271.7 \times 10^3/[(18.8 + 2(15.4 + 15.2)1.732) \times 9.9]$$

$$= 219.9 \text{ N/mm}^2$$

The average shear stress in the column web

$$f_q = R_o/(Dt) = 271.7 \times 10^3/(307.8 \times 9.9) = 89.2 \text{ N/mm}^2$$

The axial stress in the column web

$f_a = N/A = 650 \times 10^3/(12.33 \times 10^3) = 52.7$ N/mm^2

The bending stress at the root radius of the column web

$f_{bc} = Ve(d/2)/I$

$\qquad = 270 \times 10^3(600 + 307.8/2) \times (246.5/2)/(222.02 \times 10^6)$

$\qquad = 113.0$ N/mm^2

Inserting these values in the equivalent stress formula Cl. 14c BS 449

$f_e = [(f_{bc} + f_a)^2 + f_b^2 - (f_{bc} + f_a)f_b + 3f_q^2]^{1/2}$

$\qquad = [(113.0 + 52.7)^2 + 219.9^2 - (113.0 + 52.7) \times 219.9 \ldots\ldots$

$\qquad \ldots\ldots + 3 \times 89.2^2]^{1/2}$

$\qquad = 251.5 < 350$ N/mm^2

The allowable equivalent stress of 350 N/mm^2 is the value given in Table 1 BS 449 (Appendix).

Example 4.6.2   An alternative solution assuming rotation about axis G–G is as follows.

The traditional method of determining the forces and the size of a weld required for this situation is to assume rotation about the centroidal weld axis G–G and to ignore the frictional resistance.

The second moment of area of the weld group about the axis G–G for a unit size weld is

$I_{wG} = 2d_w^3/12 + 4B(d_f/2)^2$

$\qquad = 2(834.9 - 2 \times 18.8)^3/12 + 4 \times 291.6[(834.9 - 18.8)/2]^2$

$\qquad = (84.47 + 194.21) \times 10^6 = 278.68 \times 10^6$ mm^4

The maximum force per unit length on a weld element in the x direction from rotation about the G–G axis is

$F_{wx} = (Ve)(d_f/2)/I_{wG}$

$\qquad = (270 \times 600)[(834.9 - 18.8)/2]/(278.68 \times 10^6)$

$\qquad = 0.2372$ kN/mm

Ignoring friction forces the average force per unit length on a weld element in the y direction from vertical movement is

$$F_{wy} = V/L_w = 270/[4 \times 291.6 + 2 \times (834.9 - 2 \times 18.8)]$$

$$= 0.0978 \text{ kN/mm}$$

The maximum resultant vector force per unit length on a weld element is

$$F_{wR} = [F_{wx}^2 + F_{wy}^2]^{1/2} = [0.2372^2 + 0.0978^2]^{1/2} = 0.2566 \text{ kN/mm}$$

The 6 mm fillet weld specified previously is adequate. Calculations associated with web buckling and crushing are similar to those shown in Example 4.6.1.

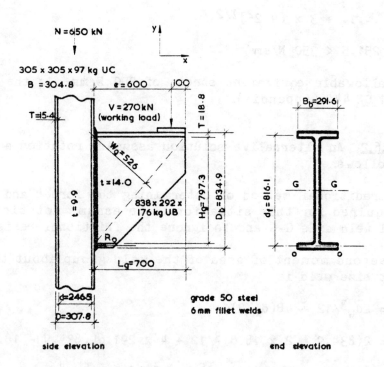

Fig. 4.24. Rigid end-bearing column bracket.

Example 4.7 Rigid beam-to-column connection Design the beam-to-column connection shown loaded at working load in Fig. 4.25. Use an end plate welded to the beam, HSFG bolts, and grade 50 steel.

The maximum tensile force in two of the top bolts is determined by assuming rotation about the axis 0-0 at the mid-thickness of the beam flange.

The second moment of area of bolts of unit area about the 0-0 axis is

$$I_o = 2(d_3{}^2 + d_2{}^2 + d_1{}^2)$$

$$= 2(54.05^2 + 254.05^2 + 339.05^2) = 364.8 \times 10^3 \text{mm}^4$$

The maximum tensile force in the bolt furthest from the axis 0-0 is

$$F_b(\text{max}) = My/I_o = 65 \times 10^3 \times 339.05/(364.8 \times 10^3) = 60.41 \text{ kN}$$

The additional prying force in a bolt associated with the cantilever portion of the end plate is determined from Equations (4.15) and (4.16). Assuming M20 HSFG bolts, a plate thickness of 20 mm, and dimensions as shown in Fig. 4.25.

$$k_e = E_p w_p t_p{}^3 g_p / (6 a_p b_p{}^2 A_b E_b)$$

$$= 210 \times 10^3 \times 100 \times 20^3 \times 23.7/(6 \times 45 \times 35^2 \times 245 \times 210 \times 10^3)$$

$$= 0.2340$$

The prying force

$$Q_{be} = \frac{F_e(1 - k_e) + k_e P_{bs}}{2(a_p/b_p) + (2/3)(a_p/b_p)^2 + k_e}$$

$$= \frac{60.41(1 - 0.2340) + 0.2340 \times 144}{2(45/35) + (2/3)(45/35)^2 + 0.2340} = 20.47 \text{ kN}$$

The total maximum tensile force on a bolt $F_{bt} = F_b(\text{max}) + Q_{be} = 60.41 + 20.47 = 80.88$ kN. The allowable tensile force on a M20 HSFG general grade bolt at working load is 84.7 kN as shown in Table 4.9.

The thickness of the end plate is related to calculations of bending stress for a cantilever. At section a-a Fig. 4.25(b), applying M = pZ

$$2[F_{bt} b_p - Q_{be}(a_p + b_p)] = p_b B_p t_p{}^2/6$$

$$2[80.88 \times 35 - 20.47 \times 80]10^3 = 230 \times 200 t_p{}^2/6$$

$$t_p = 17.64 < 20 \text{ mm therefore satisfactory.}$$

At the bolt line, section b-b Fig. 4.25(b), applying M = pZ

$$2Q_{be}a_p = p_p(B_p - 2d_h)t_p^2/6$$

$$2 \times 20.47 \times 10^3 \times 45 = 230(200 - 2 \times 22)t_p^2/6$$

$t_p = 17.55 < 20$ mm therefore satisfactory.

At the equivalent cantilever for the end plate, section c-c Fig. 4.25(c)

$$w_p = d_h + 2b_p = 22 + 85 - 7.7 = 99.3 \text{ mm.}$$

This value of w_p is approximately the same as the value of w_p used in the calculation for section a-a, but because the tensile force on the bolt is less the plate thickness is less. The 20 mm thick plate is satisfactory.

The frictional shear resistance of the HSFG bolts is reduced by the tensile forces produced by the bending moment. Developing Equation (4.9) the frictional shear resistance of the three levels of M20 HSFG bolts, reduced by the tensile force for each layer, is

$$n_b P_v = n_{bl}\frac{P_{bs}}{1.4}\left[n_1 - \frac{F_b(\max)}{(P_{bs}/1.7)}\left[1 + \frac{d_2}{d_1} + \frac{d_3}{d_1}\right]\right]$$

$$= 2\frac{144}{1.4}\left[3 - \frac{60.41}{(144/1.7)}\left[1 + \frac{254.05}{339.05} + \frac{54.05}{339.05}\right]\right] = 337.11 \text{ kN}$$

The frictional shear resistance is greater than the applied shear force of 40 kN and is therefore satisfactory.

The web of the column may be crushed or buckled by the reaction R_O at the hinge O. The magnitude of R_O is the sum of the tensile forces in the bolts from the applied bending moment

$$R_O = 2F_{b(\max)}(1 + d_2/d_1 + d_3/d_1)$$

$$= 2 \times 60.41(1 + 254.05/339.05 + 54.05/339.05) = 230.61 \text{ kN}$$

Check the bearing resistance of the column web as shown in Section 2.6.2. From the Structural Steel Tables the column component continuous over the bearing is $C_1 = 360$ kN. This value is greater than the value of R_O and is therefore satisfactory.

The column web is also subject to bearing, bending and shear stresses and these are combined as follows.

The bearing stress at the root radius of the column web from R_O is

$$f_b = R_o/[(L_b + t_p + t_p' + 2(T + r)\cot 30^\circ)t]$$

$$= 230.6 \times 10^3/[(13.7 + 20 + 9.1 + 2(20.5 + 10.2)1.732) \times 13]$$

$$= 230.6 \times 10^3/(149.1 \times 13) = 119.0 \text{ N/mm}^2$$

The bending stress at the root radius of the column

$$f_{bc} = M(d/2)/I = 32.5 \times 10^6(160.8/2)/(94.62 \times 10^6)$$

$$= 27.6 \text{ N/mm}^2$$

The axial stress in the column is

$$f_a = N/A = 300 \times 10^3/(11.01 \times 10^3) = 27.2 \text{ N/mm}^2$$

The average shear stress in the web of the column

$$f_q = R_o/(Dt) = 230.61 \times 10^3/(222.3 \times 13) = 79.8 \text{ N/mm}^2$$

The equivalent stress from Cl. 14d BS 449

$$f_e = [(f_{bc} + f_a)^2 + f_b^2 - (f_{bc} + f_a)f_b + 3f_q^2]^{1/2}$$

$$= [(27.6 + 27.2)^2 + 119.0^2 - (27.6 + 27.2)119.0 + 3 \times 79.8^2]^{1/2}$$

$f_e = 172.5 < 320 \text{ N/mm}^2$ the limit given in Table 1 BS 449 (Appendix).

Check the buckling resistance of the column web. From the Structural Steel Tables the column component, continuous over the bearing, $C_1 = 586$ kN. This value is greater than $R_o = 230.61$ kN, and therefore no stiffeners are required.

The size of the fillet weld connecting the web to the end plate is obtained as follows.

Check whether slip occurs from Equation (4.26), i.e. whether $(\mu_s/\gamma)M/(Vd_{wr}) < 1$. First calculate

$$\frac{d_{wr}}{d_f} = \frac{2/3 + 2b_w/d_f}{1 + 2b_w/d_f} = \frac{2/3 + 2 \times 166.8/297.2}{1 + 2 \times 166.8/297.2} = 0.843$$

$$(\mu_s/\gamma)M/Vd_{wr} = (0.45/1.4)65 \times 10^3/(40 \times 0.843 \times 297.2)$$

$= 2.085 > 1$, therefore no slip occurs, and the force applied to the weld is only from the bending moment.

The second moment of area of the weld group about axis 0-0, for unit size welds, is

$$I_o \cong 2(d_f^3/3 + Bd_f^2)$$

$$= 2(297.2^3/3 + 166.8 \times 297.2^2) = 46.97 \times 10^6 \text{mm}^4.$$

The maximum force per unit length applied to the fillet weld is

$$F_{wx} = My/I_o = 65 \times 10^3 \times 297.2/(46.97 \times 10^6) = 0.4113 \text{ kN/mm}.$$

A 6 mm fillet weld with grade 50 steel will resist a force of 0.672 kN/mm as shown in Table 4.3, and is therefore satisfactory.

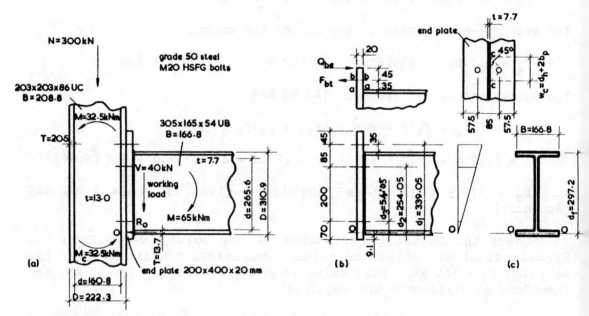

Fig. 4·25. Rigid beam-to-column connection.

4.23.2 Rigid continuous beam-to-beam connections

Where more efficient use of steel beams is required, e.g. in a steel grillage, or where deflection of the beams needs to be controlled, then continuous beam-to-beam connections are used as shown in Fig. 4.26. This has the effect of inducing a hogging bending moment at the connection and thus reducing the sagging bending moment at mid-span. Generally ($\mu M_R/Vd_r$) > 1, which means that slip does not occur, and rotation takes place about the bottom of the connection. By taking moments of forces about the bottom flange of the beam the size of the flange plate and the number of HSFG bolts may be determined.

The bottom cleat shown in Fig. 4.26(b) is a convenient support for erection purposes. An alternative method of fastening is to use two web cleats as shown in Fig. 4.26(c), and a further alternative is to use a welded end plate as shown in Fig. 4.26(a).

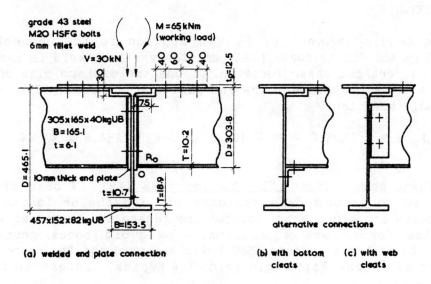

Fig. 4.26. Rigid beam-to-beam connections.

Example 4.8   Rigid continuous beam-to-beam connection   Design the welded end plate connection shown in Fig. 4.26.   Use grade 43 steel and HSFG bolts.

Assuming that the connection rotates about axis O-O shown in Fig. 4.26 and ignoring the tensile resistance of the bolts in the web, then the tensile force in the flange cover plate is

$$F_f = M/D = 65 \times 10^3/303.8 = 214 \text{ kN}$$

The thickness of the flange cover plate, assuming M20 HSFG bolts is

$$t_p = F_f/[(B - 2d_h)p_t] = 214 \times 10^3/[(165 - 2 \times 22)155] = 11.4 \text{ mm}$$

Use 12.5 mm thick grade 43 steel plate.

The number of M20 HSFG bolts required is

$$n_b = F_f/P_{bv} = 214/46.3 = 4.62. \quad \text{Use 6 M20 HSFG bolts.}$$

The magnitude of the reaction R_o at the hinge is equal to the force in the flange F_f = 214 kN.   The frictional resistance at R_o is

$$(\mu/\gamma)R_o = (0.45/1.4)214 = 68.8 > 30 \text{ kN, therefore no slip}$$

occurs.

Use 4 M20 HSFG bolts in the web connection to prevent lateral and vertical movement. The shear resistance of the four bolts is 4 x 46.3 = 185.2 > 30 kN. Use a 10 mm grade 43 steel end plate the thickness chosen to avoid crumpling and tearing associated with the failure of HSFG bolts.

The bending moment of 65 kNm applied to the connection is resisted by the flange cover plate and the vertical weld is assumed to resist the vertical shear force only. Use the minimum size of fillet weld, i.e. 6 mm, to connect the end plate to the beam. The vertical shear resistance is

$$2d_w F_w = 2 \times (303.8 - 30 - 10.2) \times 0.483 = 254.6 > 30 \text{ kN}.$$

4.23.3 Rigid beam splices (Cl. 27c and g BS 449) A beam splice is introduced to accommodate standard bar lengths or to facilitate construction and transport. Splices are generally located at sections where the forces are a minimum. To avoid local geometrical deformation of the structure HSFG bolts are used and the connection is designed as a 'non-slip' rigid joint. A typical connection is shown in Fig. 4.27.

The web splice is designed to shear force and the web bending moment as recommended in Cl. 27g BS 449. The shear force and bending are combined to form a shear force acting at an equivalent eccentricity of

$$e' = e + M_w/V \tag{4.37}$$

where M_w is the bending moment resisted by the web of the beam, V is the shear force at the splice and e is the eccentricity of the shear force with respect to one side of the web connection.

The flange splice is designed to resist the bending taken by the flanges of the beam. According to Cl. 27c BS 449 the cross sectional area of the cover plates should be 5% greater than the flange element spliced, and the strength of the fasteners on each side of the splice should be 5% greater than the element spliced. Flange plates placed either side of the flange subject the bolts to double shear forces.

Example 4.9 Rigid beam splice Determine the size of plate and number of HSFG bolts required for a beam splice shown in Fig. 4.27. Use grade 50 steel throughout.

The second moment of area of the web of the 533 x 210 x 92 UB

$$I_w = t(D - 2T)^3/12 = 10.2(533.1 - 2 \times 15.6)^3/12$$

$$= 107.5 \times 10^6 \text{mm}^4$$

From Section Tables (Appendix) the gross second moment of area of the complete section $I = 553.53 \times 10^6 \text{mm}^4$. In the elastic range of behaviour the bending moment resisted by the web is proportional to the second moment of area

$$M_w = (I_w/I)M = (107.5 \times 10^6/553.53 \times 10^6)300 = 58.26 \text{ kNm}.$$

The force to be resisted by the flange splice

$$F_f = (M - M_w)/(D - T) = (300 - 58.26)10^3/(533.1 - 15.6)$$

$$= 467.1 \text{ kN}$$

The number of M20 HSFG general grade bolts required at working load in double shear in the flange is

$$n_b = F_f/(2P_{bq}) = 467.1/(2 \times 46.3) = 5.04. \quad \text{Use 6 bolts.}$$

The thickness of the flange cover plates is related to the tensile resistance of a cross section through the bolt holes

$$t_p = \frac{F_f}{(B - 2d_h + 2w_p - 2d_h)p_p} = \frac{467.1 \times 10^3}{(209.3 - 2 \times 22 + 2 \times 70 - 2 \times 22)230} = 7.77 \text{ mm}$$

Flange plates which are 12.5 mm thick are necessary to satisfy the BS 449 requirement that the area of the flange plates should be 5% greater than the beam flange. Check as follows.

Area of flange plates $= (209.3 - 2 \times 22 + 2 \times 70 - 2 \times 22)12.5$

$$= 3266 \text{ mm}^2$$

Area of flange of beam $= (209.3 - 2 \times 22)15.6 = 2579 \text{ mm}^2$

(Area flange plates)/(area of beam flange) $= (3266/2579)$

$$= 1.266 > 1.05, \text{ therefore satisfactory.}$$

The total number of bolts in the web is 30, i.e. 15 on each side of the joint as shown in Fig. 4.27. The properties of the bolt group relevant to the x-x and y-y axes as shown in Fig. 4.27(b) are:

$$I_{bx} = 6(80^2 + 160^2) = 192 \times 10^3 \text{ mm}^4$$

$$I_{by} = 10 \times 80^2 = 64 \times 10^3 \text{ mm}^4$$

$$I_{bz} = I_{bx} + I_{by} = (192 + 64)10^3 = 256 \times 10^3 \ mm^4$$

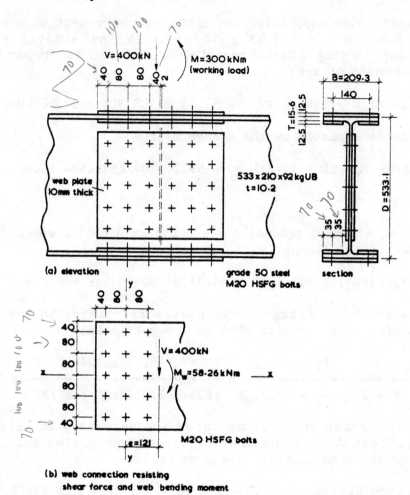

Fig. 4·27. Rigid beam splice.

The eccentricity of the shear force V to the joint is 121 mm as shown in Fig. 4.27(b), but this is increased by the web bending moment M_W to an equivalent eccentricity of

$$e' = e + M_W/V = 121 + 58.26 \times 10^3/400 = 266.65 \ mm$$

The maximum vector shear force in the x direction acting on a bolt furthest from the centroid of the bolt group is

$$F_{bx} = Ve'y_n/I_z = 400 \times 266.65 \times 160/(256 \times 10^3) = 66.66 \ kN$$

The maximum vector shear force acting on the same bolt in the y direction is

$$F_{by} = V/n_b + Ve'x_n/I_{bz} = 400/15 + 400 \times 266.65 \times 80/(256 \times 10^3)$$

$$= 60.00 \text{ kN}$$

The resultant maximum vector shear force acting on the same bolt is

$$F_{br} = [F_{bx}^2 + F_{by}^2]^{1/2} = [66.66^2 + 60.00^2]^{1/2} = 89.69 \text{ kN}$$

The double shear strength of an M20 HSFG general grade bolt from Table 4.9 is 2 x 46.3 = 92.6 > 89.69 kN, which is satisfactory.

The thickness of the web cover plates should be half the bolt diameter, or 10 mm, whichever is the lesser value. Use 10 mm thick web cover plates.

4.23.4 Rigid column splices The column splices shown in Fig. 4.28 are used for reasons of economy, convenience and to accommodate standard bar lengths. The ends of the columns are generally machined, and it is therefore assumed that part of the load is transferred by direct end bearing and part via the bolts. If a rigid connection is required, and/or the bending moment is large, then HSFG bolts should be used. Grade 4.6 or 8.8 bolts, however, may be satisfactory for a column which supports predominantly an axial load and where the flange plates are not in tension.

The simplest type of splice is shown in Fig. 4.28(a) which connects column sections of the same serial size, i.e. sections having approximately the same breadth and depth. The difference in depth is taken up by packing and the flanges and webs are connected using plates. This is similar to a beam splice, and the design method is comparable, but the column incorporates an axial force.

A more complicated connection is necessary when there is a change in serial size as shown in Fig. 4.28(b). A machined plate is introduced between the ends of the columns to transfer the load because the columns are of different size. The thickness of this plate is generally chosen to disperse the load at 45° from the outside edge of the flanges of one column to that of the other, as shown in Fig. 4.28(b). The plate is kept in position by angle cleats bolted or welded to the column webs and site bolted to the plate. The flange plate connections are similar to the simple splice, but thicker packings and longer bolts are required.

A further alternative column splice, using two end plates, is shown in Fig. 4.28(c). This method may be used for columns of the same or different serial size. An end plate with bolt holes is welded to the end of each column, and if the columns are of different serial size, the total plate thickness must be sufficient to disperse the load at 45° from the outside edge of one column to that of the other.

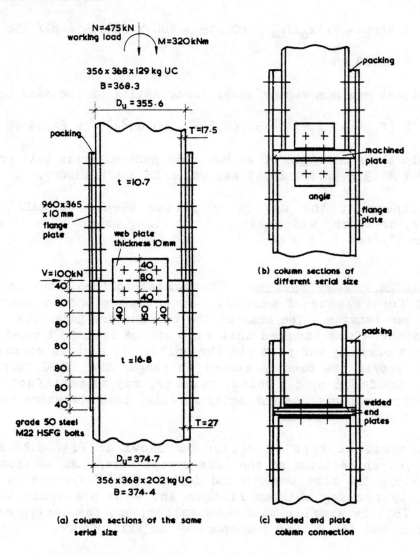

N=475 kN
working load
M=320 kNm

356 x 368 x 129 kg UC
B=368.3

D_u = 355.6

packing

T=17.5

t =10.7

960x365
x 10 mm
flange
plate

web plate
thickness 10 mm

V=100kN

40
40
80
80
80
80
80
40

40
80
40

40
50
40

t =16.8

grade 50 steel
M22 HSFG bolts

T = 27

D_u= 374.7

356 x 368 x 202 kg UC
B=374.4

(a) column sections of the same
serial size

packing

machined
plate

angle

flange
plate

(b) column sections of
different serial size

packing

welded
end
plates

(c) welded end plate
column connection

Fig. 4·28. Rigid splices in steel columns.

The design process for a column splice is first to check the position of the line of action of the equivalent eccentric load in relation to the kern of the column cross section, for the most adverse condition of loading.

If $M/N < Z/A$ which is quite common, then the load acts within the kern and there is no tensile force in the flange plate, and the thickness of the plate is then arbitrarily taken as half the thickness of the upper column flange, or 10 mm, whichever is the greater value. The width of the flange plate is chosen to fit the smaller column, and the length on either side of the joint is made equal to the width of the larger column, or 230 mm, whichever is the greater value.

If M/N > Z/A then the load acts outside the kern and there is a tensile force in the flange plates. The thickness of the flange plates is determined by taking moments of forces about the column flange.

$$M - ND/2 \simeq B_p t_p p_p D \qquad (4.38)$$

Any transverse horizontal shear force V will be resisted partly by the friction between the column ends and the machined plate, and partly by the column web connection.

The reactive force R at the point of contact of the ends of the column is obtained by taking moments of forces about the tension flange plates

$$M + N(D/2) = RD \qquad (4.39)$$

If slip does not occur the frictional resistance is greater than the horizontal shear force applied to the column, i.e.

$$(\mu_s/ \gamma)R > V \qquad (4.40)$$

Combining Equations (4.39) and (4.40) to eliminate R

$$(\mu_s/ \gamma)[M/D + N/2] > V \qquad (4.41)$$

If slip does not occur the web connection resists the difference between the horizontal force applied to the column and the frictional resistance, i.e. $V - (\mu_s/ \gamma)[M/D + N/2]$.

Traditionally the frictional resistance is ignored because contact between column ends is not assumed, as recommended in BS 5135, and the web connection is designed to resist all of the horizontal force.

Example 4.10 Rigid column splice connecting two sections of the same serial size   Determine the size of bolts, flange and web plates for the column splice shown loaded at working load in Fig. 4.28(a). The loads shown are the minimum axial load and maximum bending moment.

When column sections are of the same serial size it is possible to connect them directly with web and flange plates.

$$M/N = 320 \times 10^6/(475 \times 10^3) = 673.7 \text{ mm}$$

$$Z/A = 2264 \times 10^3/(16.49 \times 10^3) = 137.3 \text{ mm}$$

$M/N > Z/A$ therefore there is tension in the flanges, and flange plates are required to resist this tension at the splice. The ends of the columns will be in contact and rotation will take place about an axis near the outer edge of the flange of the upper column. The thickness of the flange plates may be obtained by taking moments of forces about this axis of rotation.

$$M - N(D/2) \simeq (B_p - 2d_h)t_p p_p D$$

$$320 \times 10^6 - 475 \times 10^3 (355.6/2)$$

$$= (365 - 2 \times 24) \times t_p \times 230 \times 355.6$$

$$t_p = 9.09 \text{ mm}$$

This value is less than half the flange thickness of the upper column and also less than 10 mm. Use 10 mm thick plate.

The number of bolts in the flange plates may also be found by taking moments of forces about the same axis

$$M - N(D/2) = n_b P_{bq} D$$

$$320 \times 10^6 - 475 \times 10^3 (355.6/2) = n_b \times 56.9 \times 10^3 \times 355.6$$

$$n_b = 11.64$$

Use 12 M22 HSFG bolts in single shear.

The length of the lap is not sufficient to reduce the shear strength of the connection as calculated in Equation (4.22).

The horizontal shear force on the column is resisted by the friction at the point of contact of the flange, i.e. at the axis of rotation. Check whether the frictional resistance between column ends in contact is sufficient to resist the horizontal shear force applied to the column using Equation (4.41).

Frictional shear resistance $= (\mu_s/\Upsilon)(M/D + N/2)$

$$= (0.15/1.4)(320 \times 10^3/355.6 + 475/2)$$

$$= 121.9 > 100 \text{ kN}$$ therefore the horizontal shear on the column is resisted by friction.

Most designers ignore the frictional resistance and design the web splice to resist the entire horizontal shear force as follows.

The second moments of area of the bolt group on one side of the web connection are

$$I_{bx} = 0$$

$$I_{by} = 2 \times 75^2 = 11.25 \times 10^3 \text{ mm}^2$$

$$I_{bz} = I_{bx} + I_{by} = 11.25 \times 10^3 \text{ mm}^2$$

The shear force in the x direction on the bolt is

$$F_{bx} = V/n_b = 100/2 = 50 \text{ kN}$$

The vector force in the y direction on a bolt furthest from the centroid of a bolt group is

$$F_{by} = (Ve)x_n/I_{bz} = (100 \times 40)75/(11.25 \times 10^3) = 26.66 \text{ kN}$$

The maximum resultant vector shear force on a bolt is

$$F_{bR} = [F_{bx}^2 + F_{by}^2]^{1/2} = [50^2 + 26.66^2]^{1/2} = 56.6 \text{ kN}$$

This value is less than the shear resistance of an M22 HSFG with 2 plies, i.e. 2 x 56.9 = 113.8 kN. The friction shear value of 56.9 kN is obtained from Table 4.9.

4.23.5 <u>Rigid column-to-foundation connections</u>   Generally columns are connected to a reinforced concrete foundation with a rigid joint. This type of connection reduces sway deflections and facilitates erection because the column needs no temporary support. If the bending moment is relatively small a thick steel slab base is welded to the column as shown in Fig. 4.29(a). If the bending moment is large then to avoid using excessively thick slab bases, the base is built up as shown in Fig. 4.29(b). The number of holding bolts is four, or more, positioned effectively to resist the applied bending moment.

(i) Slab bases

The design method at working load assumes that the base plate acts as a rigid body which deforms the bedding material when the external forces are applied. If the external forces are not large the dimension of the slab base may be chosen so that there is no loss of contact between the steel slab and the bedding material. It is also necessary to limit the maximum bearing pressure on the bedding material to the allowable bearing stress p_{cb}. The stress and strain distribution across the base are assumed to be linear as shown in Fig. 4.29(a). Combining axial and bending bearing pressures on the bedding material between the concrete and steel slab using the elastic theory for the condition of no loss of contact.

$$-N/(B_pD_p) + M/(B_pD_p^2/6) > 0$$

Rearranging

$$D_p > 6M/N \tag{4.42}$$

If the maximum compressive bearing pressure on the bedding material between the steel slab and the concrete is not to exceed p_{cb} then

$$-N/(B_pD_p) - M/(B_pD_p^2/6) < -p_{cb} \tag{4.43}$$

Combining Equations (4.42) and (4.43) and rearranging

$$B_p > N^2/(3Mp_{cb}) \tag{4.44}$$

The thickness of the base plate is then determined from the bending moment associated with the greatest cantilever length A.

From simple bending theory applied to a unit width of steel plate cantilevering from the face of the column as shown in Fig. 4.29(a)

$$(A^2/2)(1 - A/D_p)p_{cb} + (A^2/3)(A/D_p)p_{cb} = p_pt_p^2/6$$

rearranging

$$(t_p/A) = [(p_{cb}/p_p)(3 - A/D_p)]^{1/2} \tag{4.45}$$

or combining Equations (4.42) and (4.45) to eliminate D_p

$$(t_p/A) = [(p_{cb}/p_p)(3 - NA/6M)]^{1/2} \tag{4.45a}$$

The allowable bending stress in the steel slab base $p_p = 185$ N/mm^2 as given in Table 2, BS 449 (Appendix). A value of the allowable bearing stress on the bedding material often used in design is $0.25f_{cu}$. The holding down bolts may not be subject to tensile forces in slab bases but may have to resist the horizontal shear forces acting on the base. The minimum bolt diameter is 16 mm but generally 20 mm are used so that the bolts are capable of resisting the forces applied during erection.

   In other slab bases the ratio of bending moment to axial load is larger, necessitating a greater length of base plate D_p. If a slab base is used the length of projection A becomes large, resulting in a thick base plate. This can be avoided by limiting the value of D_p and allowing loss of contact. The holding down bolts resist all the tensile forces resulting from the bending moment. The analysis of the distribution of stresses on the base in the linear elastic range of behaviour is as follows.

Assume a linear strain distribution across the base as shown in Fig. 4.29(a), then from similar strain triangles

$$\varepsilon_c/x = \varepsilon_s/(d - x) \qquad (4.46)$$

Assuming linear stress-strain relationships for steel and bedding material

$$E_s = p_{bt}/\varepsilon_s \qquad (4.47)$$

$$E_c = p_{cb}/\varepsilon_c \qquad (4.48)$$

Combining Equations (4.46) to (4.48) to form an equation for the depth of compression zone x

$$x/d = 1/[1 + (p_{bt}/mp_{cb})] \qquad (4.49)$$

where the modular ratio $m = E_s/E_c$

Taking moments of forces about the tensile bolts

$$0.5B_p x p_{cb}(d - x/3) - N(d - 0.5D_p) - M = 0$$

rearranging

$$0.5B_p p_{cb}(x/d)(d/D_p)^2[1 - x/(3d)]D_p^2 - N(d/D_p - 0.5)D_p - M = 0 \qquad (4.50)$$

This equation is used to determine the value of D_p.

Resolving forces vertically

$$0.5B_p \times p_{cb} - A_b f_{bt} - N = 0 \qquad (4.51)$$

This equation is used to determine the area of the bolts A_b.

Equations (4.49), (4.50) and (4.51) are only valid providing the allowable stresses for the bedding material, p_{cb}, and the steel bolt, p_{bt}, coexist. If, however, values of D_p and A_b are chosen which are greater than the values obtained from Equations (4.50) and (4.51), then the bearing stress on the bedding material and the tensile stress in the bolt will be less than the allowable values.

(ii) Built-up bases

In situations where the ratio of bending moment to axial load is large it may not be practical to use a slab base because the thickness, and/or the bedding material bearing pressure, would be excessive. The alternative is a built-up base which may also be designed so that there are only compressive stresses on the bedding material and

Equations (4.42) to (4.45) are therefore applicable. Generally however the ratio of M/N is large and the length of the base obtained from Equation (4.42) is excessive. When this occurs, a practical value of D_p is adopted and the method of design is as shown previously for a slab base. In addition, calculations are required to determine the thickness of the gusset plates to prevent buckling, as shown in Section 4.18.

Example 4.11 Rigid built-up column-to-base connection Determine the dimensions and plate sizes for the built-up base at working load shown in Fig. 4.29(b). Use grade 43 steel.

Length of base required, if the holding down bolts are not in tension, is from Equation (4.42)

$$D_p > 6M/N = 6 \times 160/120 = 8.0 \text{ m.}$$ Excessive, use a built-up base.

The depth of the compression zone in relation to d is obtained from Equation (4.49), assuming $p_{bt} = 120$ N/mm^2, $p_{cb} = 0.25f_{cu} = 0.25 \times 20 = 5$ N/mm^2, and m = E_s/E_c = 15.

$$x/d = 1/[1 + (p_{bt}/mp_{cb})] \{ 1/[1 + 120/(15 \times 5)] = 0.3846$$

As shown in Fig. 4.29(b) the breadth of the base plate B_p must accommodate (column + gussets + welds + washer + edge distances).

$$B_p = 208.8 + 2 \times 12.5 + 2 \times 10 + 66 + 2 \times 68 = 455.8 \text{ use } 460 \text{ mm.}$$

The depth of the base plate is obtained from Equation (4.50)

$$0.5B_pp_{cb}(x/d)(d/D_p)^2[1 - x/(3d)]D_p^2 - N(d/D_p - 0.5)D_p - M = 0$$

$$0.5 \times 460 \times 5 \times 0.3846 \times 0.9^2(1 - 0.3846/3)D_p^2 - \ldots\ldots$$

$$\ldots\ldots 120 \times 10^3(0.9 - 0.5)D_p - 160 \times 10^6 = 0$$

$$312.3D_p^2 - 48 \times 10^3D_p - 160 \times 10^6 = 0$$

$$D_p = 796.7 \text{ mm}$$

Area required for bolts in tension is obtained from Equation (4.51)

$$0.5B_p(x/d)\overset{d}{d}p_{cb} - N = A_bp_{bt}$$

$$0.5 \times 460 \times 0.3846 \times 0.9 \times 796.7 \times 5 - 120 \times 10^3 = A_b120$$

$$A_b = 1643 \text{ mm}^2, \ 2 - M36 = 2 \times 817 = 1634 \text{ mm}^2$$

The thickness of the base is related to the cantilever projection A shown in Fig. 4.29(b). From the simple theory of bending for unit width of plate

$\quad M = p_p t_p^2/6$

$\quad 5 \times 111^2/2 = 185 t_p^2/6$

$\quad t_p = 31.6$ mm

The thickness of the base plate when related to the bolt force acting on the effective width of plate b_e as shown in Fig. 4.29(b) is

$\quad M = p_{pb} b_e t_p^2/6$

$\quad (1643/2)120 \times 43 = 185(36 + 2 \times 43) t_p^2/6$

rearranging

$\quad t_p = 33.6$ mm

Use 460 x 800 x 35 mm, grade 43 steel base plate and 4-M36 grade 4.6 holding down bolts.

The gusset plate thickness is determined as follows. Each gusset plate is subjected to a resultant force F_g which is the sum of the bearing pressures beneath the base as shown in Fig. 4.29(b).

$\quad F_g = 0.5 p_{cb}(x/d)d(0.5 B_p)$

$\quad\quad = 0.5 \times 5(0.3846 \times 0.9)796.7(0.5 \times 455.8) \times 10^{-3} = 157.1$ kN

The theoretical length of the gusset plate

$\quad L_g = 0.5(D_p - D) = 0.5(796.7 - 222.3) = 287.2$ mm

The distance

$\quad s_g = L_g - (x/d)(d/3) = 287.2 - (0.3846 \times 0.9)(796.7/3)$

$\quad\quad = 195.3$ mm

If L_g is made equal to H_g then the width of the gusset plate from Equation (4.20) is

$\quad W_g = L_g/[(L_g/H_g)^2 + 1]^{1/2} = 287.2/(1 + 1)^{1/2} = 203.1$ mm

The thickness of the gusset plate from Equation (4.19) is

$\quad t_g = 2 F_g s_g/(p_g W_g^2) + W_g/80$

$\quad\quad = 2 \times 157.1 \times 10^3 \times 195.3/(155 \times 203.1^2) + 203.1/80 = 12.14$ mm

Check the slenderness ratio of the gusset plate from Equation (4.21)

$$l_g/r_g = 2 \sqrt{3}W_g/t_g = 2 \sqrt{3} \times 203.1/12.14$$

$$= 58.0 < 185 \text{ the limit of the application of the theory.}$$

Use 12.5 mm thick gusset plates grade 43 steel.

The size of the fillet weld connecting the gusset plate to the base plate is obtained as follows. The weld group is shown in Fig. 4.29(b), and if the plates in contact are not machined then the welds are assumed to resist all of the applied forces and bending moments. Rotation is assumed to take place about the axis G-G and the properties of the weld group of unit size are:-

$$L_w = 2D_w = 2 \times 800 = 1600 \text{ mm}$$

$$I_{wG} = 2D_w^3/12 = 2(800)^3/12 = 85.33 \times 10^6 \text{ mm}^4$$

The force per unit length applied to the welds in the y direction (perpendicular to the base plate) is

$$F_{wy} = N/L_w + M(d_w/2)/I_{wG}$$

$$= 120/1600 + 160 \times 10^3(800/2)/(85.33 \times 10^6) = 0.8250 \text{ kN/mm}$$

The force per unit length applied to the welds in the z direction (parallel to the base plate) is

$$F_{wz} = H/L_w = 30/1600 = 0.0188 \text{ kN/mm}$$

The resultant vector force per unit length is

$$F_{wR} = [F_{wy}^2 + F_{wz}^2]^{1/2} = [0.8250^2 + 0.0188^2]^{1/2} = 0.8252 \text{ kN/mm}$$

Use a 12 mm fillet weld which for grade 43 steel has an allowable strength of 0.966 kN/mm as shown in Table 4.3.

Alternatively if the plates in contact are machined then rotation is assumed to be about the axis 0-0 shown in Fig. 4.29(b). The magnitude of the reaction R_0 is determined by taking moments of forces about the resultant force in the weld

$$-R_0 \times 2D_p/3 + PD_w/6 + M = 0$$

$$-R_0 \times 2 \times 800/3 + 120 \times 800/6 + 160 \times 10^3 = 0$$

$$R_0 = 330 \text{ kN}$$

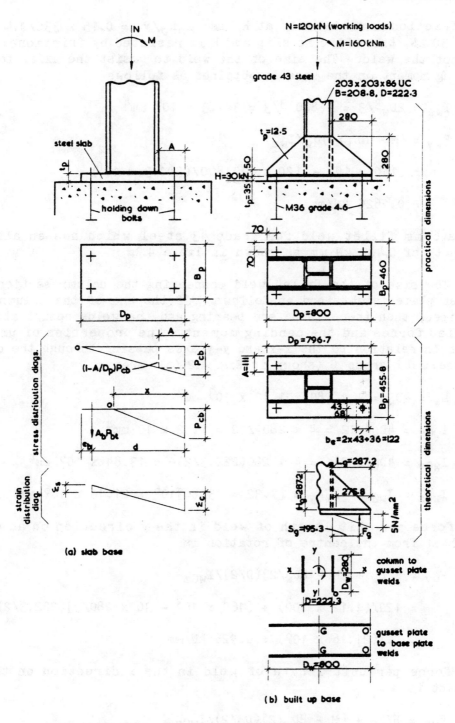

Fig. 4.29. Rigid column-to-foundation connection.

The frictional resistance at R_O is $\mu_s R_o / \gamma = 0.15 \times 330/1.4 = 35.36$ $> H = 30$ kN, therefore no slip and H is resisted by frictional forces and not the weld. The size of the weld to resist the axial force and bending moment on the base is obtained as follows

$$I_{wo} = 2D_w^3/3 = 2(800)^3/3 = 341.3 \times 10^6 \text{ mm}^4$$

$$F_{wy} = (M - ND_w/2)D_w/I_{wo}$$

$$= (160 \times 10^3 - 120 \times 400)800/(341.3 \times 10^6)$$

$$= 0.2625 \text{ kN/mm}$$

Use a 6 mm fillet weld for grade 43 steel which has an allowable strength of 0.483 kN mm as shown in Table 4.3.

The size of the fillet weld connecting the column section to the gusset plate is obtained as follows. If the end of the column is not machined, then there is no end bearing and the welds resist all of the applied forces and the bending moment. The properties of unit size welds in relation to the x-x and y-y axes passing through the centroid of the weld group G (shown in Fig. 4.29(b)) are

$$L_w = 4D_w = 4 \times 280 = 1.12 \times 10^3 \text{ mm}^2$$

$$I_{wGx} = 4D_w^3/12 = 4 \times 280^3/12 = 7.32 \times 10^6 \text{mm}^4$$

$$I_{wGy} = 4D_w(D/2)^2 = 4 \times 280(222.3/2)^2 = 13.84 \times 10^6 \text{ mm}^4$$

$$I_{wGz} = I_{wGx} + I_{wGy} = [7.32 + 13.84]10^6 = 21.16 \times 10^6 \text{ mm}^4$$

The force per unit length of weld in the y direction on an element furthest from the centre of rotation is

$$F_{wy} = N/L_w + [M - HD_w/2](D/2)/I_{wGz}$$

$$= 120/(1.12 \times 10^3) + [160 \times 10^3 - 30 \times 280/2](222.3/2)/....$$

$$.... (21.16 \times 10^6) = 0.926 \text{ kN/mm}$$

The force per unit length of weld in the x direction on the same element is

$$F_{wx} = H/L_w + [M - HD_w/2](D_w/2)/I_{wGz}$$

$$= 30/(1.12 \times 10^3) + [160 \times 10^3 - 30 \times 280/2](280/2)/...$$

$$.... (21.16 \times 10^6) = 1.058 \text{ kN/mm}$$

The resultant vector force on this weld element is

$$F_{wR} = [F_{wx}^2 + F_{wy}^2]^{1/2} = [1.058^2 + 0.926^2]^{1/2} = 1.406 \text{ kN/mm}$$

Use 18 mm fillet weld for grade 43 steel which has an allowable strength of 1.449 kN/mm as shown in Table 4.3.

Alternatively if the end of the column and the base plates are machined and in contact, rotation is assumed to take place about the axis O-O. The weld group properties are:-

$$I_{wox} = 4D_w^3/3 = 4(280)^3/3 = 29.27 \times 10^6 \text{ mm}^4$$

$$I_{woy} = 2D_wD^2 = 2 \times 280(222.3)^2 = 27.67 \times 10^6 \text{ mm}^4$$

$$I_{woz} = I_{wox} + I_{woy} = [29.27 + 27.67]10^6 = 56.94 \times 10^6 \text{ mm}^4$$

The force per unit length of weld in the y direction on an element furthest from the centre of rotation is

$$F_{wy} = [M - N(D/2)]D/I_{woz}$$

$$= [160 \times 10^3 - 120(222.3/2)]222.3/(56.94 \times 10^6)$$

$$= 0.5726 \text{ kN/mm}$$

The force per unit length of weld in the x direction on the same element assuming that the horizontal force H is resisted by friction

$$F_{wx} = [M - N(D/2)]D_w/I_{woz}$$

$$= [160 \times 10^3 - 120(222.3/2)]280/56.94 \times 10^6$$

$$= 0.7212 \text{ kN/mm}$$

$$F_{wR} = [F_{wx}^2 + F_{wy}^2]^{1/2} = [0.7212^2 + 0.5726^2]^{1/2} = 0.9209 \text{ kN/mm}$$

Use 12 mm fillet weld for grade 43 steel which has an allowable strength of 0.966 kN/mm as shown in Table 4.3.

__Example 4.12   Rigid connection between RHS sections__   Design the connection between the RHS sections shown in Fig. 4.30 which is joint C for the Vierendeel girder design in Chapter 6. The forces shown in Fig. 4.30 are from the computer analysis assuming elastic behaviour. Use grade 43 steel.

The connection between the vertical member and the horizontal member is a butt weld which necessitates a 45° chamfer on the end of the vertical member. Rotation is assumed to take place about the axis

0-0 of the weld as shown in Fig. 4.30. The axis 0-0 is assumed to be a stiff bearing because the intersecting members are of the same width. The effective length of the weld furthest from the axis 0-0 is obtained from Equation (4.1) and values given in Table 4.1

$$b_{we} = 2t + CT = 2 \times 10 + 5 \times 10 = 70 \text{ mm}$$

The position of the resultant force R_o for the weld is obtained from taking moments of forces about the resultant R_o.

$$d_{wr}/d_f = (2/3 + b_{we}/d_f)/(1 + b_{we}/d_f)$$

$$= (2/3 + 70/240)/(1 + 70/240) = 0.742$$

The magnitude of the reaction force R_o acting on the stiff bearing is obtained by taking moment of forces about R

$$R_o = [M + N(d_{wr} - d_w/2)]/d_{wr}$$

$$= [54.9 \times 10^3 + 44.5(0.742 \times 240 - 240/2)]/(0.742 \times 240)$$

$$= 322.8 \text{ kN}$$

The frictional resistance at the stiff bearing is (μ_s/γ)R_o = (0.45/1.4)322.8 = 103.8 > 27.4 kN, which is the horizontal shear force acting on the weld.

The weld group is subject only to forces in the vertical direction. The second moment of area of the weld group for unit size welds shown in Fig. 4.30 about the axis 0-0 is

$$I_{wo} = 2d_w^3/3 + b_{we}d_w^2$$

$$= 2(240)^3/3 + 70 \times (240)^2 = 13.25 \times 10^6 \text{ mm}^4$$

The moment applied about axis 0-0 is

$$M' = M - Nd_w/2 = 54.9 - 44.5 \times 240 \times 10^{-3}/2 = 49.6 \text{ kNm}$$

The force per unit length on the weld furthest from the centre of rotation is

$$F_{wy} = M'd_w/I_{wo} = 49.6 \times 10^3 \times 240/(13.25 \times 10^6) = 0.898 \text{ kN/mm}$$

From Table 4.3 the resistance of a 10 mm butt weld is 0.805 kN/mm for grade 43 steel. This is the maximum size that can be used on the side welds as shown in Fig. 4.30. To increase the strength of this weld group use 15 mm fillet welds at the ends. This increases I_{wo} to 15.26 $\times$ 10^6 mm^4, and the maximum force per unit length of weld is reduced to 0.78 kN/mm. Details of this type of connection are given in BS 5135.

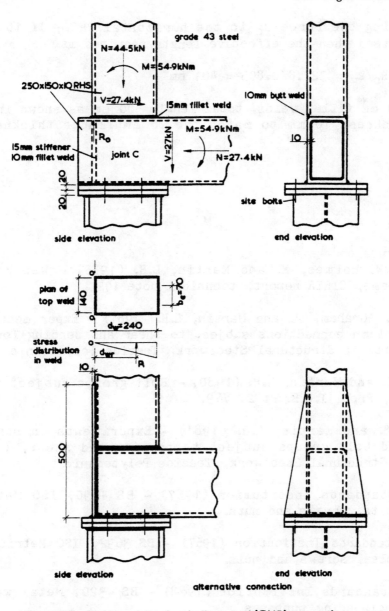

Fig. 4.30. Rigid rectangular hollow steel (RHS) connection.

The force R_O = 322.8 kN may crush the two webs of the RHS section. The crushing strength may be increased by welding a 15 mm thick stiffener in the end of the section as shown in Fig. 4.30. The allowable crushing strength of the stiffener

$$P_s = b_s t_s p_b = (150 - 20)15 \times 190 \times 10^{-3} = 370.5 > 322.8 \text{ kN}$$

A further alternative is to weld a gusset plate to the section as shown in Fig. 4.30. The welds on the vertical RHS must be capable of transferring the force R_O to the horizontal RHS. If 10 mm fillet welds are used then the effective length required is

$$L_W = R_O/P_W = 322.8/0.805 = 401 \text{ mm}$$

Use two 10 mm fillet welds, total length 500 mm as shown in Fig. 4.30. Plate thickness 10 mm to match the 10 mm flange thickness of the RHS.

REFERENCES

Astill, A.W. Holmes, M. and Martin, L.H. (1980) - Web buckling of steel I beams, CIRIA report, technical note 102.

Bahia, C.S., Graham, J. and Martin, L.H. (1981) - Experiments on rigid beam to column connections subject to shear and bending forces, Conf. Proc., Joints in Structural Steelwork, Teeside Polytechnic.

Bahia, C.S. and Martin, L.H. (1980) - Bolt groups subject to torsion and shear, Proc. I.C.E. Pt 2, V69.

Bahia, C.S. and Martin, L.H. (1981) - Experiments on stressed and unstressed bolt groups subject to torsion and shear, Conf. Proc. Joints in Structural Steelwork, Teeside Polytechnic.

British Standards Institution (1967) - BS 4190, ISO Metric black hexagon bolts, screws and nuts.

British Standards Institution (1967) - BS 3692, ISO Metric precision hexagon bolts, screws and nuts.

British Standards Institution (1968) - BS 4320, Metal washers for general engineering purposes.

British Standards Institution (1969) - BS 449, The use of structural steel in building, Pt. 2.

British Standards Institution (1979) - BS 4604 Pt 1, The use of high strength friction grip bolts in structural steelwork - general grade.

British Standards Institution (1973) - BS 4395, High Strength friction grip bolts and associated nuts and washers.

British Standards Institution (1974) - BS 5135, Metal-arc welding of carbon and carbon manganese steels.

British Standards Institution (1976) - BS 639, Covered electrodes for the manual metal-arc welding of carbon and carbon manganese steels.

British Standards Institution (1979) - BS 4360, Specification for weldable structural steels.

British Standards Institution (1981) - BS 3643, ISO Metric screw threads.

Biggs, M.S.A.B., Crofts, M.R., Higgs, J.D., Martin, L.H. and Tzogius, A. (1981) - Failure of fillet weld connections subject to static loading, Proc. Conf. on Joints in Steelwork, Teeside Polytechnic.

Butler, L.J., Pal, S. and Kulak, G.L. (1972) - Eccentrically loaded welded connections, Proc. A.S.C.E. (Struct. Div), V98.

Chesson, E. Jr., Faustino, N.L. and Munse, W.H. (1965) - High strength bolts subject to torsion and shear, A.S.C.E (Struct), V91, ST5.

Clarke, A. (1970) - The strength of fillet welded connection, M.Sc. thesis, Imperial College, University of London.

Clarke, P.J. (1971) - Basis of design for fillet welded joints under static loading, Conf. Proc., Welding Institution, Improving Welding Design Paper 10, V1.

Crawford, S.F. and Kulak, C.L. (1971) - Eccentrically loaded bolted connections, A.S.C.E(Struct), V97, ST3.

Douty, R.T. and McGuire, W. (1965) - High strength bolted moment connections, A.S.C.E.(Struct), V91, St2.

Elzen, L.W.A. (1966) - Welding seams in beam-to-column connections without the use of stiffening plates, report 6-66-2, I.I.W. document XV-213-66.

European Convention for Structural Steelwork (1981) - European recommendations for steel construction, Construction Press.

Farrar, J.C.M. and Dolby, R.E. (1972) - Lamellar tearing in welded steel fabrication, Welding Institute, Cambridge, England.

Fisher, J.W. and Struik, J.H.A. (1974) - Guide to Design Criteria for Bolted and Riveted Joints, John Wiley and Sons.

Gourd, L.M. (1980) - Principles of Welding Technology, Edward Arnold.

Holmes, M. and Martin L.H. (1983) - Analysis and Design of Structural Connections, Ellis Horwood Ltd.

International Institute of Welding (1964) - Calculation formula for welded connections subject to static loads, Welding in the World, V2.

International Institute of Welding (1976) - Design rules for arc welded connections in steel subject to static loads, Welding in the World, V14.

Johnson, L.G. (1959) - Tests on welded connections between I sections, British Welding Journal, V6.

Kato, B. and Morita, K. (1974) - Strength of transverse fillet welded joints, Welding Research.

Ligtenberg, F.K. (1968) - International test series, final report, Stevin laboratory, Technological University of Delft, Doc XV-242-68.

Mann, A.P. and Morris, L.J. (1979) - Limit state design of extended plate connections, A.S.C.E.(Struct), V105, ST3.

Martin, L.H. (1979) - Methods for limit state design of triangular steel gusset plates, Building and Environment.

Martin, L.H. and Robinson, S. (1981) - Experiments to investigate parameters associated with the failure of gusset plates, International Conference on Joints in Steelwork, Teeside Polytechnic.

Packer, J.A. and Morris, L.J. (1977) - A limit state design method for the tension region of bolted beam to column connections, Struct. Eng. V55, 10.

Rolloos, A. (1969) - The effective weld length of beam-to-column connections without stiffening plates, final report, I.I.W. document XV-276-69.

Salmon, C.G., Buettner, D.D. and O'Sheridan, T.C. (1964) - Laboratory investigation of unstiffened triangular bracket plates, Proc. A.S.C.E.(Struct), V90.

Structural Steel Handbook (1978) - Properties and safe load tables, British Constructional Steelwork Association.

Timoshenko, S. (1959) - Theory of Plates and Shells, McGraw Hill.

5   Frames and Framing

5.1   INTRODUCTION

This chapter falls into three main sections:-
   i) Structural framing.
  ii) Methods of analysis for frames.
 iii) Frame design.
Section (i) considers the interaction of members to resist the applied
forces in a structure.   Section (ii) considers methods which assume
either linear elastic (i.e. Hookean) behaviour or plastic collapse
behaviour.   Section (iii) considers the design of a single bay pitched
roof portal frame.

   Previous chapters have laid out design methods for beam elements
(Chapter 2), column elements (Chapter 3) and for connections (Chapter
4).   It remains to consider the interaction of these design methods on
the structure as a whole and the ability of the structure to withstand
all the possible applied loading.

5.2   STRUCTURAL FRAMING

5.2.1   <u>Structural form simple model</u>

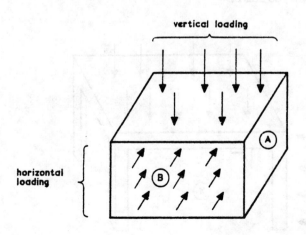

Fig. 5·I. Loading on a simple box structure.

If the simple box like structure shown in Fig. 5.1 be considered, the side panels A and B carry vertical loading from the roof and in-plane horizontal forces from each of the other sides and out of plane horizontal forces from wind action on that side panel.

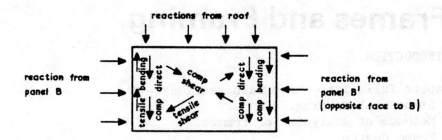

Fig. 5·2. Force distribution on panel A.

The reactions from the roof are, vertical due to dead and live load and shear due to wind loads. The reactions from each of the adjacent side panels are in-plane axial forces. The resultant force distribution due to these reactions and the wind loads applied directly to panel A is complex, but is of a form similar to that drawn in Fig. 5.2.

In practice no structure is as simple as this, but the use of this model, where the side walls are now replaced by lightweight cladding, is illustrated in Fig. 5.3., where the corner columns now take all the reactive loads from the side panels. It should be immediately apparent that the forces must be transferred from the roof to the columns by bending. This moment will be increased by the loading on the column, and there will be resultant problems in designing suitable connection to transfer this moment between the roofing system and the column.

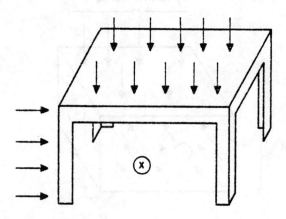

Fig. 5·3. Structural model.

In order to simplify the problem the horizontal loading on the column has been replaced by a single equivalent force at the top of the column. Then if face X in Fig. 5.3 be considered the loading is as shown in Fig. 5.4(a).

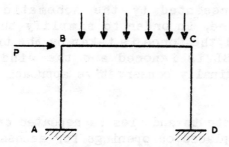

Fig. 5·4.(a) Diagrammatic view of panel X - wind force P at B.

The moment at B will be substantially reduced if a tie is put in between A and C (Fig. 5.4(b)).

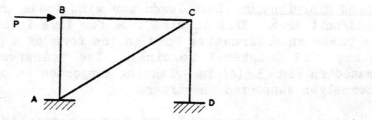

Fig. 5·4.(b) Wind force P at B resisted by tie AC.

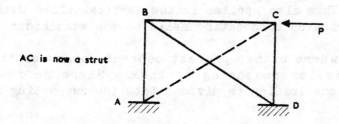

AC is now a strut

Fig. 5·4.(c) Wind force P at C resisted by tie BD.

The wind is not unidirectional but may also blow the other way round, thus the effect of P must be considered at C. AC now becomes a strut and, therefore, substantially weaker, so a further tie is placed from B to D (Fig. 5.5(c)).

The structure represented by the schematic frame ABCD is hyperstatic. In practice, in order to simplify the analysis, the strut is ignored and all the force is taken by the tie (i.e. with the wind blowing from B, BD is ignored and the wind from C, AC is ignored.) This is a marginally conservative approach but one which is generally accepted.

If this pattern of struts and ties be repeated on all four faces, severe restrictions are placed on openings for access or lighting and would produce a structure which was aesthetically unpleasant.

Very few structures are as simple as this initial model. Most structures are multi-bay in both directions and it is these which are next considered.

5.2.2 <u>Load distribution</u> Dead loads and wind loads are usually quoted as a load/unit area. This is also true for live loads except that in certain cases an alternative load in the form of a point load needs evaluating (CP3 Chapter V Loading). The transfer of loading is illustrated in Fig. 5.5(a) for cladding supported on purlins which are then themselves supported on rafters.

The cladding is assumed to be simply supported on the purlins, thus producing a total line load on the purlins equal to half the load on each panel of cladding. The purlins are also taken as simply supported on the rafters giving equal reactions on the rafters (i.e. half the line load). This also applies in the vertical plane with the wind loads being distributed to sheeting rails to the stanchions.

For floor units where either precast concrete beams or timber joists are used a similar reasoning applies. Where an in situ concrete slab is used the loading is divided into the supporting beams as in Fig. 5.5(b).

The alternative point load in live loading is usually only critical for short spans and is only to be used for local effects. It must be situated in the worst place for the effect being considered, i.e. at mid-span for the bending moment in the purlin or joist or at the support for shear in the joist. The uniformly distributed load is used for the loading on the rafter or stanchion.

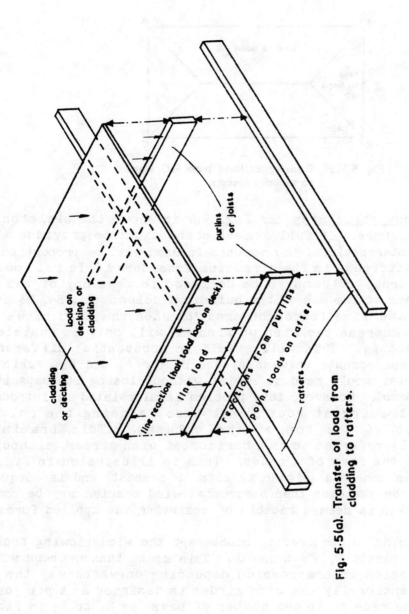

Fig. 5.5(a). Transfer of load from cladding to rafters.

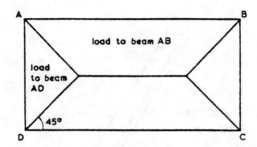

**Fig. 5·5(b). Transfer of load from RC slab
to support beams.**

5.2.3 <u>Multibay structures</u>   In Fig. 5.6 is shown the skeleton of a
multibay structure. If full moment connections be provided at all
column–beam intersections, the moments induced will be proportional to
the bending stiffness in the direction considered. It is, however,
rare for internal columns to be designed to take any of the wind
forces and thus at the end of the building, columns B_1, B_2, B_3 and B_4
will resist the wind forces on area 'b' when the wind blows from
direction P, whereas the wind on area 'a' will only be resisted by
columns A_1 and A_4. This will result in substantial differential
horizontal displacements between A_1 and B_1 (or A_4 and B_4), which the
roofing material would need to resist without losing its capacity to
remain leak proof. However, this problem is alleviated by introducing
horizontal bracing at roof level thus reducing the relative
displacement at the top of the columns. This bracing is
conventionally referred to as a horizontal wind girder, although in
fact it is in the form of a truss. This is illustrated in Fig. 5.7.
If the roof is concrete (either in situ or precast) and is adequately
connected to the skeleton then horizontal wind bracing may be omitted
as the roof slab is deemed capable of resisting the applied forces.

Full bracing is required to counteract the wind blowing from all
4 directions, namely P, P', Q and Q'. This means that members will be
in either tension or compression depending on which way the wind
blows. Conventionally the wind girder is designed as a pin jointed
truss, and the case of an odd number of bays, as B_1 to B_4 in Fig. 5.7
the truss is hyperstatic. It is conventional to use a similar
approach to design as in section 5.2.1, whereby compression members
are ignored. Alternatively an analysis may be used taking account of
all members, but this is more complex.

The loads to be resisted by the wind girder comprise

i) the horizontal forces due to the wind, and
ii) 2.5% of the vertical loading (dead and imposed) acting on the
   roof (Cl. 26e, BS 449).

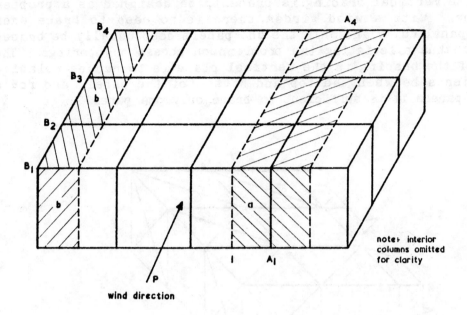

Fig. 5·6. Multibay structure under the action of wind.

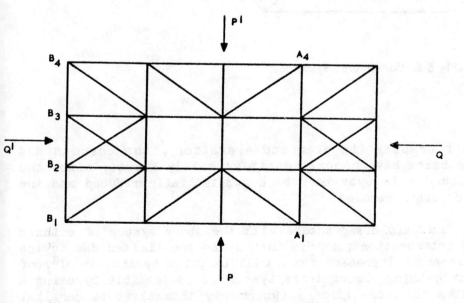

Fig. 5·7. Plan of roof showing full wind bracing.

These loads are applied at panel points and half the load will act at B_1 as at A_1 as the load distribution is taken such that the total load on a panel is divided equally between the two supporting columns. (For further details on truss design, Chapter 6 should be consulted). Note that the vertical columns may, where a wind girder or adequate vertical bracing is present, be designed as a propped cantilever. With a wind girder there is no need to brace each vertical panel fully and only the end panels need usually be braced (Fig. 5.8), thus alleviating the problems on access or openings. The purpose of the bracing in the vertical plane is to reduce relative displacements between the top and bottom of the panel, and for a series of panels it is sufficient to brace only one panel.

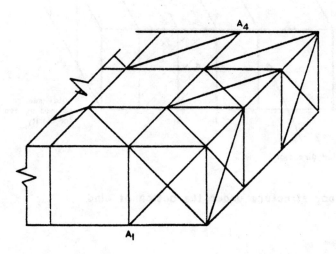

Fig. 5.8. Corner bracing detail.

With full bracing (in plan and elevation), the columns and therefore the bases have become sensibly axially loaded, since the moments induced due to sway have been substantially reduced and the need for rigid joints reduced.

However, the main disadvantage with the above system of columns at every beam intersection point is that spans are limited due to the use of rolled beams. Increased spans will be given by the use of roof trusses. Also by using a roof truss system it is possible by using a secondary system of cross girders (which may themselves be parallel chord trusses) which support the ends of the main trusses to eliminate some of the interior columns thus derestricting the use of floor space.

5.2.4 <u>Truss systems</u> In general truss systems are one way spanning in that the loads from the roof trusses which span in one direction are transferred to secondary members which span in a direction at right angles to the main trusses before the loads are transferred to the stanchions. In exceptional circumstances the truss system can be made to span in both directions, thus becoming a space frame. This will necessitate a three dimensional analysis which in general will entail computer methods. A space system is rarely economic unless the area to be covered is roughly square. This type of system is generally used for prestige structures and is not in common use.

A typical layout for a roof truss system is shown in Fig. 5.9(a).

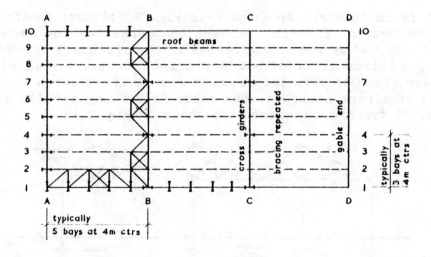

Fig. 5.9(a)   Typical layout of wind bracing for a trussed roof
              system with cross girders.

As the interior roof trusses are only supported at every third point, secondary trusses are necessary in the vertical plane on B-B and C-C. The effect of these is to reduce the number of interior columns. Horizontal bracing is required along the lines A-A and D-D to ensure sufficient resistance to horizontal loading. This has been shown for clarity on line B-B. The additional stanchions on A-B, B-C, C-D (both sides) are to support the cladding to the structure and will be of a smaller size than the main columns. A cross-section through 3-3 is shown in Fig. 5.9(b).

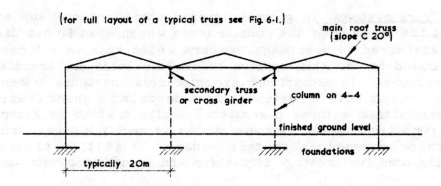

(for full layout of a typical truss see Fig. 6·1.)

main roof truss
(slope C 20°)

secondary truss
or cross girder

column on 4-4

finished ground level

foundations

typically 20m

Fig. 5·9(b) Section on 3-3 through Fig. 5·9(a).

It is immediately apparent from Fig. 5.9(b) that benefits gained by the removal of interior columns are offset by the reduction in headroom caused by the introduction of the secondary trusses. This led to the introduction of the 'Northlight' truss, so called because the steep slope of the roof which is glazed faced north to avoid the effects of direct sunlight, the other face being sheeted in normal fashion. A typical arrangement is shown in Fig. 5.10.

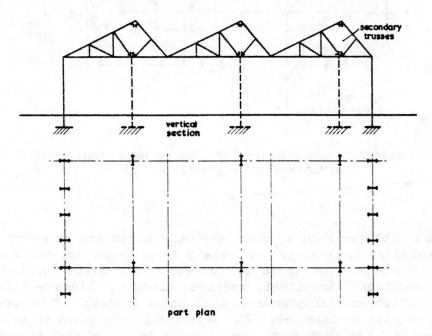

secondary
trusses

vertical
section

part plan

Fig. 5·10.   Details of northlight roof truss.

The secondary trusses now support the ridge and not the valley of the roof. Thus the secondary trusses are now contained entirely within the main trusses, saving headroom.

5.2.5 Differential settlement The effect of differential settlement on a rigidly jointed frame is to induce large moments at the joints. Thus, joints which effectively behave as pinned, and only transmit axial forces must be used in this case. A common solution for roofing is that of the cantilever-suspended span shown in Fig. 5.11(a).

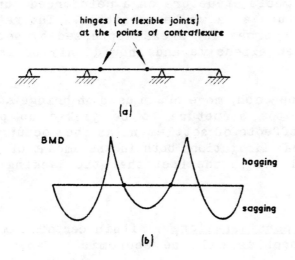

Fig. 5·11. Cantilever-suspended span arrangement.

The bending moment diagram is shown in Fig. 5.11(b). This system has the advantage of being isostatic and thus any settlement that takes place will not affect the bending moments in the roof. A disadvantage with this type of structure is ensuring an adequate watertight yet flexible joint at the connections between the cantilever and the suspended span.

If secondary trusses are used when differential settlement is likely to occur they must be simply supported on the columns, as in Fig. 5.12, to avoid the introduction of moments at A or B.

The joints at B and D must be effectively pinned and the members AA' and C'C shown dotted should be omitted to make the truss isostatic. (In normal conditions the use of members AA' and C'C will reduce the effective lengths of the stanchions.)

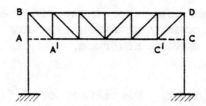

Fig. 5·12. Typical arrangement of a secondary truss.

An alternative approach where differential settlement is expected
is to place the whole structure on a reinforced concrete raft and
allow the structure as a whole to settle. The raft must then be
designed to span across any areas depressed by settlement. This
solution is often expensive and should only be used on very bad
ground.

A further approach, more often used on bridge structures, is to
allow parts of the structure to be jacked up periodically to
counteract the effects of settlement as they occur. This solution
does have practical limitations both in the amount of jacking that can
be accommodated and in the fact that the jacking points must be
accessible.

5.2.6 <u>Economic considerations</u>   Within certain limits a particular
solution to a problem will be economic.   Table 5.1 gives some
guidelines.

Rolled beams	0 - 10 m
Castellated beams	10 - 20 m
Plate girders	15 - 30 m
Roof trusses	20 - 30 m (exceptionally up to 60 m)
Pitched roof portal frames	15 m and above

Table 5.1 Economic construction methods

The main drawback with truss sytems is the large wasted space
above the bottom boom which needs heating. This is one reason for the
preference of shallow, pitched roof portal frames which are considered
later in this Chapter; another is that the erection time is much
reduced. For pitched roof portal frames grade 55 steel is not used,
as it presents instability problems when used with plastic design
methods.

5.2.7 <u>Shear walls</u> In the structures so far considered, the cladding makes little or no contribution to the stiffness or stability of structures, so in certain cases it is beneficial to use brickwork or masonry to clad the structure.   This is often required for architectural reasons, but also has the advantage, that provided the brickwork cladding is adequately tied to the skeletal frame, bracing in the vertical plane may be omitted. The brickwork panels then act as shear walls, which if they are to function correctly must be free of large openings. The brickwork must be adequately tied to the skeletal frame both to transmit the diagonal shear forces which would normally be taken by the diagonal members in the bracing and to transfer the wind loading normal to the cladding to the stanchions and hence to the foundations. For the design of brickwork infill panels the reader is referred to Curtin et al or Hendry et al.

5.2.8 <u>Erection forces</u> It is the designer's responsibility to ensure that the structure has adequate stability during construction. It is not sufficient merely to ensure stability of the completed structure. The need for temporary bracing during construction cannot be emphasised too much. During erection steelwork frequently will have different end support conditions from those assumed in the completed structure and thus will probably have a reduced load carrying capacity. This needs special consideration where parts of the structure are to be used to facilitate erection of other sections by having lifting gear attached. The designer is also responsible for ensuring that the sections of a structure needed to be positioned by crane are adequate under the forces induced whilst erection is taking place. It is also possible that members which in the completed structure are in tension may be subjected to compression during erection of part of the structure. This needs careful consideration since the maximum force which may be carried in compression is lower than that in tension. Where the need for temporary propping of the structure is envisaged during erection, it is necessary to ensure the adequacy of the structure under these propping forces during erection.

   In the case where a concrete flooring system has been assumed to provide adequate lateral support to the compression flange of a beam in the completed structure, this will not be present during construction and it will be necessary to provide adequate lateral bracing to the compression flange whilst the concrete is being poured. The soffit shutter should not be relied on to provide this lateral restraint.

   It must be reiterated that it is the designer's responsibility to ensure the adequacy of the structure during erection.

5.3   DESIGN PHILOSOPHY

Two design philosophies are in use:-

i)   Elastic
     Here the material is deemed to behave in a linear elastic or
     Hookean fashion such that the stresses are kept below yield by a
     suitable factor of safety.

ii) Plastic
     Here the material is deemed to behave accordng to an idealized
     stress-strain curve (as shown in Fig. 1.4), where the elastic
     behaviour is ignored and the member assumed to possess a large
     rotational capacity at its yield moment. At present this method
     is limited to structures or structural elements in which flexure
     is predominant, e.g. rigid jointed frames. A suitable load
     factor must be aplied to the loads on the structure.

5.4   METHODS OF DESIGN FOR FRAMES

5.4.1 Elastic methods Cl. 9b BS 449 lays down three methods that may
be used for frame (or member) design:-
     i)   Simple
     ii) Rigid
     iii) Semi-rigid
     Each of these will now be discussed in turn.

i) Simple Design Method This assumes the beams to be simply
supported, that is the beam-column connection allows full rotation.
The loading applied to the column is vertical shear applied usually
100 mm from the column face. For the stanchion design, the effective
length of the column is established by considering the rotational
restraint offered to the column by the beam and its connection. It
will at first sight appear that this method is irrational, but it
usually produces a safe solution, remembering that other design
criteria are also involved, e.g. deflection and that all horizontal
forces must be taken by suitable bracing.

     This method also implies the use of web cleats for the connection
in addition to flange cleats, although this often is not satisfied.
Alternative forms of connection will enforce some transfer of moment
from the beam to the stanchion.

     Simple design may be thought of as a lower bound method of
design. A typical usage of simple design is in triangulated roof
trusses, an example of which is given in Chapter 6.

ii) Rigid Design Method This assumes full fixity between the beam
and column, i.e. there is no relative rotation between the beam and

the column. This method allows a more accurate approach to be used to establish effective lengths of beams and columns using a similar method to that adopted in reinforced concrete in that the effective lengths are calculated from coefficients which depend on the relative stiffness of the beams and stanchions framing in at a joint, and also enables use to be made of plastic design methods, which can lead to greater economy.

The most effective rigid connection is when the beam is welded directly to the column, which may however require web stiffeners. This generally means the use of site welding which is not British practice. A less rigid beam-to-column connection consists of an end plate welded directly to the end of the beam. This end plate is then bolted to the column on site as shown in Fig. 4.23.

Rigid design may be thought of as an upper bound method of design.

iii) <u>Semi-rigid Design Method</u>   Here allowance is made for the relative rotation between the beam end and the stanchion. This means that the moment-rotation characteristic of the joint detail being used must be known. A typical schematic for the moment-rotation characteristic is drawn in Fig. 5.20(a).

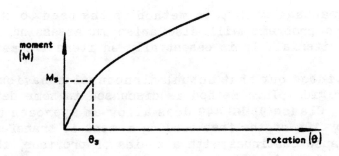

Fig. 5·13.(a) Schematic characteristic for joint moment-rotation relationship.

If a beam span L with a flexural rigidity of EI carrying a uniformly distributed load w, is considered, under support moments M_s (Fig. 5.20(b)).

Then $\theta_s = \dfrac{1}{2EI} \left[\dfrac{wl^3}{12} - M_s l \right]$ (5.1)

where θ_s and M_s must be corresponding values, e.g. from Fig. 5.13(a).

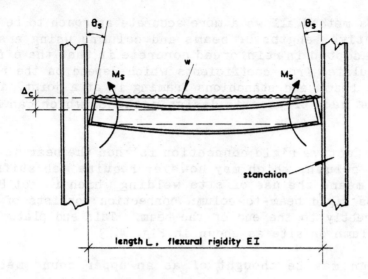

Fig. 5·13.(b) Forces and displacements on a semi-rigid beam system.

The central deflection is then given by

$$\Delta_c = \frac{5}{384} \frac{wl^4}{EI} - \frac{M_s l^2}{8EI} \qquad (5.1a)$$

The main disadvantage with this method is the need to know the joint performance and problems will also arise in assessing failure and serviceability criteria. It is essentially an iterative method.

It should be pointed out that actual structural behaviour will correspond to semi-rigid. This method is discussed in some detail by Roberts and Taylor. Clause 9b BS 449 does allow an approach to semi rigid design, whereby 10% of the free bending moment is transferred to the columns. The clause continues with a series of provisos, the last of which states that the column must be encased in concrete. Concrete encasement is now virtually obsolete, thus the approximate method for semi-rigid design advocated by BS 449 is seldom capable of being used.

5.4.2 **Plastic methods** The concept of plastic design has already been met for continuous beams in Chapter 2. It is equally applicable to frames, as is demonstrated in Examples 5.4 and 6.4. However, in frames there may be problems due to elastic instability (discussed in Example 5.4) and the effect of axial loads on the plastic moment. In the first the restraints have to be positioned to prevent elastic buckling before the member can generate its full plastic moment and in the second the reduction of the plastic section modulus due to axial load needs to be found and the plastic moment reduced accordingly. The effects of shear are not usually very significant.

5.5   METHODS OF ANALYSIS OF FRAMES

5.5.1   <u>Elastic</u>. It is expected that the student will be extensively familiar with elastic methods of structural analysis. For isostatic pin jointed frames Bow's notation, which is a graphical approach, or tension coefficients, resolution at joints or method of sections may be used. For hyperstatic pin jointed frames strain energy or matrix methods are used. For rigid jointed frames moment distribution, slope deflection equations, strain energy or matrix methods may be used. It is frequently possible to use approximate methods, one such is illustrated in the example on the Vierendeel girder (Example 5.3). For certain standard frames with standard loadings design charts are available, the best example of these are those by Kleinlogel for portal frames.

A series of references to standard texts on structural analysis are to be found at the end of this chapter.

5.5.2   <u>Plastic</u>. The use of plastic analysis for continuous beams has been covered in Chapter 2, and it now remains to extend the approach to cover frames.

It will be useful to start by stating the theorems of plastic collapse. Three conditions must be satisfied at collapse:-

1) Equilibrium. The moment field in the structure must be in equilibrium with the applied loading.
2) Yield. At all points in the structure the numerical value of the moment at any point must be less than or equal to the yield moment at that point.
3) Mechanism. Enough plastic hinges must occur to give total (or, in special cases, partial) collapse in the structure.

5.5.3   <u>Lower bound (static) theorem</u>. If there exists a moment field such that only the conditions of equilibrium and yield are satisfied, then the load set W_1 producing this moment field is less than the collapse load, W_c, i.e. $W_1 < W_c$.

5.5.4   <u>Upper bound (kinematic) theorem</u>. If there exists a moment field such that only the conditions of equilibrium and mechanism are satisfied, then the load set W_u producing this moment field is greater than the collapse load, W_c, i.e. $W_u > W_c$.

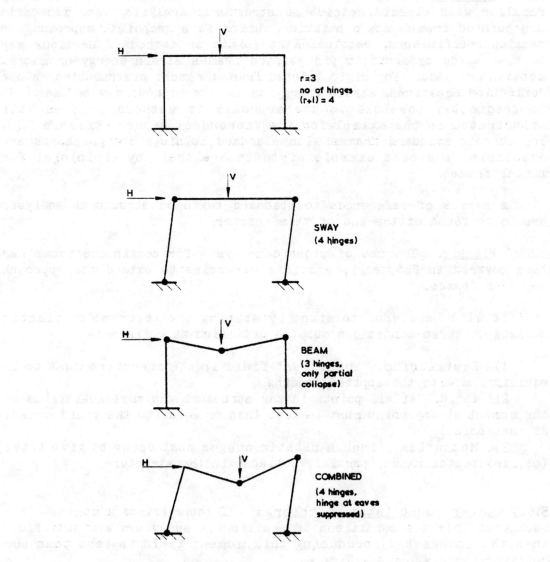

Fig. 5.14(a). Collapse modes of a single bay portal frame.

5.5.5 <u>Uniqueness theorem</u> From 5.5.3 and 5.5.4, $W_u > W_c > W_l$
W_c must satisfy all three conditions, since if the load set W_u is
adjusted until yield is satisifed or the load set W_l is adjusted until
a mechanism occurs then the same result must be reached.

   Full proofs of these theorems will be found in Horne or Neal.

5.5.6 <u>Number of hinges</u> If the structure has 'r' redundancies then,
in general r + 1 hinges will cause a mechanism to form. It is to be
stressed that less than r + 1 hinges may form to give partial collapse
(Fig. 5.14(a)).

5.5.7 <u>Types of mechanisms</u> In general mechanisms may be split into
two categories:-
   i)   Independent mechanisms, and
   ii) Combined mechanisms.

i) Independent mechanisms. These are generally defined as simple,
more basic mechanisms whose virtual work equations may be written down
easily. Typical examples are the beam collapse and sway illustrated
in Fig. 5.14. A further very useful mechanism (although no work
equation can be written down for it) is the joint rotation,
illustrated in Fig. 5.14(b).

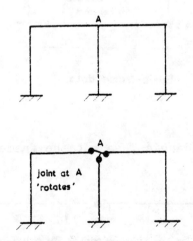

Fig. 5.14(b). Joint rotation.

If the maximum number of hinge positions (or the points at which it is required to know the bending moment) is h and the number of redundancies is r then the number of independent mechanisms is (h - r).

ii) Combined mechanisms. These are obtained by superimposing two or more independent mechanisms in such a way to eliminate hinge rotations at one or more hinge positions. This has the effect of reducing the virtual work done on the rotation at the hinges with no effect on the external virtual work done by the loads, thus reducing the load factor. There is no simple method of determining the number of combined mechanisms.

<u>Example 5.1 Flat roof portal frame</u>   This example has been worked in more detail than is usually necessary to give a solution in order to illustrate the use of the upper and lower bound theorems.

Consider the frame in Fig. 5.15. The plastic moment (M_p) is uniform throughout the structure.

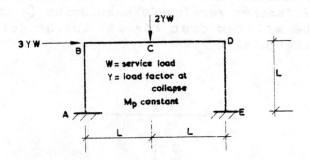

Fig. 5·15.   Basic frame data.

There are 3 redundancies, so cut the frame at C

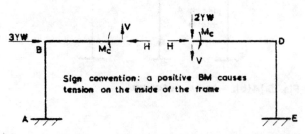

Fig. 5·16.   Frame cut at C showing redundant forces.

The values of BM at A, B, C, D, E need only be considered since the variation of bending moments in the structure is linear.

$$M_A = M_C + HL - VL - 3\gamma WL \qquad (5.2)$$

$$M_B = M_C + VL \qquad (5.3)$$

$$M_D = M_C - VL - 2\ WL \qquad (5.4)$$

$$M_E = M_C + HL - VL - 2\gamma WL \qquad (5.5)$$

Eliminating V and H from Equations (5.2) to (5.5)

$$3\gamma WL = M_B - M_A + M_E - M_D \qquad (5.6)$$

$$2\gamma WL = 2M_C - M_B - M_D \qquad (5.7)$$

These last two equations represent the overall conditions of equilibrium of the structure and must always be obeyed.

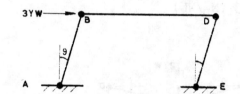

Fig. 5.17. Sidesway mechanism.

Hinge A causes compression inside so $M_A = -M_p$
Hinge B causes tension inside so $M_B = M_p$
Hinge D causes compression inside so $M_D = -M_p$
Hinge E causes tension inside so $M_E = M_p$

Note alternation of sign as the hinges are alternately closing and opening.

Use Equation (5.6) $$\gamma = -\frac{4\ M_p}{3\ WL} \qquad (5.8)$$

Equation (5.8) could have been obtained by considering virtual work on frame,

Work done at hinges $M_p\theta + M_p\theta + M_p\theta + M_p\theta = 4M_p\theta$
Work done by load $3\gamma W.L\theta$ $\qquad\qquad\qquad = 3\gamma WL\ \theta$

$$\text{or} \quad \gamma = \frac{4}{3} \frac{M_p}{WL}$$

This solution does not consider the value of M_C so substitute values for M_A, M_B, M_D, M_E into Equation (5.7) and it is then found that $M_C = 4/3 \, M_p$ which is not allowed (yield), so the value of $\gamma = 4/3 \, (M_p/WL)$ is an upper bound

i.e. $\gamma_c < 4/3 \, (M_p/WL)$

γ_c is the actual collapse load factor satisfying yield, equilibrium and mechanism.

Since $M_C > M_p$ the sway mechanism cannot be the correct mode of failure. If γ is reduced such that $M_C = M_p$ then $\gamma = 1$ (pro rata) but the remaining BMs are less than M_p and only 1 hinge forms, so the system is in equilibrium and is statically admissable and no BM is greater than M_p but no mechanism forms, so a lower bound value of γ is M_p/WL

$$M_p/WL < \quad_c < 4/3 \, (M_p/WL) \tag{5.9}$$

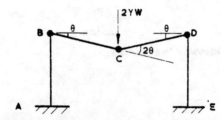

Fig. 5·18. Beam collapse mechanism.

By virtual work:

$$M_p\theta + M_p.2\theta + M_p.\theta = 2\gamma W.L$$

or $\quad \gamma = 2M_p/WL$

this value of γ is greater than the previously obtained upper bound solution, and therefore may be disregarded (Equation 5.9).

Note this mechanism indicates nothing about M_A and M_E, so using Equation (5.6) with $M_B = -M_p$, $M_D = -M_p$ and $\gamma = 2M_p/WL$,

$$6M_p = M_E - M_A \tag{5.10}$$

Thus M_E and M_A cannot be uniquely determined as only 3 hinges have formed, and no solution to Equation (5.10) can be found such that $M_E \leqslant M_p$ and $M_A \leqslant M_p$

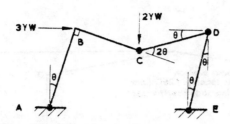

Fig. 5.19. Combined mechanism.

Using virtual work $M_p\theta + M_p.2\theta + M_p.2\theta + M_p\theta$

$$= 3\gamma W.L + 2\gamma W.L$$

or $\gamma = 1.2 (M_p/WL)$ (5.11)

The bending moment at B has not been determined, so using Equation (5.7), $M_B = 0.6M_p$.

M_B is less than M_p and is therefore satisfactory. The value of γ satisfies Equation (5.9).

No bending moment is greater than M_p (yield)
$\gamma = 1.2 (M_p/WL)$ (equilibrium)
and 4 hinges have formed (mechanism).

Therefore the value from Equation (5.11) is the actual collapse load factor, i.e. $\gamma_c = 1.2 (M_p/WL)$.

Example 5.2 Pitched roof portal frame   Consider the frame which is detailed in Fig. 5.20.

1) Plastic analysis

Using a similar approach to that already adopted for a flat roof portal frame (Example 6.1) where a combined mechanism was shown to be critical it may be demonstrated that the critical collapse mode for a pitched roof portal frame under vertical loading is as Fig. 5.21, (with hinges forming at B, C, D, E). In practice however a hinge will not form at the ridge but usually at the first purlin point either side of the ridge, (Fig. 5.22). The reason for this is that the ridge is effectively stiffened by the presence of a joint connecting the two rafters, thus locally increasing the plastic moment.

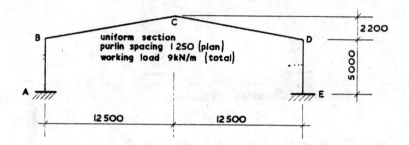

Fig. 5·20   **General arrangement of a pitched roof portal frame.**

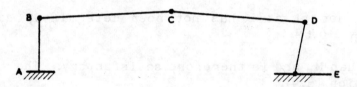

Fig. 5·21.   **Theoretical collapse mode with a ridge hinge.**

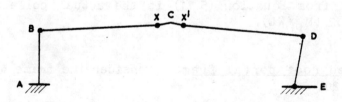

Fig. 5·22.   **Actual collapse mode.**

However with large spans, it is possible that the hinges will form at the second purlin points either side of the ridge.

Using a load factor of 1.7

Purlin loading = 1.7 x 9 x 1.25 = 19.13 kN

To determine the redundancies cut the frame at C

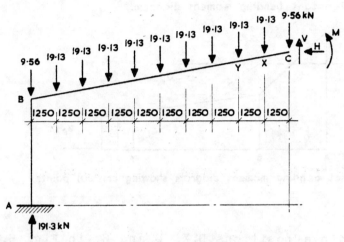

Fig. 5·23. Applied forces acting on half the frame.

The bending moment diagrams for the free bending moments, the reactant moments and the net bending moments may now be drawn (Figs. 5.24, 5.25 and 5.26) for half the frame.

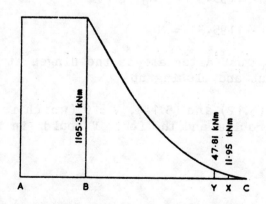

Fig. 5·24. Free bending moment diagram for half the frame.

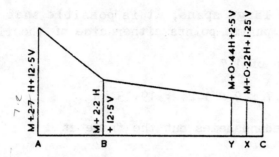

Fig. 5·25. Reactant bending moment diagram.

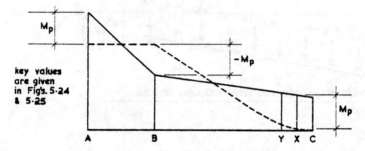

Fig. 5·26. Net bending moment diagram showing critical points.

Considering hinge positions B, X, D and E (in Fig. 5.26) the following equations may be written for the bending moments at these points:-

$$M + 2.2H + 12.5V - 1195.31 = -M_p \qquad (5.12)$$

$$M + 0.22H + 1.25V - 11.90 = M_p \qquad (5.13)$$

$$M + 2.2H - 12.5V - 1195.31 = -M_p \qquad (5.14)$$

$$M + 7.2H - 12.5V - 1195.31 = M_p \qquad (5.15)$$

The signs for M_p must alternate as the hinges at B, X, D and E are alternately opening out and closing up.

From Equations (5.12) and (5.14), V = 0 (which is expected from the symmetry of the problem and the force V could, in this case, have been omitted).

Solving Equations (5.12 to 5.15) gives

$$M = 398.7 \text{ kNm}$$
$$H = 169.5 \text{ kN}$$
$$M_p = 423.8 \text{ kNm}$$
$$V = 0$$

The bending moment at Y is

$$0.44H + 2.5V + M - 47.81 = 425.47 \text{ kNm} \qquad \text{Redo calcs}$$

which is greater than M_p, i.e. the ridge hinge must form at Y not X.
Equation (5.13) therefore becomes

$$M + 0.44H + 2.5V - 47.81 = M_p$$

and hence

$$M = 397.6 \text{ kNm}$$
$$H = 169.7 \text{ kN}$$
$$M_p = 424.4 \text{ kNm}$$
$$V = 0$$

2) Moment distribution

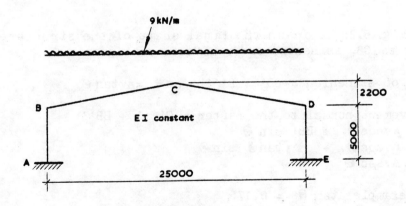

Fig. 5·27. Pitched roof portal.

Only half the frame need be considered owing to symmetry. Two
distributions need to be carried out - a no sway distribution and a
sway distribution owing to the ridge deflecting vertically downwards.

Length of BC = $[12.5^2 + 2.2^2]^{1/2}$ = 12.692 m

Distribution factors:-

B   BA (1/5)/(1/5 + 1/12.692) = 0.72

      BC (1/12.962)/(1/5 + 1/12.692) = 0.28

No sway distribution:-

Fixed end moments (FEM) in the rafter = $wl^2/12$ =

    $9 \times 12.5^2/12 = 117.2$ kNm

	A		B		C
	AB		BA	BC	CB
DF's	1.0		0.72	0.28	1.0
FEM's				-117.2	117.2
	42.2	←	84.4	32.8 →	16.4
Total	42.2		84.4	-84.4	133.6

Table 5.2   No sway distribution for the pitched roof portal

        In Fig. 5.28 is drawn the final shape of the structure. C'B''' is
parallel to CB, as second order effects are ignored.

Movement of stanchion = BB' = B'''B tan θ = Δv tan θ

Total movement normal to the rafter = C'C'' + BB''
    =   Δv cos θ + BB' sin θ
    =   Δv cos θ + Δv tan θ sin θ
    =   Δv sec θ

In this example   tan θ = 0.176
                  cos θ = 0.985

so calculating FEM's

CB   M = $-6EI\delta/L^2$ = $-6EI(\Delta v)/(12.692^2 \times 0.985)$ = $-0.0378EI\,\Delta v$

AB   M = $-6EI\delta/L^2$ = $-6EI(-0.176\,\Delta v)/5^2$ = $0.0422EI\,\Delta v$

let M_{AB} = 200 units and M_{CB} = 200(-378)/422 = -179 units. The sway
distribution is given in Table 5.3.

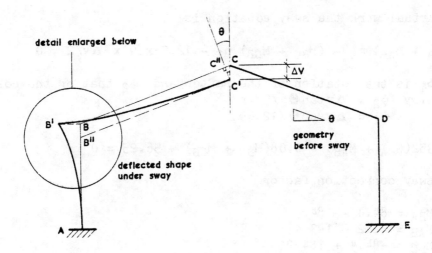

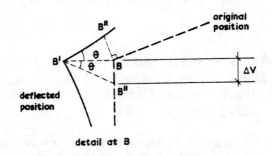

detail at B

Fig. 5·28. Sway movements for a drop at the ridge of ΔV.

	A		B		C
	AB	BA	BC		CB
DF's	1.0	0.72	0.28		1.0
FEM's	-200.0	-200.0	179.0		179.0
	7.6 ←	+15.1	5.9 →		2.9
Total	-193.4	-184.9	184.9		181.9

Table 5.3 Sway distribution for pitched roof portal

Note that M_{CB} and M_{AB} must have opposite signs since C and A are deflecting in the same direction relative to B.

Using virtual work the sway equation is

$$(M_{BA} + M_{AB})\phi_2 - (M_{BC} + M_{CB})\phi_1 - 12.5 \times 9 \times \Delta v/2 = 0$$

where ϕ_1 is the rotation of the rafter and ϕ_2 that of the column
By geometry $\phi_2 = \Delta v \tan\Theta /5$ and
$\phi_1 = \Delta v \sec\Theta /12.692$

Or $0.0352(M_{BA} + M_{AB}) - 0.08(M_{BC} + M_{CB}) - 56.25 = 0$ (5.16)

If a = sway correction factor

then $M_{BA} = 84.4 - 184.9$
$M_{AB} = 42.2 - 193.4$
$M_{BC} = -84.4 + 184.9$ (5.17)
$M_{CB} = 133.6 + 181.9$

Substituting Equations (5.17) into (5.16) gives $a = -1.31$

Final moments:-

$M_{BA} = 84.4 + 242.2 = 326.6$ kNm
$M_{AB} = 42.2 + 253.4 = 295.6$ kNm
$M_{BC} = -84.4 - 242.2 = -326.6$ kNm
$M_{CB} = 133.6 - 238.3 = -104.7$ kNm

In Table 5.4 is given comparative results from moment distribution, Kleinlogel tables and a plane frame analysis. The results from the computer analysis (plane frame) are given in Fig. 5.29.

Moment	Moment Distribution	Kleinlogel	Plane frame Computer Analysis	
M_{AB}	295.6	293.8	289.4	
M_{BA}	326.6	326.1	327.3	
M_{BC}	-326.6	-326.1	-327.3	
M_{CB}	104.7	104.3	104.3	kNm

Table 5.4 Comparison of bending moments obtained by different methods of analysis

The slight differences in Table 5.4 are due to the effect of axial deformations which are taken into account in the computer analysis.

UNIVERSITY OF ASTON IN BIRMINGHAM
DEPARTMENT OF CIVIL ENGINEERING
PLANE FRAME ANALYSIS

JOB REFERENCE pitportal

FRAME DATA

JOINT DATA

JOINT	XCOOR	YCOOR	RESTRAINTS
1	0	0	XYR
2	0	5	0
3	12.5	7.2	0
4	25	5	0
5	25	0	XYR

SECTION DATA

SECTION TYPE	MOMENT OF INERTIA	X SECTION AREA	YOUNGS MODULUS
a	3.6215e-04	.01045	200000000

(continued over)

MEMBER DATA

MEMBER	JOINT NO END I	JOINT NO END J	MEMBER TYPE	SECTION TYPE	LENGTH
1	1	2	0	3	5
2	2	3	0	3	12.6921236
3	3	4	0	3	12.6921236
4	4	5	0	3	5

LOAD CASE 1

MEMBER LOADS

MEMBER NO	LOAD W	DISTANCE a	DISTANCE b	LOAD DIR'N
2	-9	0	12.5	Y
3	-9	0	12.5	Y

(continued over)

JOINT DISP

JOINT NO	X DISP	Y DISP	ROTATION
1	0	0	0
2	-.0145042925	-2.71179002e-04	-1.29783999e-03
3	2.55610807e-06	-.0872963585	-2.64738966e-06
4	.0145101258	-2.67098518e-04	1.29550668e-03
5	0	0	0

MEMBER FORCES AND MOMENTS

MEMBER NO	JOINT NO	AXIAL FORCE	SHEAR FORCE	MOMENT
1	1	113.352823	-123.413019	-289.732038
	2	-113.352823	123.413019	-327.333058
2	2	141.193002	90.2450731	327.333057
	3	-121.692719	20.551989	104.300199
3	3	121.397066	22.2318123	-104.300199
	4	-140.897349	88.5652498	-327.299257
4	4	111.64718	123.413019	327.299258
	5	-111.64718	-123.413019	289.765838

Fig. 5.29. Plane frame analysis results for the portal frame.

<u>Example 5.3 Vierendeel girder</u>   The Vierendeel girder is a non-
triangulated fabricated girder, generally with parallel top and bottom
chords.   Since it is non-triangulated the joints will have to transmit
moments and shears as well as axial forces.   Thus it is ideally suited
to welded construction.   A Vierendeel girder has fewer members than a
triangulated truss although fabrication costs may well be higher since
the joint details will need more care.   The Vierendeel girder is
ideally suited to fabrication from rolled hollow sections, as these
are extrememly effective in bending (high radius of gyration and
section modulus), which is the manner in which the Vierendeel girder
carries the imposed loading.   The Vierendeel girder has the advantage
over the conventional triangulated trussed girder in that there are no
diagonals to cause problems with siting of windows etc.   Consider the
4 bay girder shown in Fig. 5.30, together with the imposed working
loads.   The section is uniform throughout.

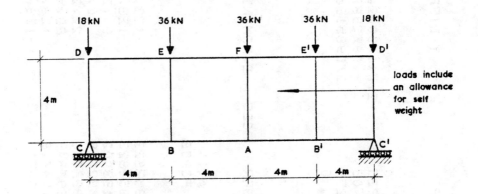

Fig. 5·30. Diagram of Vierendeel girder showing imposed loads.

Due to symmetry the girder may be considered as a pair of
cantilevers back to back loaded by the loads at D and E and the
reaction at C.   This substitute frame may be used both for plastic
analysis and elastic analysis.

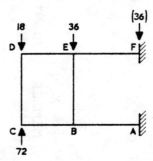

Fig. 5·31.. Diagram of frame required for analysis.

1) Plastic analysis

For the frame shown in Fig. 5.31 the bending moment is required to be known at 10 points and there are 6 redundancies thus there are 4 independent mechanisms, joint rotations at E and B and two sway modes. The joint rotations will only be involved in the combined mechanisms.

Using a load factor of 1.7 the loading at C, D and E becomes 122.4, 30.6 and 71.2 kN respectively. The method of virtual work has been used to relate the plastic moment M_p to the loading.

Independent mechanisms:-

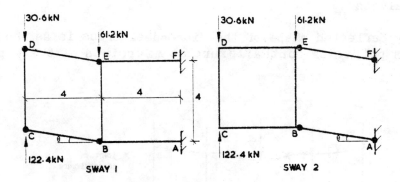

Fig. 5.32. Plastic collapse - sway modes.

Work equation:-

$(122.4 - 30.6)4\theta = 4M_p\theta$

or $M_p = 91.8$ kNm

Work equation:-

$(122.4 - 30.6 - 61.2)4\theta = 4M_p\theta$

$M_p = 30.6$ kNm

There are two possible combined mechanisms involving joint rotations.

Work equation:-
$6M_p\theta = (122.4 - 30.6)4\theta$

$M_p = 61.2$ kNm

Work equation:-
$6M_p\theta = (122.4 - 30.6)8\theta$
$\quad\quad\quad\quad - 61.2 \times 4\theta$

$M_p = 81.6$ kNm

Thus the design value for Mp is 91.8 kNm.

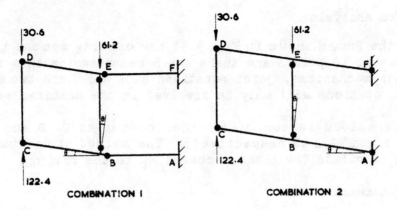

Fig. 5·33. Plastic collapse - combined modes.

2)   Elastic analysis

The likely deflected shape of the Vierendeel truss is as shown in Fig. 5.34, with points of contraflexure at midpoints of the top and bottom chords.

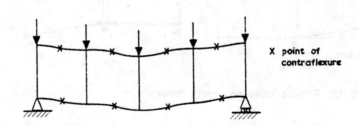

Fig. 5·34. Deflected shape of the Vierendeel girder.

Similar such points will occur at approximately mid-height of the verticals to maintain equilibrium at the joints.   If sufficient such points are considered the frame may be considered as isostatic and a simple approximate analysis be performed, the results of which are given in Table 5.5.   This approach is often used to calculate the effect of wind loads on tall structures.

In Fig. 5.35 is given the results of a plane frame computer analysis.   The approximate analysis is carried out replacing the point of contraflexure by fictitious hinges and then splitting the frame at

```
UNIVERSITY OF ASTON IN BIRMINGHAM
   DEPARTMENT OF CIVIL ENGINEERING
        PLANE FRAME ANALYSIS

    JOB REFERENCE   virendeel3

              FRAME DATA
```

JOINT DATA

JOINT	XCOOR	YCOOR	RESTRAINTS
			XY0
1	0	0	
2	0	4	0
3	4	0	0
4	4	4	0
5	8	0	0
6	12	0	0
7	8	4	0
8	12	4	0
9	16	0	Y0
10	16	4	0

(continued over)

SECTION DATA

SECTION TYPE	MOMENT OF INERTIA	X SECTION AREA	YOUNGS MODULUS
a	5.167e-05	6.11e-03	200000000

MEMBER DATA

MEMBER	JOINT NO END I	JOINT NO END J	MEMBER TYPE	SECTION TYPE	LENGTH
1	1	2	0	3	4
2	2	4	0	3	4
3	1	3	0	3	4
4	3	4	0	3	4
5	4	7	0	3	4
6	3	5	0	3	4
7	5	7	0	3	4
8	7	8	0	3	4
9	5	6	0	3	4
10	6	8	0	3	4
11	8	10	0	3	4
12	6	9	0	3	4
13	9	10	0	3	4

(continued over)

LOAD CASE 1

JOINT LOADS

JOINT NO	APPLIED MOMENT	FORCE IN Y DIR	FORCE IN X DIR
2	0	-18	0
4	0	-36	0
7	0	-36	0
8	0	-36	0
10	0	-18	0

JOINT DISP

JOINT NO	X DISP	Y DISP	ROTATION
1	0	0	-3.688506635e-03
2	5.58085653e-04	-1.47199903e-04	-3.6696381e-03
3	8.97979984e-05	-,0293657863	-4.01924046e-03
4	4.68287654e-04	-,0294248865	-4.00975431e-03
5	2.79042776e-04	-,0420397814	1.482050004e-10
6	4.68287549e-04	-,0293657849	4.01924043e-03
7	2.79042876e-04	-,0420985826	1.47601043e-10
8	8.97981034e-05	-,0294248636	4.00975428e-03
9	5.58085543e-04	0	3.68850611e-03
10	1.09535396e-10	-1.47199897e-04	3.66963786e-03

(continued over)

MEMBER FORCES AND MOMENTS

MEMBER NO	JOINT NO	AXIAL FORCE	SHEAR FORCE	MOMENT
1	1	44.9695704	-27.4332886	-54.9153233
	2	-44.9695704	27.4332886	-54.817831
2	2	-27.4332885	26.9695704	54.8178311
	4	-27.4332885	-26.9695704	53.0604506
3	1	-27.4332885	27.0304359	54.9153233
	3	27.4332885	-27.0304359	53.2064201
4	3	18.0485493	-30.380991	-60.7864894
	4	-18.0485493	30.380991	-60.7374744
5	4	57.8142798	9.01811222	7.67702379
	7	-57.8142798	-9.01811222	28.3954251
6	3	-57.8142795	8.98188876	7.5800694
	5	57.8142795	-8.98188876	28.3474856
7	5	17.9637643	-1.34019319e-06	2.6817833 7e-06
	7	-17.9637643	-1.34019319e-06	2.6789894 e-06
8	7	57.814278	-9.01811392	-28.3954277
	8	-57.814278	9.01811392	-7.67702805
9	8	-57.8142783	-8.98189044	-28.3474881
	5	57.8142783	8.98189044	-7.58007358
10	6	18.0485453	30.3809911	60.7864897
	8	-18.0485453	-30.3809911	60.7374748
11	8	27.4332871	-26.9695688	-53.0604467
	10	-27.4332871	26.9695688	-54.8178283
12	6	-27.433287	-27.0304342	-53.2064162
	9	27.433287	27.0304342	-54.9153205
13	9	44.9695685	27.4332871	54.9153204
	10	-44.9695685	-27.4332871	54.8178281

Fig. 5.35. Plane frame analysis for the Vierendeel girder.

these hinge points and inserting the shear forces and axial loads at the hinges. These axial forces and shear forces may then be obtained by simple statics and thus the moments at the joints can be calculated.

Member	End	APPROXIMATE ANALYSIS			MOMENT DISTRIBUTION			COMPUTER ANALYSIS		
		MOMENT	SHEAR	AXIAL	MOMENT	SHEAR	AXIAL	MOMENT	SHEAR	AXIAL
AB	A	18.0	9.0	+63.0	28.0	9.0	57.5	28.3	9.0	+57.8
	B	18.0	-9.0	+63.0	7.8	-9.0	57.5	7.6	-9.0	+57.8
BC	B	54.0	-27.0	-27.0	53.2	-27.1	27.4	53.2	-27.0	+27.4
	C	54.0	27.0	+27.0	55.2	27.1	27.4	54.9	27.0	+27.4
CD	C	-54.0	-27.0	-45.0	-54.9	-27.4	-45.0	-54.9	-27.4	-44.5
	D	54.0	27.0	-45.0	-54.9	27.4	-45.0	-54.8	27.4	-44.5
DE	D	54.0	27.0	-27.0	54.8	27.0	-27.4	54.8	27.0	-27.4
	E	54.0	-27.0	-27.0	53.2	-27.0	-27.0	53.1	-27.0	-27.4
EF	E	18.0	9.0	-63.0	8.0	9.0	-57.8	7.7	9.0	-57.8
	F	18.0	-9.0	-63.0	28.1	-9.0	-57.8	28.4	-9.0	-57.8
EB	E	-72.0	36.0	-18.0	-60.8	30.4	-18.0	-60.7	30.4	-18.0
	B	-72.0	-36.0	-18.0	-60.8	-30.4	-18.0	-60.7	-30.4	-18.0
FA	F	0	0	-18.0	0	0	-18.0	0	0	-18.0
	A	0	0	-18.0	0	0	-18.0	0	0	-18.0
		kNm	kN	kN	kNm	kN	kN	kNm	kN	kN

+ve moment tension on outside face
+ve axial force tension

Table 5.5   Forces in the Vierendeel girder

Moment distribution:-

There is no primary distribution as all the forces act at the joints, but two sway distributions -

SWAY1

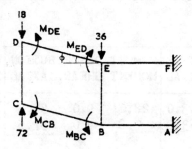

Fig. 5·36. Sway I.

FEMs: $M_{BC} = M_{CB} = M_{DE} = M_{ED} = -100$ units.

Using virtual work to obtain a sway equilibrium equation:-

$$4 \times 54\phi + (M_{BC} + M_{CB} + M_{DE} + M_{ED})\phi = 0$$

or $M_{BC} + M_{CB} + M_{DE} + M_{ED} + 216 = 0$ (5.18)

SWAY2

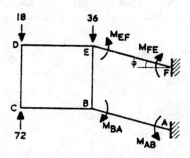

Fig. 5·37. Sway 2.

FEMs: $M_{AB} = M_{BA} = M_{EF} = M_{FE} = -100$ units

Again using virtual work to find the sway equilibrium equation:-
$$(72 \times 4 \quad - 18 \times 4 \quad - 36 \times 4 \quad)\phi + (M_{AB} + M_{BA} + M_{EF} + M_{FE})\phi = 0$$

or

$$M_{AB} + M_{BA} + M_{EF} + M_{FE} + 72 = 0 \tag{5.19}$$

Distribution factors:-

```
at  C       BC,CD       1/2
at  B       AB,BC,BE    1/3
```

	A		B		C		D		E		F	
	AB	BA	BE	BC	CB	CD	DC	DE	ED	EB	EF	FE
DF'S	1.0	1/3	1/3	1/2	1/2	1/2	1/2	1/2	1/3	1/3	1/3	1.0
SWAY 1 FEM'S				-100	-100			-100	-100			
	16.7	33.3	33.3	33.3	16.7	25.0	50	50	25.0	16.7		
			9.7	14.6	29.1	29.1	14.5	9.7	19.4	19.4	19.4	9.7
	-4.0	-8.1	-8.1	-8.1	-4.0	-6.0	-12.1	-12.1	-6.0	-4.0		
			1.7	2.5	5.0	5.0	2.5	1.7	3.3	3.3	3.3	1.7
	-0.7	-1.4	-1.4	-1.4	-0.7	-1.0	-2.1	-2.1	-1.1	-0.7		
			0.3	0.4	0.9	0.9	0.4	0.3	0.6	0.6	0.6	0.3
	-0.1	-0.2	-0.2	-0.2	-0.1	-0.1	-0.3	-0.3	-0.2	-0.1		
									0.1	0.1	0.1	
Totals	11.9	23.6	35.3	-58.9	-53.1	52.9	52.9	-52.8	-58.9	35.3	23.4	11.7
SWAY 2 FEM'S	-100	-100									-100	-100
			16.7					16.7	33.3	33.3	33.3	16.7
	13.9	27.8	27.8	27.8	13.9	-4.2	-8.4	-8.4	-4.2	13.9		
			-1.6	-2.4	-4.9	-4.9	-2.4	-1.6	-3.2	-3.2	-3.2	-1.6
	0.7	1.3	1.3	1.3	0.7	1.0	2.0	2.0	1.0	0.7		
			-0.3	-0.4	-0.9	-0.9	-0.5	-0.3	-0.6	-0.6	-0.6	-0.3
	0.1	0.2	0.2	0.2	0.1	0.2	0.4	0.4	0.2	0.1		
					-0.1	-0.1			-0.1	-0.1	-0.1	
Totals	-85.3	-70.7	44.1	26.5	8.8	-8.9	-8.9	8.8	26.4	44.1	-70.6	-85.2
1 (SWAY 1)	13.3	26.4	39.5	-66.0	-59.5	59.2	59.2	-59.1	-66.0	39.5	26.2	13.1
2 (SWAY 2)	-41.3	-34.2	21.3	12.8	4.3	-4.3	-4.3	4.3	12.8	21.3	-34.2	-41.2
FINAL MOMENTS	-28.0	-7.8	60.8	-53.2	-55.2	54.9	54.9	-54.8	53.2	60.8	-8.0	-28.1

Table 5.6   Moment distribution table for the Vierendeel girder

If a_1 = sway correction factor for the first sway mode and a_2 is that for the second mode, then Equation (5.18) becomes

$$-223.7\ a_1 + 70.5\ a_2 + 216 = 0 \qquad\qquad (5.20)$$

and Equation (5.19) becomes

$$70.6\ a_1 - 311.8\ a_2 + 72 = 0 \qquad\qquad (5.21)$$

Thus $a_2 = 0.484$ and $a_1 = 1.12$, giving the final moments in Table 5.6.

<u>Example 5.4 Design of a pitched roof portal frame</u>   The most common method used is to design at collapse under factored loads and then check the stability of the frames under elastic conditions.

The details of the frame, of uniform section, are shown in Fig. 5.21.

Purlin spacing 1.25 m
Frame spacing 6 m
Grade 43 steel is to be used
Total load (including dead load) 1.5 kN/m^2
Load factor (γ) = 1.7

The plastic analysis for the frame has been undertaken in Example 5.5.2 where the plastic collapse moment M_p was found as

$$M_p = 424.4 \text{ kNm}$$
$$Z_{px} = M_p/Y_s = (424.4 \times 10^3)/250 = 1698 \text{ cm}^3$$

From rolled steel section property tables (see Appendix) a 457 x 152 x 82 UB has a plastic section modulus of 1800 cm^3.

Actual load factor = design load factor x (actual Z_p/calculated Z_p)
$$= 1.7 \times (1800/1698) = 1.80$$

<u>Reduction of plastic moment of resistance due to the axial load</u> The theory covering this has been dealt with in Chapter 2.   The vertical reaction at the base of the stanchion is 191.3 kN (Fig. 5.23).

The axial stress at collapse, p, is given by

$$p = N/A = (191.3 \times 10)/10.45 \text{ N/mm}^2$$
$$= 18.3 \text{ N/mm}^2$$
$$n = p/Y_s = 18.3/250 = 0.073$$

$$n < 0.436 \text{ so } Z_p = 1800 - 2558 \, n^2 \text{ (from the BCSA section property tables)}$$

$$= 1800 - 2558(0.073)^2$$

$$= 1786.4 \text{ cm}^3$$

i.e. a reduction of less than 1.0% in the plastic section modulus.
∴ neglect.

Stanchion Stability. It is necessary to ensure that the column can generate its full plastic moment capacity without elastic instabilities occurring due to premature buckling. It may thus be necessary to introduce additional restraints to eliminate the possibility of instability. This is covered by design charts in BCSA Publication No 23, 1969, although Morris indicates that these design charts are conservative. From Table C (Morris and Randall) in the Appendix the torsion constant, T, for a 457 x 152 x 82 UB is 309 N/mm^2.

The parameter T is given by $T = AGJ/Z_x^2$ where A is the cross-sectional area, GJ the torsional rigidity, and Z_x the elastic section modulus about the x-x (major axis) and has units of N/mm^2. A full discussion of the background theory to the column stability curves is given by Horne. The limiting slenderness ratio curves are obtained by considering the interaction of strut buckling and torsional buckling. In the latter, the warping resistance is ignored, underestimating the critical moment, and thus in general the solution is conservative.

The charts in BCSA Publication No 23 were originally prepared in imperial units and interpolation is strictly necessary in metric units, although it is sufficiently accurate to use the lower valued chart as will be done in this example.

The chart for $T = 275 \text{ N/mm}^2$ is reproduced in Fig. 5.38.

The ultimate axial stress in the column is 18.3 N/mm^2 and the ratio of end moments, β, is -1 giving a maximum slenderness ratio of 145. The actual slenderness ratio is 5000/33.1 = 151 which is greater than the allowable. A restraint at mid height ($\beta = 0$) will be found to be satisfactory so that allowable slenderness ratio for $\beta = 0$ and p = 18.3 is 105.

It has been suggested by Morris and Randall, that sufficient restraint at the eaves hinge is given by a member whose plastic moment of resistance about the major axis is the same as that of the stanchion about the minor axis, so

Z_{py}, stanchion = 235.4 cm^3 or
(M_p) minor axis = 250 x 235.4/10^3 = 58.85 kNm.

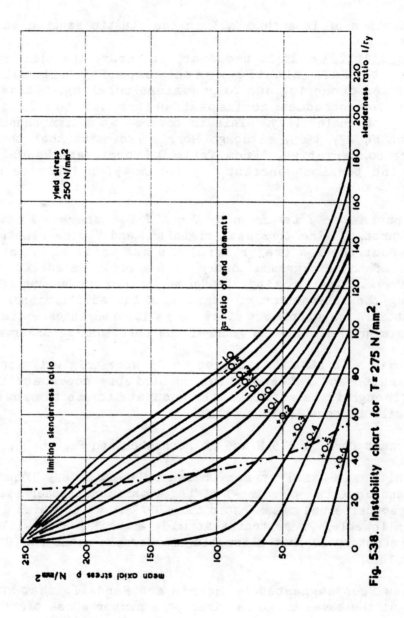

Fig. 5.38. Instability chart for T = 275 N/mm².

Assuming two members framing in then,

(M_p) restraint = 0.5 x 58.85 kNm, i.e. a
203 x 133 x 25 UB (Z_p = 259.8) may be used.

The connection must be designed to take this moment.

__Rafter stability__ It is necessary to ensure that the rafter can
generate the full plastic moment without elastic instability occuring.
The bending moment diagram for the rafter at collapse is plotted in
Fig. 5.39.

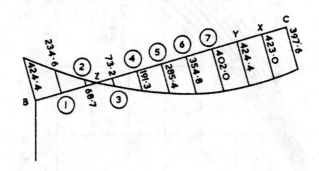

Fig. 5·39. Net bending moment diagram for the rafter.

To ensure adequate stability conditions are placed on the values
of the slenderness ratios of the rafter,
  i) from the eaves restraint to the first purlin point, (i.e length
     B1 in Fig. 5.39), and
 ii) from the eaves restraint to purlin beyond the point of
     contraflexure of the rafter, (length BZ in Fig. 5.39).

Condition 1:

ratio of end moments, β = 234.6/424.4 = 0.55, this is greater than the
critical value of 0.5 from chart 1 Morris and Randall. The maximum
spacing of 1.70 m is given in Table B in the same reference.   The
actual spacing on the slope is 1.25/cos10° = 1.27 m which is less than
the maximum.

Condition 2:

β = -73.2/424.4 = -0.17, T = 309 N/mm² as before. From chart 1,
reproduced in Fig. 5.38 from Morris and Randall, the allowable
slenderness ratio is 124.   Thus the maximum spacing is (124 x
33.1)/1000 = 4.104 m, and the actual spacing 3 x 1.27 or 3.81 m on the
slope.

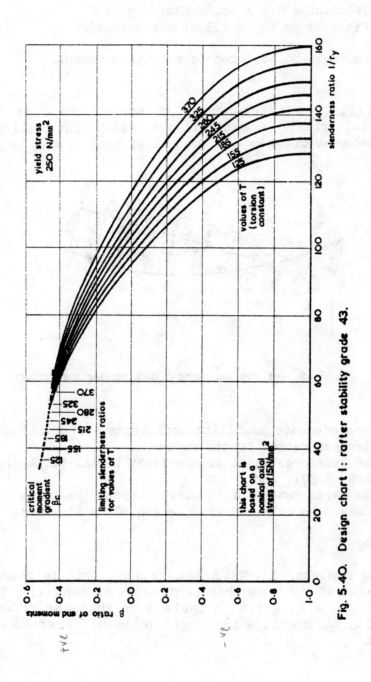

Fig. 5.40.   Design chart I: rafter stability grade 43.

<u>Deflections</u> It has tacitly been assumed in the calculations that no significant geometry changes have occurred. If significant geometry changes do occur, then the effective load factor of the structure will be much reduced. This is because the critical load causing elastic buckling will be reduced which leads to a reduction in the ability to sustain full plastic action. It is also necessary to check the lateral deflection under working loads at the eaves for any detrimental effect on the sheeting. The lateral deflection is limited to height/325 by Cl. 31b BS 449.

i) Changes in geometry

Design charts to assess these are given in BCSA Publication No 23 and chart 1/29 is reproduced in Fig. 5.41.

For grade 43 steel the effective slenderness ratio is given by (1/61) x (span/member depth) i.e. a value of (1/61) x (25/0.465) or 0.88 is obtained for the frame under design. The span/height to eaves ratio is 25/5 or 5. Chart 1/29 then gives a 1% reduction in the load factor due to geometry changes. In certain cases this reduction can be such to cause the value of the load factor to go below 1.70.

ii) Working load deflections

Either the approximate methods given in BCSA Publication No 23 may be used or an elastic analysis undertaken. It is instructive to compare the answers:
   a) Plane Frame Analysis
      The output is given in Fig. 5.29 from which the eaves
      deflection is 14.5 mm.
   b) Using the Area Moment theorems on the moment distribution
      results given in Example 5.2 the deflection is 15.2 mm.
   c) BCSA Publication No 23 method
      The relevant chart is reproduced in Fig. 5.42

      Slope, ϕ = 10°, h/l = 5 thus D = 0.41
      $D = (\delta_1 \times 10^6)/(h_1 \times (L/d) \times (Y_s/\gamma))$
      L/d = 25/0.465 = 53.8: h_1 = 5.0
      Y_s/γ = 250/1.80 = 138.9 N/mm^2 (the value of γ to be used
                                   is the actual collapse load factor)
         δ_1 = 15.3 mm

The limit given in Cl. 31b BS 449 is height/325 or 5/325 = 15.4 mm.

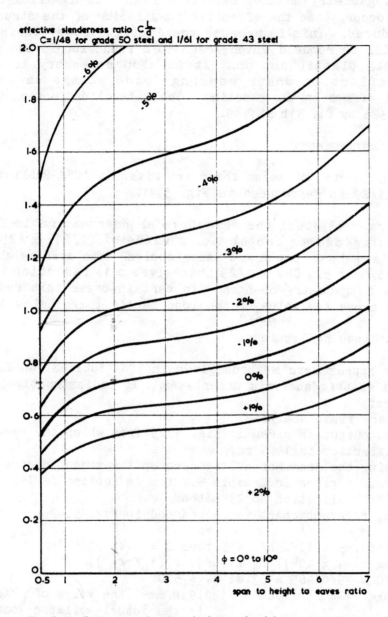

effective slenderness ratio $C\frac{L}{d}$

C=1/48 for grade 50 steel and 1/61 for grade 43 steel

-6%

-5%

-4%

-3%

-2%

-1%

0%

+1%

+2%

ϕ = 0° to 10°

span to height to eaves ratio

Fig. 5.41. Percentage load variations - fixed base frames.

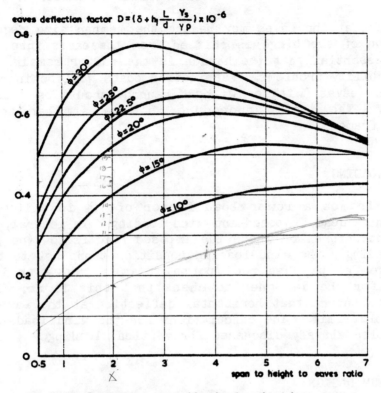

eaves deflection factor $D = (\delta + h_1 \frac{L}{d} \cdot \frac{Y_s}{Y\rho}) \times 10^{-6}$

Fig. 5·42. Deflections at working loads - fixed base frames

5.6   HAUNCHES

A haunch at the ridge will reduce the deflections under working load
and will facilitate siting the bolts for the ridge joint which is to
be designed in accordance with methods given in Chaper 4.

A haunch at the eaves may reduce the member size required since
the position of the hinge is shifted from the eaves into the column,
but it is essential that the haunch in the rafter remains elastic to
avoid instability problems. Fabrication costs will be increased. The
joint at the eaves (with or without haunch) should be designed using
Chapter 4.   The design procedure is very similar to that for
unhaunched frames.

5.7   WIND LOADING

On portal frames a lower load factor of 1.4 is allowed for the
combination of dead, imposed and wind loading. In general, except for
very tall structures or where the imposed load is low (no possibility
of snow loading), the wind loading condition tends not to be critical
at collapse for pitched roof portal frames. However it is still
necessary for the designer to check that this is so. It is also
necessary to ensure that horizontal deflection at the eaves is below
the allowable limit. The calculations for the wind load case follow
the principles already discussed for vertical loading.

5.8   OPTIMUM DESIGN

Two common forms of optimum design are used:-
   i)   minimum weight
   ii) minimum cost

Minimum weight design

Minimum weight design may either be performed by an elastic analysis
or a plastic analysis.

Elastic approach

The constraints usually used in this method are maximum allowable
stresses with no allowance for reduction due to buckling effects.
However consideration, especially if sway is involved, must be given
to maximum deflections. Such an approach involves empirically
relating the relevant section properties of rolled sections to the
cross sectional area. No account is usually taken of fabrication
problems which may ensue at beam-column connections due to varying
sized members framing in.

Plastic approach

Here it is only necessary to set up a relationship between plastic moment and weight. Considering first geometrically similar sections, when the weight is proportional to d^2 and the plastic section modulus to d^3 or

$$W = k(M_p)^{0.67}$$

where W is the weight/unit run, M_p the plastic moment and k a constant depending on steel grade.

If Z_p is plotted against mass/unit length for rolled Universal beams a function of the type $W = k(Z_p)^{0.6}$ is a reasonable fit. However over a reasonable range this may be approximated by a linear function where a and b are constants

$$W_i = a + bM_{pi} \tag{5.22}$$

where W_i is the mass/unit length of the ith member and M_{pi} its plastic moment. Thus the total weight of the structure is given by

$$W_T = \Sigma(W_i)L_i \tag{5.23}$$

or $\quad W_T = \Sigma(a + bMp_i)L_i$

$$= a\Sigma L_i + b\ \Sigma(L_iMp_i) \tag{5.24}$$

The first term in Equation (5.24) is geometry dependent and so its value is fixed and only the second term must be optimised, i.e. $\Sigma(L_iMp_i)$ must be minimized.

The method used is to set up, with no restrictions on the relative values of the plastic moments, a complete set of equations governing collapse of the structure and then to minimize $\Sigma(L_iMp_i)$ with respect to these solutions.

For other than very simple structures, the method demands linear programming techniques and is best suited to computer methods.

Minimum cost designs

The process is basically similar to minimum weight except that additional factors are considered. Minimum weight design is a simplified minimum cost design where only the cost of the materials for the basic frame are considered important.

Minimum cost design must take into account the following
   i) Cost of materials for the structural frame
   ii) Cost of cladding (including purlins and sheeting rails)

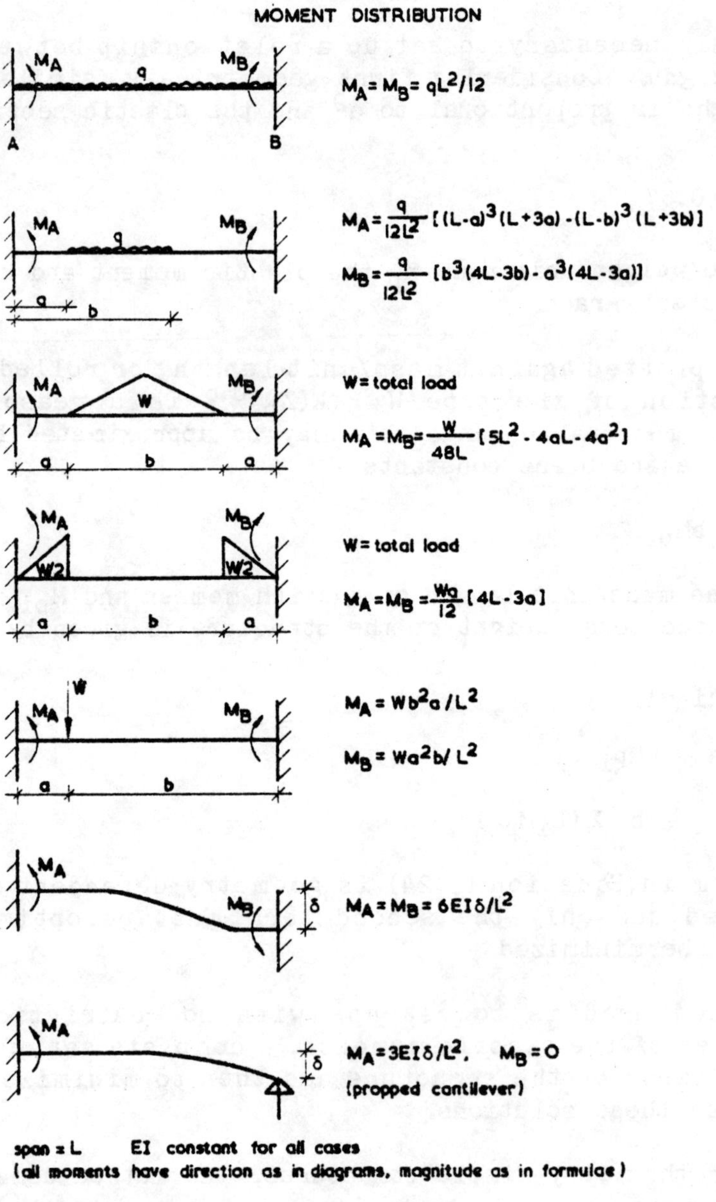
MOMENT DISTRIBUTION
M_A
M_B
q
A
B
$M_A = M_B = qL^2/12$
a
b
$M_A = \frac{q}{12L^2}[(L-a)^3(L+3a) - (L-b)^3(L+3b)]$
$M_B = \frac{q}{12L^2}[b^3(4L-3b) - a^3(4L-3a)]$
W
W = total load
$M_A = M_B = \frac{W}{48L}[5L^2 - 4aL - 4a^2]$
W/2
W = total load
$M_A = M_B = \frac{Wa}{12}[4L - 3a]$
W
$M_A = Wb^2a/L^2$
$M_B = Wa^2b/L^2$
δ
$M_A = M_B = 6EI\delta/L^2$
$M_A = 3EI\delta/L^2$, $M_B = 0$
(propped cantilever)
span = L EI constant for all cases
(all moments have direction as in diagrams, magnitude as in formulae)

Fig. 5·43. Fixed end moments for standard cases.

   iii) Cost of foundations
   iv) Fabrication and erection costs
    v) Costs of painting, fire protection and maintenance.

This whole process is an extremely complex example of linear programming and has only been carried out in a limited form for some design studies.

5.9 FIXED END MOMENTS (MOMENT DISTRIBUTION)

In Fig. 5.43 are given the fixed end moments for some standard cases of loading. For other loading cases reference should be made to a standard text or evaluated from first principles.

REFERENCES

Bray, K.H.M., Croxton, P.C.L. & Martin, L.H. (1976) - Matrix Analysis of Structures, Arnold.

Constrado (1979) - Plastic Design Supplement.

Coates, R.C., Couttie, M.G. & Kong, F.K. (1972) - Structural Analysis, Nelson.

Curtin, W., et al (1982) - Structural Masonry Designers' Manual, Granada.

Hendry, A. et al (1981) - An Introduction to Load Bearing Brickwork Design, Ellis Horwood.

Horne, M.R. (1978) - Continuous construction and special problems of single storey frames, Lecture 9 - The background to the new British standard code for structural steelwork, Constrado, Imperial College.

Horne, M.R. & Morris, L.J. (1981) - Plastic Design of Low Rise Frames, Constrado Monograph, Granada.

Kleinlogel, Rahmenformeln, Wilhelm Ernst & Sohn, Berlin. (Quoted in Steel Designers Manual, Crosy Lockwood & Staples, 1972).

Livesley, R.K. (1964) - Matrix Methods of Structural Analysis, Pergamon Press.

Morris, L.J. (December 1981) - A commentary on portal frame design. The Structural Engineer, Vol. 59A No 12, pp 394-403.

Morris, L.J. & Randall, A.L. (1979) - Plastic Design, Constrado Publication.

Moy, S.S.J. (1981) – Plastic Methods for Steel and Concrete Structures, Macmillan.

Neal, B.G. (1965) – The Plastic Methods of Structural Analysis, Chapman Hall.

Roberts, E.H. (1980) – Semi-Rigid Design Using the Variable Stiffness Method of Column Design. Joints in Structural Steelwork, Eds J.H. Howlett, W.M. Jenkins and R. Stainsby, Pentech Press.

Taylor, J.C. (1980) – Semi-Rigid Beam Connections: Effects on Column Design: 1320 Code Method, Joints in Structural Steelwork, Eds. J.H. Howlett, W.M. Jenkins and R. Stainsby, Pentech Press.

6   Trussed Parts of Complete Structures

6.1   INTRODUCTION

This chapter is concerned with the design of members needing to be
fabricated from standard sections, e.g. angles, structural tees, 'I'
sections or other section types illustrated in Fig. 1.1. Such members
are needed where spans are in excess of those capable of being spanned
by rolled beams or in the case of columns where the forces are too
great to be sustained by standard stanchions. Such members are
usually fabricated off site and are then bought onto site for
erection, thus minimising construction time. Some form of splice may
be necessary for large trusses due to possible transportation
problems. The design of such trussed members may often be
conveniently isolated from the design of the complete structure. The
form of trussed members, as is pointed out in Chapter 5, is often
dictated by external requirements such as drainage, lighting etc.
Trusses are normally economic on weight but their fabrication costs
will be high, thus partially offsetting the saving in weight. A
typical roof truss is illustrated in Fig. 6.1.

In this chapter is set out design examples on a triangulated roof
truss, gusset plates and purlins, together with the design of a
Vierendeel girder. The chapter concludes with designs of compound
stanchions.

6.2   ANALYSIS OF TRIANGULATED ROOF TRUSSES

In general it is adequate for the determination of member forces,
provided the member axes are coincident at a joint, to assume pin
joints. If an alternative analysis assuming fixed joints be
performed, then in general, the moments will be low, but the
deflections will be much reduced. Welded trusses will tend to behave
as fixed joint trusses, bolted trusses are somewhere between fixed and
pinned, as some joint rotation or slip will take place. The effect of
gusset plates is, in general, ignored although it may be taken into
account in computer analyses by representing the gusset plates as
short rigid extensions to the members as in the methods set down by
Jennings and Majid, or Just. Non-coincidence of member axes will
cause substantial problems; these are discussed in relation to an

actual design by Purkiss and Croxton. It is also assumed that all the forces act at nodal points, but usually the purlin spacing is such that additional bending is applied to the rafters. This will need considering in a design, and is normally allowed for by approximate methods as shown in example 6.1.

6.3   MEMBER DESIGN IN A TRIANGULATED ROOF TRUSS

A single angle member will tend to buckle about its principal axes as discussed in Chapter 3. In contrast a pair of angles back to back, adequately connected, or a 'T' section will tend to buckle about the 'x-x' or 'y-y' axes.

The design procedure when both axial forces and bending moments are present, is similar to the design of stanchions, where the interaction formula used is

$$f_a/p_a + f_b/p_b \leqslant k \tag{6.1}$$

where f_a is the actual axial stress, f_b the actual bending stress, p_a the allowable axial stress, p_b is the allowable bending stress and k equals 1 if no wind load be present and 1.25 under wind load. This equation applies in tension or compression when the appropriate allowable stresses must be adopted. Strictly BS 449 Cl. 13 allows a 25% increase in the allowable stresses with wind loading present, with the factor k always equal to unity. The method adopted above produces the same result and needs the calculation of only one set of allowable stresses.

If there be no bending (i.e. $f_b = 0$) then Equation (6.1) reduces to $f_a \leqslant p_a$ with no wind forces or $f_a \leqslant 1.25p_a$ with wind forces.

It is normally necessary to consider three load combinations:-
  i) Dead and imposed loading acting together.
 ii) Dead, imposed and downward wind loading.
      This needs considering only if the wind increases the member forces by more than 25%.
iii) Dead and upward wind loading.
      This may cause a reversal in the member forces. On some structures without dominant openings where the wind pressures are such that uplift is small, this effect may not occur.

Gusset plates are not normally used when 'T' sections are used for the top and bottom booms, as the other members are welded directly to the stem of the 'T'. Any effects due to non-coincident neutral axes in the vertical plane are ignored, since in general the joint is detailed such that where two or more members are connected to the stub of the tee, they are connected on opposite sides.

<u>Example 6.1 Triangulated roof truss</u>   The roof truss is illustrated in Fig. 6.1, together with other relevant geometrical data. A roof slope of around 20° is normal for such a truss as it allows rain water to be shed adequately and will tend to optimise the forces in the frame. Assume the truss to be fabricated from grade 43 steel and to be situated in the Oxford area.

Dead load is taken from CP3 Ch V, Part 1.

Loading	kg/m^2	kN/m^2 (on plan)
Truss self-weight (assumed)	18	0.176
Roof sheets	18	0.176
Lining (insulation)	6	0.060
Purlins	10	0.098
--------	--	-----
Total	52	0.510

Snow loading (Cl. 6.2, CP3 Ch V Part 1) 0.750

Total	1.260
Truss self-weight	(0.176)

Total excluding truss self-weight	1.084

Panel dimensions (on plan) are 4.5 (truss spacing) by 2.0 (node spacing)
Panel loading – Dead 0.51 x 2 x 4.5 = 4.6 kN
               Live 0.75 x 2 x 4.5 = 6.8 kN

<u>Wind Loading</u> (In accordance with CP3 Ch V, Part 2).

Note all code references in this section are to CP3 Ch V, Part 2. This section may be omitted, but if the reader intends to follow it through, a copy of CP3 Ch V, Part 2 is required.

a)   Wind loading blowing at right angles to the ridge
     External pressure – 0.32q   WINDWARD SLOPE }   Table 8
                       – 0.40q   LEEWARD SLOPE   }   Roof slope 22°
                       where q is the dynamic pressure.

     Internal pressure : Appendix E allows the internal pressure coefficients to be taken as 75% of the external. Table 8 gives values of C_{pe} the external pressure coefficient for the walls. Assuming the length width ratio (L/W) to lie between 3/2 and 4, the worst values of C_{pe} for

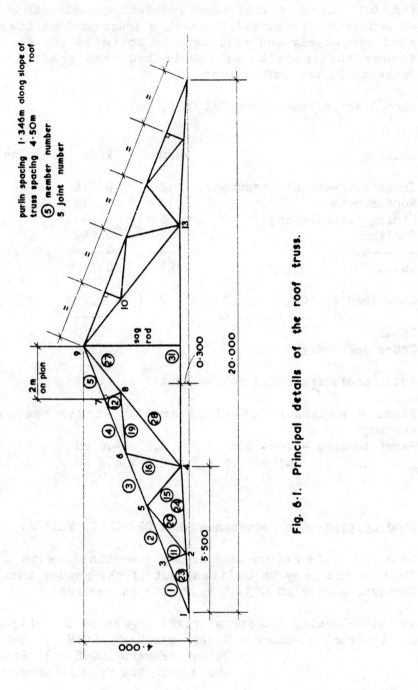

purlin spacing 1·346m along slope of roof
truss spacing 4·50m
⑤ member number
5 joint number

Fig. 6·1. Principal details of the roof truss.

the walls are 0.7q and -0.6q, giving internal pressures of 0.75 x 0.7q (or 0.53q) and -0.75 x 0.6q (or -0.45q).

Thus the worst effect will be

WINDWARD SIDE   0.32q + 0.53q = 0.85q upward
         or   0.45q - 0.32q = 0.13q downward

and LEEWARD SIDE   0.40q + 0.53q = 0.93q upward
         or   0.45q - 0.40q = 0.05q downward

b)   Wind blowing parallel to the ridge
     External pressure - 0.70q on one slope and 0.60q on the other (Table 8)
     Internal pressure - From Table 7 the worst values are 0.7q and -0.5q, i.e. internal coefficients of 0.75 x 0.7q (or 0.53q) and -0.75 x 0.5q or (-0.58q)

Thus the worst effect will be 0.70q + 0.53q = 1.23q upward on one slope, and 0.60q + 0.53q = 1.13q on the other.

Thus the wind forces to be considered in the analysis are 1.23q upward on one and 1.13q on the other, 0.13q downward on one slope and 0.05q downward on the other.

Wind forces   For Oxford the maximum 40 year gust speed is 40 m/s (Fig. 1).

Topography: Factor S_1 = 1.0   (Table 2)
Height and shelter factor: Class A (Cl. 5.5.2)
                     Ground roughness Category 2 (Cl. 5.5.1)
         S_2 = 0.93 (Table 3)
Life factor - building to have 100 year life
         s_3 = 1.05 (Fig. 2).

Design windspeed V_s = $S_1 S_2 S_2 V$ (Cl. 6)
                = 1.0 x 0.93 x 1.05 x 40
                = 39 m/s.

Dynamic pressure q = $0.613 V_s^2$ (Table 5)
                   = 0.613×39^2 = 932 N/mm^2

Thus the wind pressures are 1.23q = 114.6, 1.13q = 1053, 0.13q = 121, 0.05q = 47 N/mm^2 respectively. The panel loads are 11.1, 10.2, 1.18 and 0.46 kN respectively, using a panel area of 4.5 x 2/cos 21.8 as the wind acts normal to the slope of the roof.

For an asymetric loading case it is normally considered
sufficient that the total horizontal reaction be apportioned equally
to both supports, assuming that the columns upon which the roof truss
is supported have equal stiffnesses.

<u>Member Forces</u>   The analysis may be performed by any of the methods
mentioned in Chapter 5.

The member forces given in Table 6.1, have been calculated on the
assumption that all the joints are pinned joints, and thus the forces
are axial.

Table 6.1   Roof Truss Member Forces (in kN)

MEMBER	DEAD	LIVE	WIND DOWN	WIND UP	DESIGN COMBINATIONS		
					DEAD + SNOW	DEAD + LIVE + WIND DOWN	DEAD + WIND UP
1	-64.5	-95.4	-12.5	143.0	-159.9	-171.4	78.5
2	-62.8	-92.9	-12.5	143.0	-155.7	-168.2	80.2
3	-55.7	-82.3	- 9.5	114.5	-138.0	-147.5	58.8
4	-61.0	-90.2	-12.9	147.1	-151.2	-164.1	86.1
5	-59.3	-87.7	-12.9	147.1	-147.0	-159.9	87.8
11	- 4.3	- 6.3	- 1.2	11.3	- 10.6	- 11.8	7.0
12	- 4.3	- 6.3	- 1.2	11.3	- 10.6	- 11.8	7.0
15	- 7.2	-10.7	- 2.0	19.1	- 17.9	- 19.9	11.9
16	- 7.8	-11.5	- 2.2	20.6	- 19.3	- 21.5	12.8
19	7.1	10.4	1.9	- 18.7	17.5	19.4	-11.6*
20	6.7	9.9	1.8	- 17.7	16.6	18.4	-11.0*
23	60.0	88.7	12.1	-134.0	148.7	160.8	-74.0*
24	53.3	79.8	9.1	-105.6	133.1	142.2	-52.3*
27	31.1	45.9	8.2	- 78.9	77.0	85.2	-48.7*
28	24.0	35.5	6.2	- 61.2	59.5	65.7	-37.2*
31	31.1	45.9	4.6	- 59.7	77.0	81.6	-28.6*

NOTE: Compression -ve

The dead and snow combination will be the one to use for the
design of most members and, except for the rafters (members 1 to 5)
which need special checks, nowhere does the addition of downward wind
produce force increases of greater than 25%. A special check on the
rafter is needed since the wind load is applied in a different
direction to the dead and snow loads and increase in stress is
therefore not proportional to increase in axial load.   The forces in
the last column are those for dead load combined with the upward
forces due to the wind, the forces marked * are compressive forces
induced in members normally in tension under dead and snow load only.
This needs special attention since the allowable load on a member in
compression is less than that in tension due to buckling.

Member Design

i) Rafter (members 1, 2, 3, 4, and 5)
    Since the load is applied at the purlins which will not in general
    correspond to nodal points on the actual structure, it is
    necessary to allow for the effects of bending in the rafter. This
    will be done assuming the rafter to be a continuous beam. This
    assumption is acceptable since the rafter is very much stiffer
    than the other members which offer little rotational restraint.

    a) Dead and live load
    The self-weight of the truss may still be assumed to act at nodal
    points and thus the reduced loading of 1.084 kN/mm^2 is used.

    Purlin load = purlin spacing x cos(roof slope) x loading x

    truss spacing

    $$= 1.35 \text{ x } \cos 21.8 \text{ x } 1.084 \text{ x } 4.5 = 6.1 \text{ kN} \qquad (6.2)$$

    The equivalent structure for analysis of the rafter is shown in
    Fig. 6.2(a). The analysis was performed using moment
    distribution. The bending moment diagrams are drawn on the
    tension side of the member in all cases, and the final bending
    moment diagram is shown in Fig. 6.2(b).

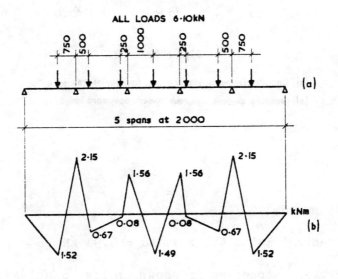

ALL LOADS 6·10kN

5 spans at 2000

Fig. 6·2.

(a) Continuous beam idealisation of rafter under imposed load

(b) Bending moment diagram under imposed loads

b)   Wind downwards

Purlin load = 1.346 x 4.5 x 0.12 = 0.73 kN

Note, here the loading is normal to the rafter.
The equivalent structure is shown in Fig. 6.3(a) and the BMD in
Fig. 6.3(b).

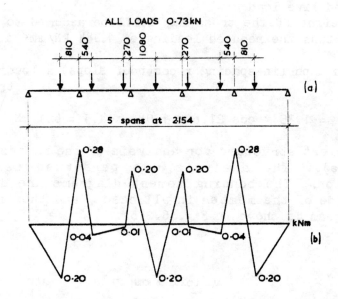

Fig. 6·3.

(a) Continuous beam idealisation for downwards wind

(b) Bending moment diagram under downward wind

c)   Wind upwards

Purlin load = 1.346 x 4.5 x 1.146 = 6.94 kN

The equivalent structure is shown in Fig. 6.4(a) and the final BMD
in Fig. 6.4(b).

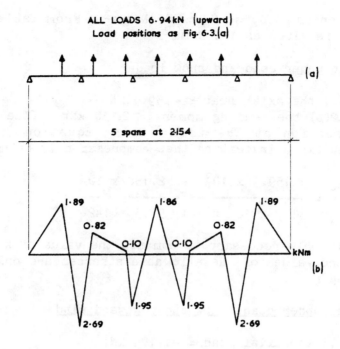

Fig. 6·4.

(a) Continuous beam idealisation for upwards wind

(b) Bending moment diagram under upwards wind

Using a 'T' section:

the effective length for buckling in plane of truss = 0.7 x 2.154 = 1.508 m and the effective length for buckling out of plane of the truss is taken as the purlin spacing = 1.346 m. Try a 165 x 152 x 27 kg structural 'T' section.

From the section property tables in the Appendix
A   = 34.2 cm^2 r_y = 3.94 cm
Z_{ex} = 51.5 cm^3 r_x = 4.31 cm

$(1/r_x)$ = 1.508/43.1 = 35.0
$(1/r_y)$ = 1346/39.4 = 34.2   Table 17a BS 449 P_c = 141 N/mm^2

The method of obtaining the allowable bending stress for a 'T' section is given in Section 2.6.2 Category 'c'.

The value of D/T required is 155.4/7.7 = 20.2 and $1/r_y$ (from above) = 34.2. From the section property tables y_t = 31.9 mm and y_c = 166.8 – 31.9 = 134.9 mm. Using the equations given at the head of Table 7 BS 449, for the coefficients of A and B, values of 2564.9 and 2398.7 are obtained respectively. C_s is given by C_s = (A + K_2B) y_c/y_t where K_2

is to be taken as -1.0, so C_s = 703 N/mm^2. From Table 8 BS 449, the value of P_{bc} is given as 142 N/mm^2.

Rafter design under dead and snow load

From Table 6.1 the axial load = -159.9 kN.
From Fig. 6.2(b) the bending moment = 2.150 kNm. (The support moment gives compression at the toe). Using Equation (6.1), with the additional suffix c indicating that compression is being considered

$$\frac{f_c}{P_c} + \frac{f_{bc}}{P_{bc}} = \frac{159.9 \times 10^3}{34.2 \times 10^2 \times 141} + \frac{2.150 \times 10^3}{51.5 \times 142} = 0.63$$

It is usually only necessary to compute the value of k to two decimal places, as the values of the allowable stresses are only known to the nearest integer.

Rafter design under dead, snow and downward wind

From Table 6.1 the axial load = -171.4 kN.
From Figs. 6.2(b) and 6.3(b) the bending moment = 2.15 + 0.28 = 2.43 kNm.
The calculation is similar to that above, with the value of the interaction constant k being calculated as 0.69 (the allowable value is 1.25).

Dead + upward wind

Dead load moment under purlin =

    1.52 x [(0.510 - 0.176)/1.084] = 0.47 kNm.

This is obtained by simple proportion as the total dead load is 0.51 kN/m^2, the self-weight of the truss is 0.176 kN/m^2 and the total applied load excluding the self-weight of the truss is 1.084 kN/m^2.

From Fig. 6.2(b) the dead load moment at first support =

    2.150 x [(0.510 - 0.176)/1.084] = 0.66 kNm

Net moment = -1.89 + 0.47 = -1.42 kNm at the first load point
             +2.69 - 0.66 = 2.03 kNm at support

Axial load = 80.2 kN (tension)

M = 2.03 kNm (support moment giving tension in the toe)

$$\frac{f_t}{P_t} + \frac{f_{bt}}{P_{bt}} = \frac{80.2 \times 10^3}{34.2 \times 10^2 \times 155} + \frac{2.029 \times 10^3}{51.5 \times 165} = 0.4 < 1.25$$

Therefore satisfactory.

ii) Members (11) and (12)

Axial load due to dead + snow = -11 kN
effective length = 0.85 x 740 for single angle
try 65 x 50 x 5 angle long leg attached.

$$(l/r_v) = (0.85 \times 740)/10.7 = 59$$

From Table 3a BS 449 $p_c = 126$ N/mm^2

$$\frac{f_c}{P_c} = \frac{11 \times 10^3}{5.54 \times 10^2 \times 126} = 0.16 \quad < 1.0 \quad \text{therefore satisfactory}$$

Dead + upward wind = 7.0 kN

For angles in tension connected by one leg, Equation (3.4) should be used.
For the 65 x 50 x 5 angle the area of the connected leg (a_1) (65 - 5/2) x 5 = 312.5 mm^2 and the unconnected leg (a_2) (50 - 5/2) x 5 = 237.5 mm^2.
The equivalent area is given by $a_1(1 + 3a_2/(3a_1 + a_2))$, i.e. a value of 502 mm^2 is obtained.

$$\frac{f_t}{P_t} = \frac{7.0 \times 10^3}{502 \times 155} = 0.09 \quad < 1.25 \quad \text{therefore satisfactory}$$

iii) Members (15) and (16)

From Table 6.1 dead + live = -19.3 kN
Use a 60 x 60 x 6 single angle
$(l/r_v) = 0.85 \times 2100/11.7 = 152.6$ from Table 17a $p_c = 39$ N/mm^2

$$\frac{f_c}{P_c} = \frac{19.3 \times 10^3}{6.91 \times 10^2 \times 39} = 0.72 \quad < 1.00$$

Downward wind causes less than a 25% increase, and the upward wind although reversing the force is of a small magnitude, therefore no further calculations are necessary in this member.

iv) <u>Members (19) and (20)</u>

From Table 6.1 the axial force due to dead + snow = 17.5 kN
Use a 50 x 50 x 5 single angle

$$\frac{f_t}{p_t} = \frac{17.5 \times 10^3}{(7/8) \times 4.80 \times 10^2 \times 155} = 0.24 \quad < 1.0 \quad \text{therefore satisfactory}$$

Dead + upward wind - 11.6 kN

$$(1/r_v) = \frac{0.85 \times 2274}{9.7} = 199 \quad \text{Table 17a BS 449} \quad p_c = 24 \text{ N/mm}^2$$

$$\frac{f_c}{p_c} = \frac{11.6 \times 10^3}{4.80 \times 10^2 \times 24} = 1.01 \quad < 1.25 \quad \text{therefore satisfactory}$$

v) <u>Members (23) and (24)</u>

Use 'T' section 165 x 152 x 20 'T' $A = 25.8 \text{ cm}^2$
$r_x = 4.30 \text{ cm}$
$r_y = 3.85 \text{ cm}$

From Table 6.1 dead + snow = 148.7 kN (tension)

$$\frac{f_t}{p_t} = \frac{148.7 \times 10^3}{25.8 \times 10^2 \times 155} = 0.37 \quad < 1.0 \quad \text{therefore satisfactory}$$

From Table 6.1 dead + upward wind −52.3 (24)
−74.0 (23)

$$(l_{23}/r_x) = \frac{0.7 \times 2300}{43.0} = 37.4$$

$$(l_{24}/r_x) = \frac{0.7 \times 3300}{43.0} = 53.7$$

$(l_{23-24}/r_y) = 5600/38.5 = 145.5$ (assuming a longitudinal tie
at joint 4)

so $p_c = 42$ N/mm^2 (Table 17a BS 449)

$$\frac{f_c}{p_c} = \frac{74.0 \times 10^3}{25.8 \times 10^2 \times 42} = 0.68 \; < \; 1.25 \quad \text{therefore satisfactory}$$

vi) **Members (27) and (28)**

Each member may buckle individually about their v-v axes, but the combined member may buckle out of plane of the truss.

From Table 6.1 dead + snow = 77.0 kN (tension)
Use 125 x 75 x 8 angle short leg attached
$a_1 = (75 - 8/2)8 = 568 \; mm^2$
$a_2 = (125 - 8/2)8 = 968 \; mm^2$

Net area = $(3a_1/(3a_1 + a_2)) \times a_2 + a_1$

$$= \frac{3 \times 568}{3 \times 568 + 968} \times 968 + 568 = 1185 \; mm^2$$

$$\frac{f_t}{p_t} = \frac{77 \times 10^3}{1185 \times 155} = 0.42 \; < \; 1.0 \quad \text{therefore satisfactory}$$

Dead + upward wind   -48.7 kN (27)
                      -37.2 kN (28)

$$(l_{28}/r_v) = \frac{0.85 \times 3500}{16.3} = 183$$

$$(l_{27}/r_v) = \frac{0.85 \times 2300}{16.3} = 120$$

$$(l_{27,28}/r_v) = 5800/40.0 = 145.6$$

From Table 17a (BS 449) $(p_c)_{28} = 28 \; N/mm^2$ $(p_c)_{27} = 43 \; N/mm^2$

$$\left(\frac{f_c}{p_c}\right)_{27} = \frac{48.7 \times 10^3}{15.5 \times 10^2 \times 43} = 0.73 \; < \; 1.25 \quad \text{therefore satisfactory}$$

$$\left(\frac{f_c}{p_c}\right)_{28} = \frac{37.2 \times 10^3}{15.5 \times 10^2 \times 28} = 0.86 \; < \; 1.25 \quad \text{therefore satisfactory}$$

vii) <u>Member (31)</u>

Use 'T' section 165 x 152 x 20 'T' (as other bottom member)
From Table 6.1 dead + snow = 77 kN

$$\frac{f_t}{p_t} = \frac{77 \times 10^3}{25.8 \times 10^2 \times 155} = 0.19 \ < 1.00 \quad \text{therefore satisfactory}$$

Dead + upward wind -30.8 kN

$$(1/r_x) = \frac{9000 \times 0.85}{4.30 \times 10} = 178$$

$$p_c = 24 \text{ N/mm}^2 \quad \text{(Table 17a)}$$

$$(1/r_y) = \frac{9000 \times 0.85}{3.85 \times 10} = 199$$

$$\frac{f_c}{p_c} = \frac{30.8 \times 10^3}{25.8 \times 10^2 \times 24} = 0.50 \ < 1.25 \quad \text{therefore satisfactory}$$

In practice a sag rod as shown in Fig. 6.1 is provided which reduces the effective length for the y-y axis.

The next stage is to design all the welds in accordance with Chapter 4. Note that it may be necessary to increase some member sizes to accomodate the correct size welds. A check also needs to be made on the deflection of the truss using methods given in Chapter 5. In this case, the visual effects of deflection will not be obvious as the bottom boom of the truss is, in effect, cambered.

An alternative to welding members directly together is to use gusset plates, the design of which is dealt with in the next section. A summary of the member sizes is given in Table 6.2.

Table 6.2   Summary of roof truss members

Member	Section
1,2,3,4,5	165 x 152 x 27 T
11,12	65 x 50 x 5 angle
15,16	60 x 60 x 6 angle
19,20	50 x 50 x 5 angle
23,24,31	165 x 152 x 20 T
27,28	125 x 75 x 8 angle

In practice a check now needs to be made on the self-weight of the truss and a comparison made with the assumed value. Adjustments will need to be made in the design if the assumed value is grossly in error.

6.4   GUSSET PLATES

Gusset plates are used to assist in the transfer of forces, and to ease fabrication, at joints in trusses fabricated from rolled steel work. Gusset plates would have been necessary if pairs of angles back to back had been used for the main members as there may not have been sufficient space for the necessary welds.

A typical joint layout is given in Fig. 6.5.

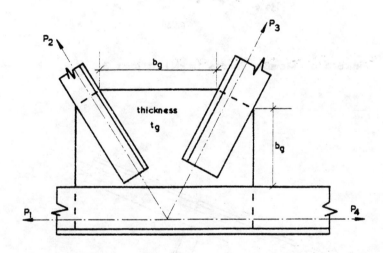

Fig. 6-5. Typical joint layout.

The overall size of the plate must be such that there is sufficient room to place and to develop the strength of all the connections (bolts or welds).

Fisher and Struik in 1974 commented that a considerable amount of work still needed to be done to evaluate gusset plate behaviour but that certain empirical rules may be used. This seems still to be true.

The thickness is determined by the condition that the equivalent stress at any point does not exceed the allowable values (see Table 1). The maximum length, b_g, of an unsupported edge is given by Cl. 12.8.2 of BS 5400 Part 3 (there being no guidance in BS 449),

$$b_g = 50\ t_g(355/Y_s)^{1/2} \qquad\qquad (6.3)$$

Y_s is the guaranteed minimum yield stress.

Example 6.2 Gusset plate design. Consider joint 3 in the truss
designed in section 6.4, the forces from which are given in Fig.
6.6(a). The roof truss was designed using T sections for the top and
bottom rafters which due to the depths of the stem avoid gussets. An
alternative would have been to have used a pair of 100 x 65 x 8 angles
with the long legs attached.

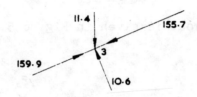

Fig. 6.6 (a). Member forces (in kN) at joint 3.

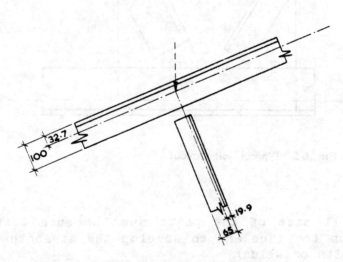

Fig. 6.6 (b). Details of members at joint 3.

Minimum weld lengths for 6 mm fillet welds

Main members : 159.9/(2 x 0.48) = 167 mm
               10.6/(2 x 0.48) = 11 mm

In Fig. 6.7 is drawn the gusset plate, the dimensions of which are determined from considerations of being able to perform the welding in a satisfactory manner.

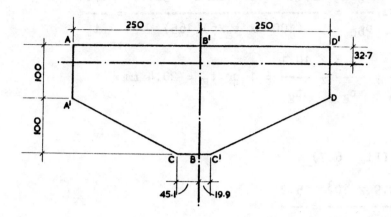

Fig. 6·7. Dimensions of the gusset plate.

From Table 1 BS 449 maximum allowable equivalent stress is 230 N/mm^2 for grade 43 and from Table 2 BS 449 p_{bc} is 165 N/mm^2 and p_c may be taken as 155 N/mm^2.

The method used to design the gusset plates is to consider
 i) the effect of bending and axial effects at a cross section using Equation (6.1), and
ii) the effect of bending, shear and axial effects to calculate an equivalent stress as in Cl. 14c BS 449.

In general if the combined effects of bending and axial effects are satisfactory, it will not normally be necessary to check for equivalent stresses.

Only the effects of dead and snow loading are considered in this example, hence the value of k in Equation (6.1) may be taken as unity. The forces are considered to be acting along the centroidal axes of the members.

Each of the critical sections will be considered in turn.

Section AA' (Fig 6.7)

$$\text{Axial stress } f_c = \frac{159.9 \times 10^3}{100t} \quad , \quad \frac{f_c}{p_c} = \frac{159.9 \times 10^3}{100t_g \times 155} = \frac{10.3}{t_g}$$

where t_g is the thickness of the gusset plate

Bending stress $\dfrac{f_{bc}}{p_{bc}} = \dfrac{159.9 \times 10^3(100/2 - 32.7)}{(100^2 \times t_g/6)(165)} = \dfrac{10.1}{t_g}$

$\dfrac{f_{bc}}{p_{bc}} + \dfrac{f_c}{p_c} = \dfrac{10.3}{t_g} + \dfrac{10.1}{t_g} = 1$ or $t_g = 20.4$ mm

Section BB' (Fig. 6.7)

$\dfrac{f_c}{p_c} = \dfrac{159.9 \times 10^3}{200 t_g \times 155} = \dfrac{5.2}{t_g}$

$\dfrac{f_{bc}}{p_{bc}} = \dfrac{159.9 \times 10^3(100 - 32.7)}{(200^2 \times t_g/6)(165)} = \dfrac{9.8}{t_g}$

$\dfrac{f_c}{p_c} + \dfrac{f_{bc}}{p_{bc}} = 1$ or $t_g = 15.0$ mm

Section CC' (Fig. 6.7)

$\dfrac{f_c}{p_c} = \dfrac{10.6 \times 10^3}{65 \times t \times 155} = \dfrac{1.05}{t_g}$

$\dfrac{f_{bc}}{p_{bc}} = \dfrac{10.6 \times 10^3 \times 12.6}{((65)^2 \times t_g/6) \times 165} = \dfrac{1.15}{t_g}$

$\dfrac{f_c}{p_c} + \dfrac{f_{bc}}{p_{bc}} = \dfrac{2.2}{t_g}$ or $t_g = 2.2$ mm

The effect of shear is negligible and therefore use a 25 mm plate. There is no need to check the equivalent stresses as the allowable value of the equivalent stresses is much higher than that for bending.

Maximum length of unsupported edge C'D $= [100^2 + (250 - 19.9)^2]^{1/2}$
$= 250$ mm

from Equation (6.3) maximum length b_g is given by

$$b_g = 50t_g(355/Y_s)^{1/2}$$

t_g = 25 mm Y_s = 250 N/mm^2 therefore b_g = 1490 mm, which is greater than 250, therefore satisfactory.

In general the gusset plate should be at least as thick as the thicknesses of the members attached to it at any section.

6.5 DESIGN OF PURLINS

The design procedure follows that for beams, except that adequate lateral restraint is deemed to be provided by the roof sheeting. Commonly used sections are pairs of angles bolted back to back, channels or proprietary cold formed 'Z' section purlins. Due to the presence of the roof sheeting, an asymetric section will tend not to bend about the principal axis but about an axis parallel to the roof slope. For the purposes of calculation, a degree of fixity may be assumed where the purlin is bolted to the roof members and the purlin may be designed for a maximum approximate sagging bending moment of $wL^2/10$ where w is the uniformly distributed applied load and L the span (Walker). Full deflection calculations or lateral buckling calculations are not required if the provisions of Cl. 45 (BS 449) are adopted. These may be summarised as follows:-

Minimum depth of purlin	L/45	
Minimum width of purlin	L/60	(6.4)
Minimum section modulus in cm^3	WL/1.8 x 10^{-3}	

where W is the total load in kN. These rules apply to the use of grade 43 steel on roof slopes less than 30°. If these conditions are not satisfied then purlins are to be designed as uncased beams using full bending stresses (Table 1 BS 449). No checks on deflections are then required.

These provisions were originally formulated to cover angle purlins, but are now also applicable to 'Z' purlins and channels. It is sufficient to consider dead and live load only. Similar considerations apply to sheeting rails on the sides of buildings.

Example 6.3 Design of a purlin. Design an angle purlin in grade 43 for the roof truss in Example 6.1.

Since the slope does not exceed 30° and the steel grade is grade 43, the approximate rules in Equation (6.4) may be used.

```
Span         4.5 m
Spacing      1.346 m  (slope distance)
Loading      1.084 kN/m²  (snow load and superimposed dead load)
```

Uniformly distributed load = 1.084 x 1.346 = 1.46 kN/m
Total load   = 1.46 x 4.5 = 6.57 kN
Using Equation (6.4):-
Minimum width = 4.5 x 10^3/60 = 75 mm
Minimum depth = 4.5 x 10^3/45 = 100 mm
Minimum Z_{ex} = (6.57 x 4.5)/1.8 = 16.43 cm^3
Maximum bending moment = 1.46 x 4.5^2/10 = 2.96 kNm

p_{bc} = 165 N/mm² so $(Z_e)_{req}$ = 2.96 x 10^3/165
$$= 17.94 \text{ cm}^3$$
which is greater than minimum. Therefore use 100 x 75 x 8
angle (Z_{ex} = 19.3 cm^3)

It is seen in this example that the criteria on minimum dimensions
determine the section size.

6.6 DESIGN OF VIERENDEEL GIRDERS

A Vierendeel girder may either be designed using the result of an
elastic analysis or a plastic analysis. A plastic analysis will in
general be more economical, although a check will still need to to be
made on the deflection of the girder.

Example 6.4 Design of a Vierendeel girder in grade 43 steel. The
truss to be designed is illustrated in Fig. 5.30. Both the plastic
and elastic analyses are made in Chapter 5 (Example 5.3).

i) Elastic Design

As the truss will tend to buckle out of plane an effective length
factor of 1.0 is used for the y-y axis.

Try a 250 x 150 x 10 RHS
     A = 75.5 cm^2 I = 6279 cm^4 Z_{ex} = 501 cm^3 r_y = 6.07 cm

l/r_y = 400/6.07 = 66

From Cl. 19d BS 449 mentioned in Section 2.5.2 the allowable bending
stress (p_{bc}) is 165 N/mm², from Table 17a the allowable axial stress
(p_c) is 120 N/mm².

From the approximate analysis given in Table 5.5 (Chapter 5) the worst
loaded section is EB, M = 72 kNm and N = 18.0kN, so

$$\frac{18 \times 10^3}{75.5 \times 10^2 \times 120} + \frac{72 \times 10^6}{501 \times 10^3 \times 165} = 0.89 < 1.00, \text{ therefore satisfactory}$$

Alternatively using either the computer analysis or the moment distribution analysis, the above factor reduces to 0.75. The approximate analysis will tend to overestimate the moments due to assumptions made on the position of the points of contraflexure. The above section size has been used for the computer analysis the results of which are given in Fig. 5.35.

The maximum deflection at joint A in Fig. 5.35 is 42.0 mm (allowable is 16000/360 = 44.4 mm), thus the section could be made lighter. It might be possible to use a 250 x 150 x 8 RHS (Z_{ex} = 413 cm^2)

ii) **Plastic Design**

From Example 5.3, M_p = 91.8 kNm so

$$Z_p = M_p/Y_s = 91.8 \times 10^6/(250 \times 10^3) = 367.2 \text{ cm}^3$$

i.e. a 250 x 150 x 6.3 mm RHS may be used (Z_p = 405 cm^3). Since the shear forces and axial forces are low, no account has been taken of the reduction in M_p due to these effects. The section used may give slightly unacceptable deflections, but it should be noted that economy is effected using plastic design methods, since a lighter section has been used.

The welded joints should be designed in accordance with the principles laid down in Chapter 4.

6.7 COMPOUND STANCHIONS

These are used where the axial loads are high. A single rolled section would be inadequate because of the tendency to lateral instability. The most common use of compound stanchions is for supporting gantries for heavy duty cranes. For light duty cranes a bracket support from a rolled stanchion is usually adequate.

Compound stanchions may be either laced or battened. These are illustrated in Fig. 6.8(a).

The battened column behaves like a Vierendeel girder; the laced like a parallel chord truss. The design of such columns is considered by Cl. 35 and 36 BS 449.

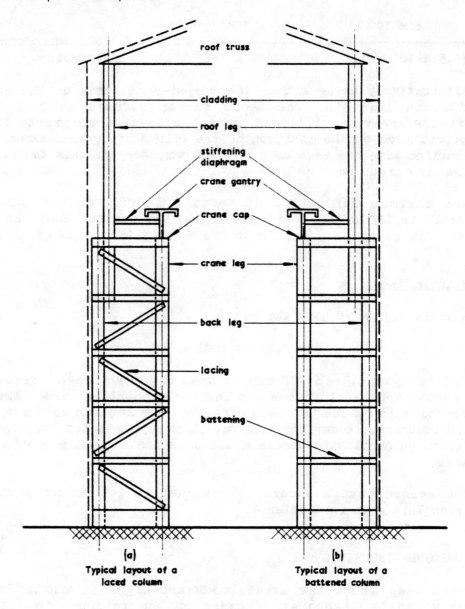

(a)
Typical layout of a
laced column

(b)
Typical layout of a
battened column

Fig. 6·8.

Clauses 35 and 36 (BS 449) are only partially relevant to the design of compound stanchions as they also apply to the situation where a compound member (e.g. angles arranged in a cruciform shape) resists direct compression. Also extensive reference is made to riveting, a method of jointing which is virtually obsolete.

Essentially the provisions laid down in Cl. 35 and 36 ensure that local buckling will not occur at unduly low stresses, and thus some restriction is laid down on allowable slenderness ratios. A further requirement is that the total horizontal loading to be resisted is the applied horizontal forces and 2.5% of the vertical loading. This provision has been met previously in connection with the design of wind bracing (see Chapter 5).

A battened column will probably contain less weight of steel work than a laced column (the examples that follow illustrate this), although the amount of welding and therefore fabrication costs will be higher. There is probably little difference in overall costs in each of the two cases.

Example 6.5 Design of a laced column. The column is illustrated in Fig. 6.9. The overall dimensions are laid down by consideration of the working areas and heights required in the structural layout. Grade 43 steel is to be used.

Loading

			kN
VERTICAL			
Roof leg	–	reaction from roof	120
		side cladding	12

			132
Crane leg	–	max. crane reaction	240
		gantry girder self-weight	35
		self-weight of leg	15

			290
Back leg	–	cladding	12
		self-weight of leg	12

			24

HORIZONTAL	
Wind load at eaves	22
Crane surge at rail level	6

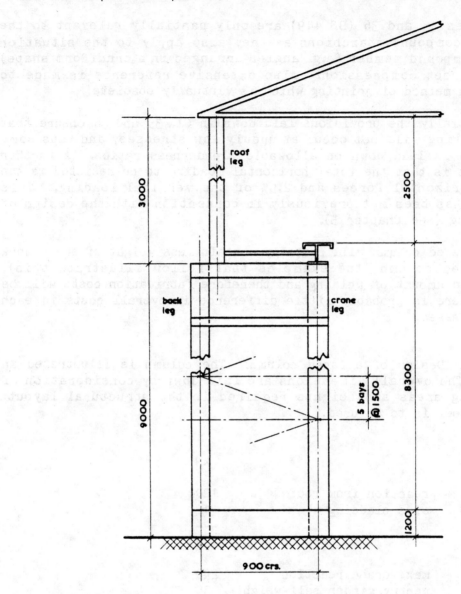

Fig. 6·9. Principal dimensions of the design example.

The moments due to horizontal loads are calculated at 1) the level of the crane cap and 2) the base, and are then treated as a pair of equal and opposite forces in the crane leg and the back leg.

1) at crane cap level
   Moment due to wind = 22 x 3   = 66 kNm
   Moment due to surge = 6 x 0.5 = 3 kNm (usually ignored due to stiffening diaphragm)

2) at base level
   Wind moment   = 22 x 12 = 264 kNm
   Surge moment   = 6 x 9.5 = 57 kNm

These may be considered as a pair of equal and opposite forces in each of the legs:-

equivalent wind force ± 264/0.9 = ± 293.3 kN
equivalent surge force ± 57/0.9 = ± 63.3 kN

Design of roof leg

Actual length 3.0 m
effective length x-x axis = 2 x 3.0 = 6.0 m (acting as a cantilever)
effective length y-y axis = 1 x 3.0 = 3.0 m (considered attached to sheeting rails)
axial load 132 kN moment 56 kNm

Try 305 x 165 x 40 UB
$A = 51.5$ cm^2, $Z_{ex} = 561.2$ cm^3, $r_x = 12.86$ cm
$r_y = 3.85$ cm, $D/T = 29.9$

$(l/r_x) = 46.7$ $P_c = 106$ N/mm^2 Table 17a (BS 449)
$(l/r_y) = 77.9$ $P_{bc} = 165$ N/mm^2 Table 3a (BS 449)

$$\frac{f_c}{P_c} + \frac{f_{bc}}{P_{bc}} = \frac{132 \times 10^3}{51.5 \times 10^2 \times 106} + \frac{66 \times 10^6}{561.2 \times 10^3 \times 165} = 0.95 < 1.25$$

therefore satisfactory.

Assume all the roof leg loading goes through to back leg.

Cl. 35e BS 449 states the angle of the bracing shall be between 40° and 70°. Cl. 36a BS 449 states that there shall be at least 3 bays. Try 5 bays of bracing.

From Fig. 6.10 actual slope = $\tan^{-1}$ ((1.50 - 0.3)/0.9))
                            = 53.1°

0.3 is an estimate of the eccentricity of the connection which will be allowed for in the design of the bracing.

Back leg

Total axial load	=	24	cladding + s.wt. of back leg
		132	roof leg
		293.3	wind
		63.3	surge

		512.6 kN	

Effective lengths - in plane of bracing 1.50 m - bracing centres
out of plane 0.85 x 9 = 7.65 allows for base
fixity and crane
girder.

Try a 406 x 140 x 46 UB

r_x = 16.29 cm, r_y = 3.02 cm,   A = 59.0 cm^2,

(l/r_x) = 7650/162.9 = 47   Table 17a p_c = 133 N/mm^2
(l/r_y) = 1510/30.2 = 50   Table 3a p_{bc} = 165 N/mm^2

$$\frac{f_c}{p_c} = \frac{512.6 \times 10^3}{59.0 \times 10^2 \times 133} = 0.65 \;<\; 1.25, \text{ therefore satisfactory}$$

Note, Cl. 35f BS 449 states that the maximum slenderness ratio shall not exceed 50 and that the slenderness ratio between bracings shall not exceed 0.7 x max (l/r) for member. This is to avoid premature buckling. This condition is satisfied by the above values.

Crane leg

Max. axial load		290	crane load
		293.3	wind
		63.3	surge

		646.6 kN	
	or	353.3 kN	- excluding wind

Moment due to crane load = crane load x D/2
D = depth of section, use 406 deep member
so crane moment = 290 x 0.406/2 = 58.9 kNm

Try a 406 x 178 x 54 UB

r_x = 16.5 cm, r_y = 3.85 cm, A = 68.4 cm^2, Z_{ex} = 925.3 cm^3,   D/T = 37

(l/r_x) = 0.85 x 9000/165 = 46 p_c = 136 N/mm^2 Table 17a BS 449
(l/r_y) = 1500/38.5 = 39 p_{bc} = 165 N/mm^2 Table 3a BS 449

No wind $\dfrac{f_c}{p_c} + \dfrac{f_{bc}}{p_{bc}} = \dfrac{353.3 \times 10^3}{68.4 \times 10^2 \times 136} = \dfrac{58.9 \times 10^6}{925.3 \times 10^3 \times 165} = 0.77$

$$< 1.00$$

with wind $\dfrac{f_c}{p_c} + \dfrac{f_{bc}}{p_{bc}} = \dfrac{646.6 \times 10^3}{68.4 \times 10^2 \times 136} = \dfrac{58.9 \times 10^6}{925.3 \times 10^3 \times 165} = 1.08$

$$< 1.25$$

both satisfactory.

Bracing

From Cl. 35a BS 449 the total horizontal load on the bracing shall be taken as the horizontal loading + 2.5% applied vertical loading i.e.

surge	6	
wind	22	
Total vertical load	132	
	290	
	240	

2.5% of	662	16.6

		44.6 kN

i.e. 22.3 kN per face.

The bracing layout is shown in Fig. 6.10.

Axial force in diagonals   = 22.3/cos θ
                           = 22.3/cos 53.4
                           = 37.4 kN

Moment due to eccentricity = 22.3 x 0.3 = 6.7 kNm divided equally between the two members, i.e. 3.35 kNm to each diagonal. This moment is induced since the centroidal axes of the inclined bracing members and the main stanchions are not coincident. Length of diagonals from Fig. 6.10,

$$l = [(1.51 - 0.3)^2 + 0.9^2]^{1/2} = 1.5 \text{ m}$$

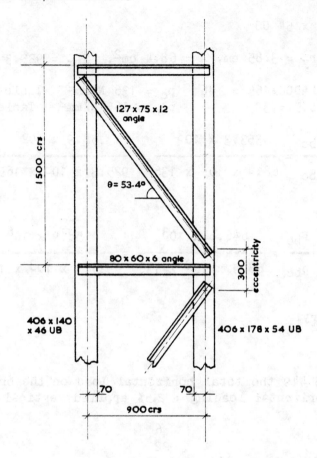

Fig. 6·10. Bracing layout for the laced column.

Try 125 x 75 x 12 angle
$A = 22.7$ cm^2
$Z_{ex} = 43.2$ cm^3
$r_v = 1.61$ cm

$l/r_v = 1500/16.1 = 93$

Note, Cl. 36b BS 449 states that the l/r ratio for the lacing members shall not exceed 140.

$p_{bc} = 165$ N/mm^2 Table 3a   BS 449
$p_c = 87$ N/mm^2 Table 17a   BS 449

$$\frac{f_c}{p_c} + \frac{f_{bc}}{p_{bc}} = \frac{37.4 \times 10^3}{22.7 \times 10^2 \times 87} + \frac{3.35 \times 10^6}{43.2 \times 10^3 \times 165} = 0.66 < 1.25$$

therefore satisfactory.

The function of the horizontal brace is to reduce the effective length of the stanchions for in plane buckling. They are not necessary to resist the applied horizontal loading.

It is therefore reasonable to consider that the horizontal braces need only take a load of 2.5% of the vertical load, i.e. 16.6 kN or 8.3 kN per face.

Try a 80 x 60 x 6 angle,

r_v = 1.29 cm,   A = 8.11 cm^2
$1/r_v$ = 900/12.9 = 70, p_c = 115 N/mm^2 Table 3a BS 449

$$f_c/p_c = (8.3 \times 10^3)/(8.11 \times 10^2 \times 115) = 0.09 < 1.25$$

therefore satisfactory.

Welds are to be designed using the principles evolved in Chapter 4.

Example 6.6 Design of a battened column. The main difference here is in the need to calculate the moments at the intersection of the vertical members and the battening. This is done by considering the column to act as a Vierendeel girder and then using an approximate analysis, assuming points of contraflexure at the mid points of panels to calculate the moments. This is further discussed in Chapter 5.

Loading as for the laced column.

Roof leg - as before
Assume a 457 x 152 x 82 UB, I_x = 36215 cm^4 for the back leg and 457 x 191 x 89 UB, I_x = 41021 cm^4 for the crane leg.
Total horizontal shear = 44.6 kN (Example 6.5)
The shear is shared between the two legs in proportion to their second moments of area.
Shear on back leg   = 44.6 x 36215/(41021 + 36215) = 20.9 kN
Shear on crane leg = 44.6 x 41021/(41021 + 36215) = 23.7 kN.

Using a 1.5 m spacing of battens, and assuming points of contra-flexure at member midpoints (as in a Vierendeel girder)
the moment in the back leg = 20.9 x 1.5/2 = 15.7 kNm
the moment in the crane leg = 23.7 x 1.5/2 = 17.8 kNm

For a 457 x 152 x 82 UB
r_y = 3.31 cm,   A = 104.5 cm^2,   D/T = 24.6
r_x = 18.62 cm, Z_{ey} = 149 cm^3

$(1/r_x)$ = 0.85 x 9000/186.2 = 41 p_c = 136 N/mm^2 Table 17a BS 449
$(1/r_y)$ = 1500/33.1   = 45 p_{bc} = 165 N/mm^2 Table 3a BS 449

$$\frac{f_c}{p_c} + \frac{f_{bc}}{p_{bc}} = \frac{512.6 \times 10^3}{104.5 \times 10^2 \times 136} + \frac{15.7 \times 10^6}{149 \times 10^3 \times 165} = 1.0 < 1.25$$

Crane leg

457 x 191 x 89 UB
r_y = 4.28 cm, A = 113.9 cm^2, Z_{ex} = 1770 cm^3,   D = 463.6 mm
r_x = 18.98 cm, D/T = 26.3, Z_{ey} = 217.4 cm^3

(l/r_y) = 1500/42.8 = 35 p_c = 138 N/mm^2 Table 17a BS 449
(l/r_x) = 0.85 x 9000/189.8 = 40 p_{bc} = 165 N/mm^2 Table 3a BS 449

Check under full loading only:-

$$(f_c/p_c) + (f_{bc}/p_{bc})xx + (f_{bc}/p_{bc})yy =$$

$$= \frac{646.6 \times 10^3}{113.9 \times 10^2 \times 138} + \frac{290 \times (463.6/2) \times 10^3}{165 \times 1770 \times 10^3} + \frac{17.8 \times 10^6}{165 \times 217.4 \times 10^3}$$

= 1.14 < 1.25, therefore satisfactory.

The case without wind load will also be found to be satisfactory.

Design of battens

Maximum axial force = 23.7 kN
Maximum moment = 23.7 x 0.9/2 = 10.7 kNm
These forces are taken by two channel sections, i.e. the forces on
each are a moment of 5.4 kNm and an axial force of 11.9 kN.

Effective length = 0.7 x 900 = 630 mm.
Try a 102 x 51 channel

r_y = 1.48 cm, A = 13.28 cm^2, Z_{ex} = 40.89 cm^3, D/T = 13.3

(l/r_y) = 630/14.8 = 43

p_{bc} = 165 N/mm^2 Table 3a p_c = 137 N/mm^2 Table 17a BS 449

$$\frac{f_c}{p_c} + \frac{f_{bc}}{p_{bc}} = \frac{11.9 \times 10^3}{13.28 \times 10^2 \times 137} + \frac{5.4 \times 10^6}{40.89 \times 10^3 \times 165} = 0.87$$

< 1.25 therefore satisfactory.

The welds are designed according to Chapter 4.

REFERENCES

BS 5400, Part 3 (1982) - Steel, Concrete and Composite Bridges, Part 3, Code of Practice for Design of Steel Bridges, BSI.

CP3, Ch V, Part 1 (1967) - Code of basic data for the design of buildings Chapter V Loading, Part 1, Dead and imposed loads BSI.

CP3, Ch V, Part 2 (1972) - Code of basic data for the design of buildings, Chapter V loading, Part 2, Wind loads, BSI.

Fisher, J.W., Struik, J.H.A. (1974) - Guide to Design Criteria for Bolted and Riveted Joints, Chapter 15, Gusset Plates. John Wiley & Sons.

Jennings, A., Majid, K.I. (1966) - The computer analysis of space-frames using sparse matrix techniques. International Conference on Space Structures, University of Surrey.

Just, D.J. (December 1978) - Deep plane frame works with stiffened joints. Journal Institution of Structural Engineers, Vol 56B, No. 4, pp 85-92.

Purkiss, J.A., Croxton, P.C.L. (1980) - Design of Eccentric Welded Connections in Rolled Hollow Sections. Joints in structural steelwork, Eds. J.H. Howlett, W.M. Jenkins and R. Stainsby, Pentech Press.

Walker, A.C. (Ed) (1975) - Design and Analysis of Cold-formed Sections, Intertext Books.

7 Fabricated Beams

7.1 INTRODUCTION

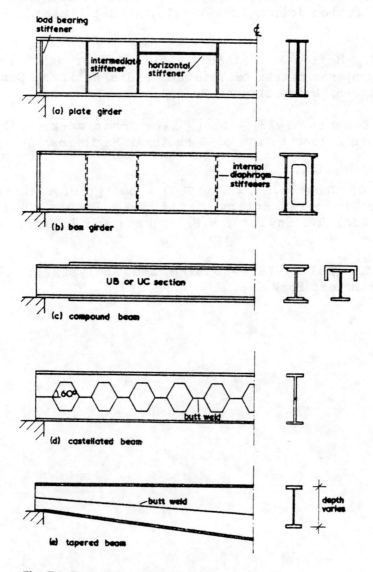

Fig. 7.1. Types of fabricated beams.

If the size of the beam required is greater than those available in the standard rolled sections then special fabricated beams are used. The cost of fabricated beams, however, is greater than that of rolled sections because of the labour of cutting, jigging, assembly and welding. Common types of fabricated beam are shown in Fig. 7.1.

Generally the same grade of steel is used throughout a fabricated beam but it is possible for the web of a plate girder to be of grade 43 steel and the flanges to be of grade 50 or 55, although this is not common in British practice.

7.2   PLATE GIRDERS

Plate girders are built-up from two flange plates and a web plate. Originally they were riveted together with angles at the web-flange joint, but modern fabrication is by continuous automatic electric welding, using a fillet weld between the web and the flange. The welds on one side of the web are done first then the beam is turned over and the welds completed on the return pass. To ensure that the web is at right angles to the flanges and that the flanges are parallel, the initial set of welds is performed with the flanges non-parallel as in Fig. 7.2.

Fig. 7.2. Fabrication of a plate girder of I section.

Finally the stiffeners are welded manually to the web and the flanges. The minimum thicknesses are 6 mm for the web plates and 15 mm for the flanges. The cross section of the girder is generally an I shape with a single web, but may be a rectangular box with twin webs. The depth of the girder is generally constant for the entire span but may be tapered for economic reasons or to meet constructional requirements, e.g. to fit a roof slope.

Originally plate girders were designed for spans which were too long for the then limited range of rolled sections. Since then the range of rolled sections has increased, and plate girders are only used to support large loads on crane gantries or bridge girders and for large span beams.

7.2.1   <u>Lateral torsional buckling of a plate girder</u>   Lateral
torsional buckling of rolled sections is discussed in Chapter 2.   The
effect is enhanced in plate web girders because of the increased
distance between flanges which decreases the stabilising effect of the
tension flange.   The following theory shows the derivation of the
critical elastic moment at which buckling occurs and from which the
allowable design stresses in bending given in BS 449 can be obtained.

a)   Girder symmetric about both axes.
Consider the lateral torsional buckling of the beam shown in Fig. 7.3.

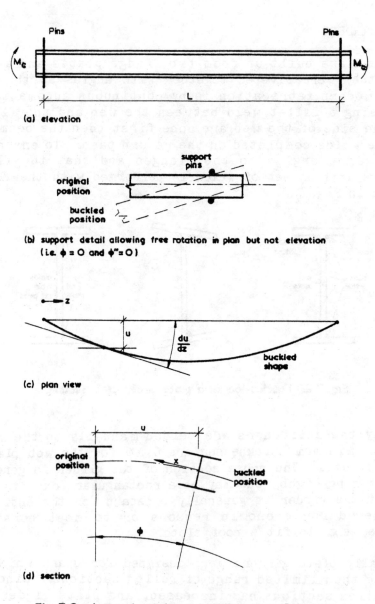

Fig. 7.3.   Lateral torsional buckling of an I beam.

The equation governing bending in a plane parallel to the flange is given by

$$EI_y d^2 u/dz^2 = -M_c \qquad (7.1)$$

where EI_y is the bending rigidity about y-y axis of the beam: the remaining symbols are defined in Fig. 7.3.

The torsion equation is given by

$$GJd\phi/dz - EI_w d^3\phi/dz^3 = M_c du/dz \qquad (7.2)$$

where the first term is due to uniform torsion (GJ is the torsional rigidity), the second term the warping effect (EI_w is the warping rigidity), and the third term the disturbing torque.

To facilitate the solution of Equations (7.1) and (7.2) the deflection, u, may be expressed in the form of a Fourier series. In practice only the first term is necessary, i.e.

$$u = \delta \sin(\pi z/l) \qquad (7.3)$$

Substituting Equation (7.3) into Equation (7.1) gives

$$u = M_c \phi / \pi^2 (EI_y/l^2) \qquad (7.4)$$

When the critical moment M_c is reached the beam just attains unstable equilibrium. If the governing torsion equation (Equation 7.2) is to be satisfied then by substituting Equations (7.3) and (7.4) into Equation (7.2) the value of the critical moment M_c is given by

$$M_c = \pi (EI_y GJ)^{1/2}/l \ \{1 + (\pi^2 EI_w/GJl^2)\}^{1/2} \qquad (7.5)$$

Equation (7.5) ignores the effect of major axis curvature. The exact solution is

$$M_c = \frac{\pi}{l} \left[\frac{EI_y GJ}{\gamma} \right]^{1/2} (1 + \frac{\pi^2 EI_w}{l^2 GJ})^{1/2} \qquad (7.6)$$

where $\gamma = 1 - I_y/I_x$

For most sections I_y is very much less than I_x (of the order of one fifteenth and one twenty fifth) and thus γ may be taken as unity and Equation (7.5) used.

For rolled sections values of I_w, J and I_y have been tabulated, otherwise for parallel-flange beams the corresponding values may be calculated approximately using the following equations:-

Warping constant:

$$I_w = I_y h^2/4 \qquad\qquad (7.7)$$

where h is the distance between flange centroids.

Torsional constant:

$$J = (2/3)k_f BT^3 + (1/3)k_w(D - 2T)t^3 \qquad\qquad (7.8)$$

where $k_f = (1 - 0.63T/B)$ and
$k_w = [1 - 0.63t/(D - 2T)]$.

To facilitate the use of Equation (7.5), certain approximations may be made, which are

$$Z_{ex} = 1.1BTD$$
$$I_y = B^3T/6$$
$$J = 0.9BT^3 \qquad\qquad (7.9)$$
$$I_w = I_y D^2/4$$
$$B = 4.2r_y$$
and $E = 2.5G = 200 \text{ kN/mm}^2$

The full derivations of Equations (7.1) to (7.9) are given by Trahair.

Thus the critical bending stress is given by

$$\sigma_{cr} = \frac{M_c}{Z_{ex}} = \frac{2.65 \times 10^6}{(l/r_y)^2} \left[1 + \frac{1}{20}\left(\frac{l\,T}{r_y D}\right)^2\right]^{1/2} \text{ N/mm}^2 \qquad\qquad (7.10)$$

Due to the sweeping nature of the assumptions made in Equations (7.9), the accuracy of Equation (7.10) is variable, and care should be taken in its use.

For beams l/r_y is the slenderness ratio of the beam and is defined in a similar fashion to that for rolled sections which are dealt with in Chapter 2. For example if the complete solution is carried out for a cantilever, l is replaced by 2l, thus reducing the critical moment by approximately 75%.

For design of plate girders to BS 449 the constant 2.65×10^6 is replaced by 1675^2 ($= 2805625$) and the critical stress C_s is defined by

$$C_s = \left[\frac{1675}{l/r_y}\right]^2 \left[1 + \frac{1}{20}\left(\frac{l\,T}{r_y D}\right)^2\right]^{1/2} \qquad\qquad (7.11)$$

Allowance for imperfections is made using a Perry-Roberston type approach, in which the maximum stress in an elastic stress-free beam occurring under simultaneous lateral instability and in plane bending is limited to the yield stress, Y_s. The design curve is obtained by considering the limiting case when D/T tends to infinity. Using a value of initial imperfection   given by

$$\eta = 0.003(1/r_y) \qquad (7.12)$$

Note this is the same as the original imperfection coefficient for column buckling (Section 3.3). Using Equation (7.12) and (7.11) the critical bending stress p_{br} is given by the solution of

$$p_{br}^3 - [Y_s + 385/(1/r_y)]p_{br}^2 - C_s[C_s + 385/(1/r_y)]p_{br} \ldots$$

$$\ldots\ldots + Y_s C_s^2 = 0 \qquad (7.13)$$

The value of C_s given by Equation (7.11) is increased by 20% when T/t < 2 and d_1/t < 85 for grade 43 steel. The corresponding values are 75 for grade 50 and 65 for grade 55 steels, since Equation (7.11) is unduly conservative for these values of T/t and d_1/t. The solutions to Equation (7.13) are given in Table 8 BS 449 (Appendix).

It is recognised that below a limiting slenderness ratio, lateral torsional buckling will not occur. This limit is set at 60. A straight line transition is used between slenderness ratios of 60 to 100 where Equation (7.13) takes over. A variable factor of safety is then used to reduce the values of p_{br} to allowable stresses p_{bc}. For l/r > 100 a safety factor of 1.7 is employed, elsewhere a factor between 1.63 and 1.52 is used, the latter at low values of slenderness as the beam here will fail by yielding in the extreme fibres.

b) Girder symmetric about the y-y axis only.
i) Second moment of area of the compression flange greater than that of the tension flange, about the y-y axis.

An additional torque is introduced when the shear centre is no longer coincident with the centroid of the cross section of the beam and consequently the effective torsional rigidity is decreased. Winter gives the solution for the critical moment, M_c, for the general case of a beam with unequal flanges as

$$M_c = \frac{\pi}{1} (EI_y GJ)^{1/2} \left[\left[1 + \frac{\pi^2 EI_w}{1^2 GJ} \right]^{1/2} + (2\lambda - 1)\frac{\pi}{1} \left[\frac{EI_w}{GJ} \right]^{1/2} \right] \qquad (7.14)$$

where λ is defined as the ratio of the second moment of area of the compression flange about the y-y axis to that of the whole girder about the y-y axis. In practice the ratio λ may be expressed as that second moment of area of the compression flange to that of both flanges, since the second moment of area of the web is negligible. Equation (7.14) reduces to Equation (7.5) for equal flanges since λ is then 0.5.

Kerensky notes that when the compression flange is larger than the tension flange Winter's solution is too conservative and that the correction factor for this case, i.e. when λ lies between 0.5 and 1.0, has been reduced by 50%. The correction factor used by BS 449 is K_2 and is plotted in Fig. 7.4.

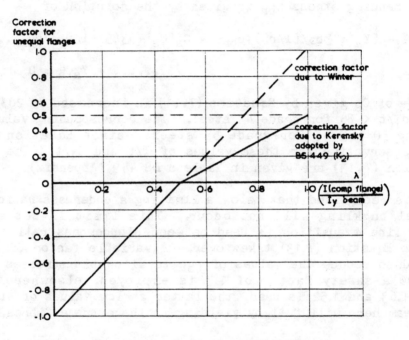

Fig. 7.4. Correction factor for unequal flanges.

Similar approximations are made for the section properties with no allowance for the effect on the elastic section modulus caused by the movement of the neutral axis, thus

$$C_s = \frac{(1675)^2}{(1/r_y)^2}\left[1 + \frac{1}{20}\left(\frac{1}{r_y}\frac{T}{D}\right)^2\right]^{1/2} + K_2\frac{(1675)^2}{(1/r_y)^2} \qquad (7.15)$$

ii) Second moment of area of tension flange larger than that of the compression flange.

In this case the effective torsional inertia increases and also a correction for the shift of the neutral axis is made so that C_s is given by

$$C_s = \left[\frac{1675}{1/r_y}\right]^2 \left[1 + \frac{1}{20}\left(\frac{1}{r_y}\frac{T}{D}\right)^2\right]^{1/2} + K_2\left[\frac{1675}{1/r_y}\right]^2 \frac{y_c}{y_t} \qquad (7.16)$$

where y_c is the distance from the neutral axis to the extreme
        compression fibre and y_t is that to the tension fibre
    P_{bc} is calculated in the same manner as symmetric girders.

    For all the calculations an effective flange thickness is used
defined as $K_1 T$. K_1 depends on the parameter N which is defined as the
ratio of the total area of both flanges at the point of least bending
moment to that of the greatest, between the points of restraint,
except that if considering the compression flanges alone gives a lower
value, then this value shall be used (Clause 20, BS 449). Flange
breadths shall not be reduced to give values of N lower than 0.25.
The values of K_1 are plotted in Fig. 7.5.

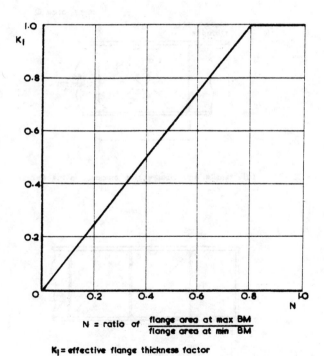

N = ratio of $\dfrac{\text{flange area at max BM}}{\text{flange area at min BM}}$

K_1 = effective flange thickness factor

Fig. 7.5. Values of effective flange thickness factor.

7.2.2   <u>Web buckling of plate girders in shear</u>   Where the
depth/thickness ratio of the web of beams is less than approximately
70, as in the standard rolled steel sections, the allowable shear

stress in the web is controlled by the Von Mises-Huber-Hencky yield criterion as described in Section 1.16.   Plate girders generally have more slender webs and the allowable shear stress in the web is controlled by web buckling.

Web buckling is produced by the 45° diagonal compressive stresses which develop from complementary shear stresses, as shown in Fig. 7.6. The buckling strength of the web depends on the d_1/t ratio and also on the spacing of the intermediate stiffeners, or in more general terms the dimensions of the panel abcd shown in Fig. 7.6.   The buckling strength also depends on the edge support conditions of the panel.   At ultimate load after buckling has occurred the overall effect of combining the web and stiffeners is to force the girder to act as a truss as shown in Fig. 7.6.   The intermediate stiffeners act as compression members and the buckled web plate acts as a diagonal tension member.   This truss analogy helps to explain the reserve of strength which exists at ultimate load, after the web has buckled.

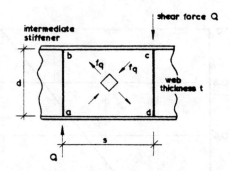

(a) tensile and compressive stresses in the web

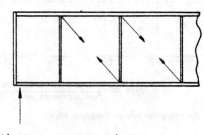

(b)   truss analogy at ultimate load

Fig. 7.6. Shear effects in plate web girders.

The first exact solution for the buckling of a rectangular plate loaded by shear forces was by Southwell and Skan. The plate tends to buckle across compression diagonals, and the number of buckles depends on the aspect ratio of the plate. The solution for the elastic critical buckling shear stress may be expressed algebraically as

$$\tau_{crit} = \pi^2 Ek/[12(1 - v^2)(d_1/t)^2] \qquad (7.17)$$

where $k \simeq 5.35 + 4(d_1/s)^2$ when $(s/d_1) > 1$ $\qquad (7.18)$

and $k \simeq 5.35(d_1/s)^2 + 4$ when $(s/d_1) < 1$ $\qquad (7.19)$

If the preceding theory is applied to the standard rolled UB and UC sections it can be shown that intermediate non-load bearing stiffeners are not required. If there are no stiffeners s is large and $k \simeq 5.35$. Also if the buckling is elastic $\tau_{crit} < Y_s/\sqrt{3}$, or in a design situation $\tau_{crit} < p_e/\sqrt{3}$ where the value of p_e is obtained from Table 1 BS 449 (Appendix). Substituting these values in Equation (7.17) and rearranging

$$d_1/t \leqslant [\pi^2 Ek/(12(1 - v^2)\tau_{crit}]^{1/2}$$

$$\leqslant [\pi^2 \times 210 \times 10^3 \times 5.35/[12(1 - 0.25^2)(p_e/\sqrt{3})]]^{1/2}$$

i.e. $d_1/t \leqslant 86.2$ for grade 43 steel, $p_e = 230$ N/mm^2

$\leqslant 73.0$ for grade 50 steel, $p_e = 320$ N/mm^2

$\leqslant 66.2$ for grade 55 steel, $p_e = 390$ N/mm^2

The corresponding d_1/t ratios given in BS 449 Cl. 27f are 85, 75 and 65. The d_1/t ratios of standard UB and UC sections are all less than these values and therefore do not require intermediate stiffeners. It should be noted that this solution to the problem assumes that the web plate is simply supported at the edges whereas in I beam sections the out of plane rotational restraint provided by the flanges is considerable and increases the buckling strength as shown by Kerensky et al.

It can be seen from Equations (7.17) to (7.19) that by increasing the aspect ratio (d_1/s) there is an increase in the elastic critical stress. In practice the aspect ratio can be changed by using vertical and horizontal stiffeners of the required size as shown in Sections 7.2.3 and 7.2.4.

Before the web plate buckles the stresses in it are due to shear only but after buckling, tensile membrane stresses produced by angular distortion react with the compressive stresses resulting from out-of-plane bending, thus inhibiting the buckle size and enhancing the shear buckling strength. The shear stresses given in BS 449 Table 12

(Appendix) are therefore not related to the elastic critical shear stresses but are calculated so that steel yields at the crest of a buckle. The shear stresses at this yielding condition are divided by a stress factor of 1.43 to produce values to be used as allowable stresses in Table 12 BS 449. A more detailed description of the derivation of these stresses is given by Kerensky et al.

7.2.3   Intermediate vertical stiffeners for plate girders (Cl. 28 BS 449) Intermediate transverse stiffeners as shown in Fig. 7.1(a) are introduced to strengthen the web of a plate girder. The web plate is restrained and supported on two opposite sides by the stiffeners and on the other two sides by the flanges of the girder. The intermediate stiffeners must have a minimum rigidity, i.e. a minimum second moment of area about an axis in the plane of the web perpendicular to the stiffener.

For an initially flat plate, subject to shear, Timoshenko derived the expression $I_s = 0.3d_1^4 t^3 / s^3$, but for plates which are not flat this value is too small and Moore recommended $I_s = 4d_1^3 t^3 / 3s^2$. Ultimate load tests have shown this value to be adequate but the value in BS 449 Cl. 28b has been increased further to $I_s = 1.5d_1^3 t^3 / s^2$. When vertical intermediate stiffeners are subject to external forces then this value is again increased to values given in Cl. 28b BS 449.

The actual width of the stiffener is equal to or less than the width of the flange and to prevent local buckling the outstand should not be greater than $12t_s$ for flats. The web of the girder interacts with the stiffener and a length of 20t on either side of the stiffener, if available, is specified in BS 449 Cl. 28a. The second moment of the actual stiffener about the centre line of the web is calculated from

$$I_s = t_s B_s^3 / 12 + 40t^4 / 12 \qquad\qquad (7.20)$$

This value must be greater than or equal to $I_s = 1.5d_1^3 t^3 / s^2$.

7.2.4 Horizontal stiffeners for plate girders (Cl. 28b BS 449) The addition of horizontal stiffeners reduces the d_1/t ratio of a web and consequently strengthens the web against buckling. This type of stiffener shown in Fig. 7.1(a) is only used in large girders where the increase in labour cost is offset by a reduction in web thickness.

Little experimental data is available for the design of this type of stiffener and according to Kerensky et al BS 449 recommendations are based on theoretical work by Bleich. The theoretical values are increased by a factor which is less than that for vertical stiffeners, because horizontal stiffeners are not required to carry increases in axial load when the critical load is exceeded. They are also further restrained by the dissimilar buckled forms of the adjacent panels.

The most effective position for a horizontal stiffener on a deep web subjected to bending and shear is at 2/5 of the distance from the compression flange to the neutral axis of the girder as specified in Cl. 28 BS 449. The stiffener itself should have a second moment of area about the centre line of the web

$$I_s > 4s_1 t^3 \tag{7.21}$$

where s_1 is the actual distance between stiffeners. A horizontal stiffener is used when the thickness of the web is less than $d_2/200$ for grade 43 steel, $d_2/180$ for grade 50 steel and $d_2/155$ for grade 55 steel. The distance d_2 is defined as twice the clear distance from the compression flange plate to the neutral axis.

The following recommendations for a second horizontal stiffener are taken from Cl. 28b BS 449. A second horizontal stiffener (single or double) should be placed on the neutral axis of the girder when the thickness of the web is less than $d_2/250$ for grade 43 steel, or $d_2/225$ for grade 50 steel, or $d_2/190$ for grade 55 steel. This stiffener should have a second moment of area about the centre line of the web

$$I_s > d_2 t^3 \tag{7.22}$$

Horizontal web stiffeners should extend between vertical stiffeners but need not be continuous over them.

7.2.5 <u>Load bearing stiffeners (Cl. 28 BS 449)</u> Where concentrated loads, or reactions, exceed the buckling or bearing strength of the web of a standard UB or UC section, as described in Section 2.6, then load bearing stiffeners should be provided. Generally stiffeners are symmetrical about the web as shown in Fig. 7.7, and are used for beams and columns. The concentrated loads for the beam-to-column connection are the forces in the flanges of the beam. One pair of stiffeners act in compression and the other pair in tension. These forces may be resisted by stiffeners as shown in Fig. 7.7.

According to Cl. 28 BS 449, a pair of compression stiffeners is designed as a strut, assuming that the section consists of a pair of stiffeners together with a length of web on each side of the centre line of the stiffeners equal, where available, to twenty times the web thickness. To avoid local buckling the outstand of a stiffener flat should not exceed $12t_s$. A section of the strut is shown in Fig. 7.7(a) and the radius of gyration about the y-y axis is

$$r_s = \left[\frac{I_s}{A_s}\right]^{1/2} = \left[\frac{(t_s B_s^3/12) + (40t^4/12)}{B_s t_s + 40t^2}\right]^{1/2} \tag{7.23}$$

The slenderness ratio is calculated assuming an effective length factor of 0.7.

$$l_s/r_s = 0.7h_s/r_s \qquad (7.24)$$

The allowable buckling stress p_c at working load is obtained from Table 17 BS 449 (Appendix) and the design load is

$$P_s = A_s p_c = (B_s t_s + 40t^2)p_c \qquad (7.25)$$

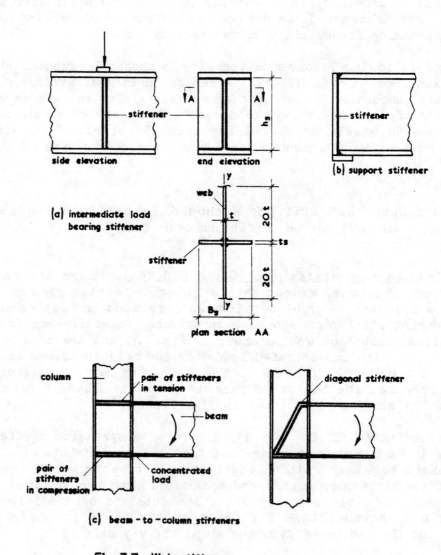

(a) intermediate load bearing stiffener

(b) support stiffener

(c) beam-to-column stiffeners

Fig. 7.7. Web stiffeners.

The stiffener is cut to fit to the web and the flanges of the beam with a corner chamfer to avoid the root radius of the beam.   The bearing stresses at working load between the stiffener and the flange of the beam should not exceed the allowable bearing stresses given in Table 9 BS 449 (Appendix).   The stiffener is welded to the web and flanges of the beam and the web should be capable of resisting the total applied load.

Load bearing stiffeners must be provided at points of concentrated load for plate girders.   The stiffener is generally the full depth of the web, symmetrical about the web and its cross sectional area includes part of the web as described previously.   The method of design is similar to that described for UB and UC sections.

At the supports, load bearing stiffeners for plate girders, which are the sole means of restraint against torsion, must in addition to being designed as a strut, be designed according to Cl. 28a BS 449 so that the second moment of area of the stiffener about the centre line of the web plate

$$I_s \geqslant (D^3 T/250)(R/W)\tag{7.26}$$

7.2.6   _Minimum weight of plate girders_   The depth of plate web girders is made as large as practicable to minimise the area of the flanges required to resist the bending moment.   If the depth is large then a large sectional area of plate is available in the web to resist the shear force and the thickness of the web plate can therefore be kept as small as possible.   Theoretical work on optimisation of the cross sectional area of girders is given in Schilling, but practical solutions need to be modified to take account of stiffeners and labour costs.   The weight of steel in a plate girder is related to the cross sectional area of the girder, which, for a girder symmetrical about the x-x and y-y axes is

$$A = 2BT + d_1 t\tag{7.27}$$

Ignoring the variation of bending stress through the thickness of the flanges, the maximum bending moment about the x-x axis is

$$M_{max} \simeq BTd_1 p_{bc} + p_{bc} t d_1^2/6\tag{7.28}$$

where p_{bc} is the maximum allowable compressive stress.

Combining Equations (7.27) and (7.28) to eliminate BT

$$A = 2M_{max}/d_1 p_{bc} + 2d_1 t/3\tag{7.29}$$

Where web buckling is likely to occur it is convenient to express A in terms of t and d_1/t and Equation (7.29) is modified to

$$A = 2M_{max}/(d_1/t)tp_{bc} + 2(d_1/t)t^2/3 \qquad (7.30)$$

For a particular value of d_1/t the optimum value of A is obtained when $dA/dt = 0$, i.e. when

$$-2M_{max}/[(d_1/t)t^2p_{bc}] + 4(d_1/t)t/3 = 0$$

from which

$$t = [3M_{max}/(2(d_1/t)^2p_{bc})]^{1/3} \qquad (7.31)$$

Combining Equations (7.30) and (7.31) to eliminate t

$$A = 3(2/3)^{1/3}(M_{max}/p_{bc})^{2/3}/(d_1/t)^{1/3} \qquad (7.32)$$

An estimate of the self-weight may be obtained by multiplying A from Equation (7.32) by the length L and the density of the steel ρ . The stiffeners increase this self-weight by approximately 10% but curtailment of the flanges reduces the weight. An approximate first estimate of the self-weight is

$$W_{sw} \simeq 3\rho L(M_{max}/p_{bc})^{2/3}/(d_1/t)^{1/3} \qquad (7.33)$$

Initially the self-weight can be omitted when calculating M_{max}, without significant error.

It is evident from an inspection of Equation (7.33) that the larger the ratio of d_1/t the smaller the self-weight. An inspection of Table 12 BS 449 (Appendix) shows that if the number of stiffeners is to be kept to a minimum to minimise labour costs, then the maximum spacing of $1.5d_1$ should be used. For grades 43 and 50 steel the maximum ratio of d_1/t that allows this is 180. For grade 55 steel the corresponding value is 150. The thickness of the web is therefore determined using Equation (7.31) with this maximum value of d_1/t. The thickness of the web plate then must be chosen from the available thicknesses of 6, 8, 10, 15 or 15 plus 5 mm increments. When one of these values has been selected then the depth d_1 can be determined from the d_1/t ratio.

At this stage of the optimisation process the shear resistance of the web involving the values of d_1 and t should be checked and compared with the external applied shear force Q, i.e.

$$Q \leqslant d_1 t p_q' \qquad (7.34)$$

The value of the allowable average shear stress p_q' is obtained from Table 12 BS 449 (Appendix) for the appropriate d_1/t ratio, assuming maximum spacing of the stiffeners. If Equation (7.34) is not satisfied then the optimisation process must be modified to include this restraint.

The thickness of the flange at the section of maximum bending moment is obtained by combining the maximum outstand ratios of B/2T given in Table 14 BS 449 (Appendix) and Equation (7.28).

$$T = \{[(M_{max}/d_1 p_{bc}) - (td_1/6)]/2(B/2T)\}^{1/2} \tag{7.35}$$

The flange thickness actually used is chosen from the available plate thicknesses and the breadth of the flange B is compared with the B/2T ratio.

Example 7.1 Welded plate girder Design a welded plate girder in grade 50 steel to carry a uniformly distributed load of 2400 kN over an effective span of 18 m as shown in Fig. 7.8. The compression flange is restrained laterally at quarter points.

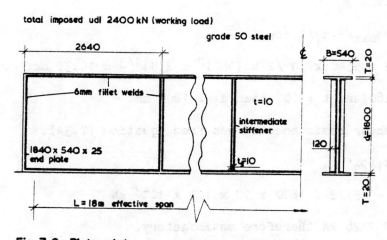

total imposed udl 2400 kN (working load)

Fig. 7·8. Plate girder.

A check on available UBs in the BCSA Structural Steel Handbook shows that the largest section, i.e. a 914 x 419 x 388 kg UB grade 55 steel will support a maximum udl of approximately 1870 kN, i.e. less than 2400 kN, therefore a plate girder is necessary.

The maximum bending moment at centre span from the superimposed load is

WL/8 = 2400 x 18/8 = 5400 kNm

From Equation (7.33) an estimate of the self-weight of the girder assuming maximum values of $d_1/t = 180$ and $p_{bc} = 230$ N/mm^2 is

$$W_{sw} \simeq 3 \rho L(M_{max}/p_{bc})^{2/3}/(d_1/t)^{1/3}$$

$$= [3(7850/10^6)18(5400 \times 10^6/230)^{2/3}/(180)^{1/3}] \times 9.81 \times 10^{-3}$$

$$= 60.42 \text{ kN.} \quad \text{Use 65 kN.}$$

Maximum bending moment at mid-span from the self-weight of girder is

$$W_{sw}L/8 = 65 \times 18/8 = 146 \text{ kNm}$$

Total bending moment at mid-span

$$M_{max} = 5400 + 146 = 5546 \text{ kNm}$$

The web thickness from Equation (7.31) assuming $d_1/t = 180$ and $p_{bc} = 230 \text{ N/mm}^2$ is

$$t = [3M_{max}/2(d_1/t)^2 p_{bc}]^{1/3}$$

$$= [3 \times 5546 \times 10^6/2 \times (180)^2 \times 230]^{1/3} = 10.37 \text{ mm.} \quad \text{Use 10 mm.}$$

If $d_1/t = 180$ and $t = 10$, then $d_1 = 1800$ mm.

Check the shear resistance of web from Equation (7.34).

$$Q < d_1 t p_q'$$

$$(2400 + 65)/2 < 1800 \times 10 \times 107 \times 10^{-3} \text{ kN}$$

1232 < 1926 kN therefore satisfactory.

The flange thickness from Equation (7.35) is

$$T = \{[(M_{max}/d_1 p_{bc}) - (t d_1/6)]/2(B/2T)\}^{1/2}$$

$$= \{[(5546 \times 10^6/1800 \times 230) - (10 \times 1800/6)]/(2 \times 14)\}^{1/2}$$

$$= 19.27 \text{ mm.} \quad \text{Use 20 mm.}$$

Width of flange for $B/2T = 14$ (see Table 14 BS 449 (Appendix)) is $B = 28 \times 19.27 = 539.6$ mm. Use $B = 540$ mm.

Check the maximum allowable bending stress.

Effective length of beam between lateral restraints is

$$l = (18 \times 10^3)/4 = 4500 \text{ mm}$$

The area of the cross section of the girder is

$$A = 2BT + d_1t = 2 \times 540 \times 20 + 1800 \times 10 = 39.6 \times 10^3 \text{ mm}^2$$

The second moment of area of the girder cross section about the y-y axis is

$$I_y = 2TB^3/12 + d_1t^3/12 = 2 \times 20 \times 540^3/12 + 1800 \times 10^3/12$$

$$= 525.03 \times 10^6 \text{ mm}^4$$

The radius of gyration of the girder about the y-y axis is

$$r_y = (I_y/A)^{1/2} = (525.03 \times 10^6/39.6 \times 10^3)^{1/2} = 115.15 \text{ mm}$$

The slenderness ratio for lateral buckling about the y-y axis is

$$l/r_y = 4500/115.15 = 39.08$$

The ratio of overall beam depth to flange thickness is

$$D/T = (d_1 + 2T)/T = (1800 + 2 \times 20)/20 = 92$$

Where flanges have equal second moments of area about the y-y axis of the girder, then from Table 7 BS 449 (Appendix), $C_s = A$ 1762 N/mm^2, and from Table 8 p_{bc} = 230 N/mm^2, which agrees with the value assumed in the initial stages of the design.

Check the actual maximum bending stress. The second moment of area about the x-x axis is

$$I_x = td_1^3/12 + 2BT^3/12 + 2BT[(d_1 + T)/2]^2$$

$$= 10 \times 1800^3/12 + 2 \times 540 \times 20^3/12 + 2 \times 540 \times 20[(1800 + 20)/2]^2$$

$$= 22.75 \times 10^9 \text{ mm}^4$$

The actual bending stress is

$$f = My/I_x = 5546 \times 10^6 \times 920/(22.75 \times 10^9) = 224.3 < 230 \text{ N/mm}^2$$

Check the bending deflection of the beam at mid-span when subject to applied loads

$$\Delta_{max} = 5WL^3/384EI_x$$

$$= 5(2400 \times 10^3) \times (18 \times 10^3)^3/(384 \times 210 \times 10^3 \times 22.75 \times 10^9)$$

$$= 38.14 \text{ mm}$$

The allowable deflection according to Cl. 15 BS 449 is

$$\text{span}/360 = 18 \times 10^3/360 = 50 \ > 38.14 \text{ mm}$$

therefore satisfactory.

The number of intermediate stiffeners spaced at $1.5d_1$, in an overall length of girder L_o is

$$n = 1 + L_o/(1.5d_1) = 1 + 18.5 \times 10^3/(1.5 \times 1800)$$

$$= 7.85 \quad \text{use 8, with a spacing of } (18.5 \times 10^3)/7 = 2642 \text{ mm.}$$

Use spacing s = 2640 mm.

If the thickness of the intermediate stiffener is equal to the web thickness $t_s = 10$ mm, and the outstand is $12t_s = 12 \times 10 = 120$ mm, then the second moment of area of the stiffener about the centre line of the web from Equation (7.20) is

$$I_s = t_s B_s^3/12 + 40t^4/12 = 10 \times 240^3/12 + 40 \times 10^4/12$$

$$= 11.55 \times 10^6 \text{ mm}^4$$

This value is greater than the minimum specified in Cl.28 BS 449 of

$$I_s = 1.5d_1^3t^3/s^2 = 1.5(1800)^3 \times 10^3/2640^2 = 1.26 \times 10^6 \text{ mm}^4$$

A load bearing stiffener is required at the support to resist the end reaction and if the stiffener is the full width of the flange there is less welding required. The outstand of the stiffener should not be greater than $12t_s$ therefore, $t_s = B/(2 \times 12) = 540/(2 \times 12) = 22.5$ mm.   Use 25 mm thick end stiffener as shown in Fig. 7.8.

Check load bearing stiffener at the support as a strut with an effective length $1 = 0.7d_1 = 0.7 \times 1800 = 1260$ mm.   The cross sectional area of the strut is

$$A_s = Bt_s + 20t^2 = 540 \times 25 + 20 \times 10^2 = 15.5 \times 10^3 \text{ mm}^2$$

The second moment of area of the strut about an x–x axis parallel to the web of the beam

$$I_x = t_s B^3/12 + 20t \times t^3/12$$

$$I_x = 25 \times 540^3/12 + 20 \times 10^4/12 = 328.05 \times 10^6 \text{ mm}^4$$

This value is greater than the minimum value specified in Cl. 28a BS 449

$$I = (D^3T/250)(R/W) = (1840^3 \times 20/250)(1/2) = 249.18 \times 10^6 \text{ mm}^4$$

The radius of gyration of the stiffener about its x-x axis is

$$r_y = [I_x/A_s]^{1/2} = [328.05 \times 10^6/15.5 \times 10^3]^{1/2} = 145.48 \text{ mm}$$

The slenderness ratio of the stiffener is

$$l/r_x = 0.7 \times 1800/145.48 = 8.66$$

From Table 17b BS 449 (Appendix) the allowable axial stress for grade 50 steel at this slenderness ratio is $p_c = 210$ N/mm^2. The actual axial stress is $f_c = R/A_s = 0.5(2400 + 65) \times 10^3/(15.5 \times 10^3) = 79.52$ N/mm^2, which is less than 210 N/mm^2 and is therefore satisfactory.

Check the self-weight of the girder originally estimated to be 65 kN. Assume a constant cross section for the girder.

$$W_{sw} = [(2BT + d_1t)L_o + 8B_st_sd_1]\rho$$

$$= [(2 \times 540 \times 20 + 1800 \times 10)18.5 \times 10^3 + 2 \times 540 \times 25 \times \ldots\ldots$$

$$\ldots\ldots 1800 + 6 \times 240 \times 10 \times 1800] \times 7850 \times 10^{-9} \times 9.81 \times 10^{-3}$$

$$= 62.16 < 65 \text{ kN therefore satisfactory.}$$

The size of the fillet welds connecting the web to the flanges is determined as follows. The horizontal shear force per unit length is given by Equation (2.18), i.e. $q_s = QA\bar{y}/I$

$$F_z = QA\bar{y}/I = Q(BT)0.5(d_1 + T)/I_x$$

$$= 0.5(2400 + 65) \times 540 \times 20 \times 0.5(1800 + 20)/(22.75 \times 10^9)$$

$$= 0.532 \text{ kN/mm}$$

From Table 4.3 the shear resistance of two continuous 6 mm fillet welds is $2 \times 0.672 = 1.344$ kN/mm for grade 50 steel which is greater than 0.532 kN/mm and is therefore satisfactory.

The size of the welds connecting the end bearing stiffener to the web is determined by assuming that all of the support reaction is resisted by these welds.

The force per unit length applied to these two welds is therefore

$$F_z = R/d_1 = 0.5(2400 + 65)/1800 = 0.684 \text{ kN/mm}$$

Two 6 mm fillet welds are again satisfactory.

Check the bearing stresses between the load bearing stiffener at the support and the flange. If it is assumed that the support reaction is transferred directly to the end of the load bearing stiffener, then the bearing stress is

$$R/(B_s t_s) = 0.5(2400 + 65) \times 10^3/(540 \times 25) = 91.3 \text{ N/mm}^2$$

This value is less than the allowable bearing stress $p_b = 260$ N/mm^2 given in Table 9 BS 449 (Appendix), and is therefore acceptable.

The size of the welds connecting an intermediate stiffener to the web is determined from the equation given in Cl. 28(b)(iv) in BS 449, which states that the force per unit length applied to the two welds on each component is

$$F_z = t^2/(8h) = 10^2/(8 \times 120) = 0.104 \text{ kN/mm.}$$

where h is the outstand of the stiffener in mm. Two 6 mm fillet welds are again satisfactory.

It is always possible to curtail the flange plates of a plate girder but for this design example it is considered that the saving in steel plate will be balanced by the extra work necessary to make the butt weld. If the span of the girder had been greater then the saving in material would have been more significant.

7.3   COMPOUND BEAMS

A compound beam is a standard beam section which is strengthened by the addition of flange plates, e.g. a UB with additional welded flange plates as shown in Fig. 7.9. Compounding has the effect of increasing the moment of resistance of a section without significantly increasing its overall depth. Compound beams are therefore used where the construction depth is limited.

The approximate self-weight of a steel I beam of constant cross section , maximum overall depth D, flange thickness T, symmetrical about the x-x and y-y axes, and of length L is obtained as follows.

$$W_{sw} = L \rho (2BT + Dt) \tag{7.36}$$

At the section of maximum bending moment where p_{bc} is the maximum allowable bending stress as defined in Cl. 19b BS 449

$$M \simeq [BTD + tD^2/6]p_{bc} \tag{7.37}$$

At the section of maximum shear

$$Q = tDp_q'\qquad(7.38)$$

Combining Equations (7.36), (7.37) and (7.38) to eliminate B,T,D and t

$$W_{SW} \simeq L\,\rho\,[2M/Dp_{bc} + 2Q/3p_q']\qquad(7.39)$$

The advantage of this equation is that the self-weight is expressed in parameters which are known at the commencement of a design problem. The values of M and Q should include the self-weight, but in many cases the self-weight is small in relation to the applied load and the consequent error is not large. An iteration process may be developed if necessary.

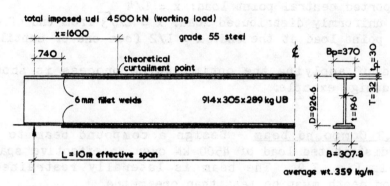

Fig. 7.9. Compound beam.

Over lengths where the additional flange plate is not required to resist the bending moment, it is often curtailed to optimise the self-weight of the beam. The optimisation process is illustrated in the following theory for a simply supported beam carrying a uniformly distributed load. The weight of steel in the compound beam, if x is the distance of the theoretical curtailment point from the end of the span, is

$$W_{SW} = [2BT + 2Bt_p(1 - 2x/L) + Dt]\,\rho\,L\qquad(7.40)$$

Applying M = pZ_e at mid-span

$$wL^2/8 \simeq p_{bc}[(BT + Bt_p)D + tD^2/6]\qquad(7.41)$$

Applying M = pZ at the theoretical curtailment point

$$wx(L - x)/2 \simeq p_{bc}[BTD + tD^2/6]\qquad(7.42)$$

Combining Equations (7.40), (7.41) and (7.42) to eliminate B, T and t_p, and rearranging

$$W_{sw}/(\rho L) = (w/Dp_{bc})(L^2/4 - Lx/2 + 2x^2 - 2x^3/L) + 2Dt/3 \qquad (7.43)$$

Differentiating the self-weight with respect to x and equating to zero to determine the value of x to give the minimum self-weight of the beam

$$-L/2 + 4x - 6x^2/L = 0$$

from which $x/L = 1/6$ $\qquad\qquad (7.44)$

Other useful results for alternative load cases and support conditions for a span of length L are:-

Simply supported central point load: $x = L/4$
Cantilever uniformly distributed load: $x = L/\sqrt{3}$ from end of cantilever
Cantilever point load at the end: $x = L/2$ from end of cantilever.

The method of applying the optimisation process is shown in the following design example.

Example 7.2 Compound beam   Design a compound beam to support a uniformly distributed load of 4500 kN over an effective span of 10 m as shown in Fig. 7.9. The beam is laterally restrained and the construction depth must be less than one metre.

To optimise on the self weight of the beam use grade 55 steel. From BCSA Structural Steel Handbook the maximum udl that can be carried by a 914 x 419 x 388 kg UB is 3498 kN, i.e. less than 4500 kN, therefore a compound beam is required.

The maximum bending moment due to the applied load is

    WL/8 = 4500 x 10/8 = 5625 kNm

Maximum shear force due to the applied load Q = W/2 = 4500/2 = 2250 kN. Approximate self-weight of the beam with a density of steel ρ = 7850 kg/m^3, p_{bc} = 280 N/mm^2 from Table 2 BS 449 (Appendix) and p_q' = 170 N/mm^2 from Table 11 BS 449 (Appendix), is from Equation (7.39)

$$W_{sw} \simeq L\rho\,[2M/Dp_{bc} + 2Q/3p_q']$$

$$= 10 \times 7850 \times 10^{-6}[2 \times 5625 \times 10^6/(1 \times 10^3 \times 280) \ldots$$

$$\ldots + 2 \times 2250 \times 10^3/(3 \times 170)] \times 9.81 \times 10^{-3}$$

$$= 37.74 \text{ kN. Use 40 kN.}$$

Maximum bending moment = WL/8 = 4540 x 10/8 = 5675 kNm

Maximum shear force = W/2 = 4540/2 = 2270 kN

Minimum thickness of web of beam based on shear resistance is

$$t = Q/Dp_q' = 2270 \times 10^3/(1 \times 10^3 \times 170) = 13.35 \text{ mm}$$

Equation (7.44) gives the curtailment position at L/6 for a simply supported beam carrying a udl for a section of optimum weight. The required elastic section modulus of the UB in determined from applying M = pZ at the theoretical curtailment point.

$$Z_{UB} = wx(L - x)/(2p_{bc})$$

$$= 454(10/6)[10 - (10/6)] \times 10^6/(2 \times 280) = 11.26 \times 10^6 \text{ mm}^3$$

Use 914 x 305 x 289 kg UB, Z = 10.891 x 10^6 mm^3, t = 19.6 > 13.35 mm.

Since this is the largest UB available in this series and because Z(actual) ≠ Z(theoretical) the revised position of the theoretical curtailment point is obtained from the following form of the M = pZ_e equation

$$wx(L - x)/2 = p_{bc}Z_{UB}$$

$$454x(10 - x)/2 = 280 \times 10.891$$

from which x = 1.6 m.

The approximate thickness of the additional flange plate is obtained from applying M = pZ at mid-span where the bending moment is a maximum.

$$WL/8 \simeq p_{bc}[Z_{UB} + Bt_pD]$$

$$5675 \times 10^6 \simeq 280[10.891 \times 10^6 + 307.8 \times t_p \times 926.6]$$

$$t_p = 32.9 \text{ mm}$$

Use t_p = 30 mm and B_p = 370 mm.

Try section shown in Fig. 7.9.   Check maximum bending stress f = My/I at mid-span.

The second moment of area of the compound section about the x-x axis is

$$I_x = I_{UB} + 2B_p t_p^3/12 + 2B_p t_p[(D + t_p)/2]^2$$

$$= 5.046 \times 10^9 + 2 \times 370 \times 30^3/12 + 2 \times 370 \times 30[(926.6+30)/2]^2$$

$$= 10.126 \times 10^9 \text{ mm}^4.$$

The maximum bending stress is

$$f = My/I_x = 5675 \times 10^6(986.6/2)/(10.126 \times 10^9) = 276.5 < 280 \text{ N/mm}^2$$

therefore satisfactory.

Determine the size of fillet welds connecting the flange plates to the UB at a section resisting the maximum shear force. The horizontal shear force per unit length is given by Equation (2.18)

$$F_z = QA\bar{y}/I_x$$

$$= (4540/2)370 \times 30 \times 478.3/(10.126 \times 10^9) = 1.19 \text{ kN/mm}$$

The shear resistance of 2 No. 6 mm fillet welds is 2 x 0.819 = 1.638 kN/mm for grade 55 steel as shown in Table 4.3.

The theoretical curtailment point of the plate is 1.6 m from the support. The flange plate is extended beyond this point so that there is sufficient length of fillet weld capable of resisting the force in the flange plate at the theoretical curtailment point.

The bending stress in the plate at the curtailment point from $f = My/I$ is

$$f_p = (wx/2)(L - x)(D/2 + t_p)/I_x$$

$$= 454(1.6/2)(10 - 1.6)10^6(926.6/2 + 30)/(10.126 \times 10^9)$$

$$= 148.6 \text{ N/mm}^2$$

The tensile force in the flange plate is approximately

$$F_p = B_p t_p f_p = 370 \times 30 \times 148.6 \times 10^{-3} = 1649.5 \text{ kN}$$

This force is resisted by the welds at the sides of length l_w and at the end of the flange plate of length B. If 6 mm fillet welds are used

$$F_p = F_{wz}(2l_w + B)$$

$$1649.5 = 0.819(2l_w + 307.8)$$

$$l_w = 853.1 \text{ mm}$$

Practical curtailment point is 1600 - 853.1 = 746.9 mm. Curtail at 740 mm from the support. This reduction in the length of the flange plate is relatively small and the resulting reduction in weight may not be considered to be cost effective.

Total weight of the compound beam is

$$W_{sw} = [m_{UB}L + B_p t_p L_p \rho]g$$

$$= [289 \times 10 + 370 \times 30(10 - 2 \times 0.74)7850 \times 10^{-6}]9.81 \times 10^{-3}$$

$$= 35.63 \text{ kN}.$$

This value is less than the 40 kN assumed in the design and is therefore satisfactory. The average mass per metre is

$$35.24 \times 10^3/(10 \times 9.81) = 359.2 \text{ kg/m}$$

Check the maximum deflection due to bending from the applied load at mid-span assuming the maximum second moment of area is constant for the entire length. The error involved in this assumption is small.

$$\Delta = 5WL^3/384EI$$

$$= 5 \times 4500 \times 10^3 \times (10 \times 10^3)^3/(384 \times 210 \times 10^3 \times .....$$

$$.....10.126 \times 10^9) = 27.6 \text{ mm.}$$

Allowable deflection is L/360 = 10 × 10³/360 = 27.8 > 27.6 mm, therefore satisfactory.

Check end support for web crushing and web buckling as shown in Section 2.6.

7.4   CASTELLATED BEAMS

Castellated beams are fabricated from UB, UC or joist sections by flame cutting the web to form castellations, and then butt welding the teeth as shown in Fig. 7.1(d). The sizes of castellations are standardised as shown in the Tables in the Appendix extracted from the Structural Steel Tables.

This type of fabricated beam is suitable for large spans and light loading where shear forces are not large. The increase in depth that results from the fabrication process increases the strength of the beam in bending, but consideration must be given to the design requirements for shear and deflection. The castellated beam is used where the construction depth is not restricted and where the holes in the web can be used for building services.

7.4.1 *Section for maximum bending stresses*   The maximum bending stress in a castellated beam occurs at one of the openings at or near to the section which is subject to the maximum bending moment. The primary bending stresses are due to the bending moment resisted by the net cross section of the castellated universal beam (CUB). The secondary bending stresses are due to the shear deformation between openings. For a simply supported beam carrying a udl the maximum compressive elastic bending stress at point 2 in section AA Fig. 7.10 at a distance x from the support is

$$f_2 = M_x/Z_c + Q_x(l_0/2)/(2Z_2)$$

$$= wx(L - x)/(2Z_c) + w(L/2 - x)(l_0/2)/(2Z_2) \qquad (7.45)$$

where Z_c and Z_2 are the elastic section moduli for the CUB and the tee section respectively. The value of x for which f_2 is a maximum is obtained by differentiating Equation (7.45) and equating to zero.

$$w(L - 2x)/2Z_c - w(l_0/2)/(2Z_2) = 0$$

rearranging $x = [L - (Z_c/Z_2)(l_0/2)]/2$ \qquad (7.46)

From inspection of the stress diagrams in Fig. 7.10 the maximum bending stress will occur either at point 1 or point 2 at a cross section. The maximum compressive bending stress at point 1 on the stalk of the tee section is

$$f_1 = M_x/Z_c' + Q_x(l_0/2)/(2Z_1)$$

$$= wx(L - x)/(2Z_c') + w(L/2 - x)(l_0/2)/(2Z_1) \qquad (7.47)$$

where Z_c' and Z_1 are the elastic section moduli for the CUB and the tee section respectively at point 1.

The value of x corresponding to Equation (7.46), for which f_1 is a maximum is

$$x = [L - (Z_c'/Z_1)(l_0/2)]/2 \qquad (7.48)$$

The solutions presented in Equations (7.45) to (7.48) are for a udl on a simply supported beam, but solutions for other loading conditions can be determined in the same way.

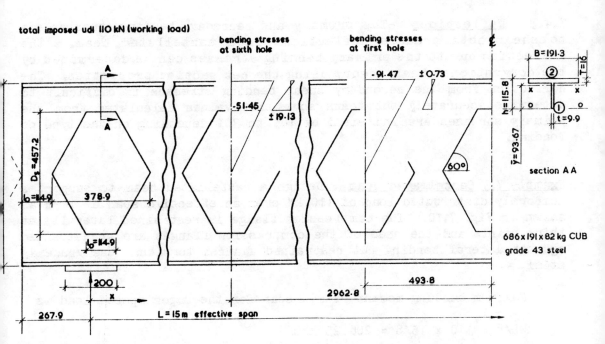

total imposed udl 110 kN (working load)

bending stresses at sixth hole

bending stresses at first hole

-91.47   ± 0.73

-51.45
± 19.13

$D_x = 457.2$

$l_o = 114.9$ 378.9

$l_o = 114.9$

200

267.9   L = 15 m effective span

2962.8   493.8

60°

B = 191.3
② T = 16
h = 115.8
$\bar{y} = 93.67$ x—x   o——o   t = 9.9

section AA

686 x 191 x 82 kg CUB
grade 43 steel

Fig. 7·10. Castellated beam.

The maximum allowable compressive bending stress is controlled by lateral buckling of the beam. Experiments on castellated UB's by Nethercot show that Table 3 values in BS 449 (Appendix) give safety factors as low as 1.3, but Tables 7 and 8 BS 449 (Appendix) generally used for plate girders give safety factors of approximately 1.7 and are therefore to be preferred.

7.4.2 Shear stresses The holes in the web increase the magnitude of the shear stresses. The simplest method of calculating the shear stress is to modify the maximum shear stress formula to allow for the reduction in area of web. The maximum shear stress at the neutral axis x-x of the CUB assuming that the maximum shear stress is proportional to the net cross sectional area of the web is

$$f_q = (QA\bar{y}/I_g t)(1.08 D_s/l_o) \qquad (7.49)$$

where I_g is the gross second moment of area of the castellated beam about the x-x axis of the cross section, D_s is the depth of the hole in the web, $1.08 D_s$ is the longitudinal spacing of the holes in the web and l_o is the length of web between holes (as shown in Fig. 7.10).

7.4.3 Deflections The primary and secondary bending stresses due to shear both produce deflections in a castellated beam. The deflection due to the primary bending stresses can be determined by the conventional beam theory using the net section properties. The deflection from the secondary shear bending stresses is difficult to determine accurately but tests show that values calculated from the primary stresses are increased by 10% to 30% depending on the type of loading.

Example 7.3 Castellated beam Design a castellated beam to support a uniformly distributed load of 110 kN over an effective span of 15 m as shown in Fig. 7.10. The compression flange is restrained laterally at third points and the ends of the compression flanges are unrestrained against lateral bending but restrained against torsion. Use grade 43 steel.

Maximum bending moment at mid-span for the superimposed load is

$$WL/8 = 110 \times 15/8 = 206.25 \text{ kNm}$$

If the maximum bending stress is applicable to grade 43 steel then from $M = pZ_e$

$$Z_{min} = M/p_{bc} = 206.25 \times 10^6/165 = 1.25 \times 10^6 \text{ mm}^3$$

and from Table 2.3 the overall depth to limit bending stresses and deflection is

$$D = L/17 = 15 \times 10^3/17 = 882 \text{ mm}$$

Using these two limits as a guide, from the Structural Steel Tables (Appendix) try 686 x 191 x 82 kg CUB. The self-weight of the beam is $82 \times 15 \times 9.81 \times 10^{-3} = 12$ kN.

The radius of gyration for the net cross section of the CUB about the y-y axis is

$$r_y = [I_{net}/A_{net}]^{1/2} = [18.7 \times 10^6/8.19 \times 10^3]^{1/2} = 47.78 \text{ mm}$$

The slenderness ratio for lateral torsional buckling is

$$l/r_y = (L/3)/r_y = (15000/3)/47.78 = 104.6$$

The ratio of overall depth to flange thickness is

$$D_c/T = 688.8/16 = 43.05$$

From BS 449 Table 7 (Appendix), $A \simeq 292$, and from Table 8, $p_{bc} \simeq 95$ N/mm².

The actual maximum bending stress in the outer fibres of the top flange is part due to the bending stress for the full depth of the castellated beam, and part due to secondary bending stresses due to shear in the tee section of the flange.

The properties of the tee section shown in Fig. 7.10 section AA are obtained as follows.

Taking moments of areas about the axis O-O to determine the position of the centroidal axis x-x for the tee section

$$BT(h - T/2) + t(h - T)^2/2 = [BT + t(h - t)]\bar{y}$$

$$191.3 \times 16(115.8 - 16/2) + 9.9(115.8 - 16)^2/2$$

$$= [191.3 \times 16 + 9.9(115.8 - 16)]\bar{y}$$

$$\bar{y} = 93.67 \text{ mm}$$

The second moment of area of the tee section about its own centroidal axis x-x is

$$I_x = t\bar{y}^3/3 + t(h - T - \bar{y})^3/3 + BT^3/12 + BT(h - \bar{y} - T/2)^2$$

$$= 9.9 \times 93.67^3/3 + 9.9(115.8 - 16 - 93.67)^3/3 \ldots\ldots$$

$$\ldots\ldots + 191.3 \times 16^3/12 + 191.3 \times 16(115.8 - 93.67 - 16/2)^2$$

$$= 3.389 \times 10^6 \text{ mm}^4$$

The elastic section moduli about the x-x axis are

$$Z_1 = I_x/\bar{y} = 3.389 \times 10^6/93.67 = 36.18 \times 10^3 \text{ mm}^3$$

$$Z_2 = I_x/(h - \bar{y}) = 3.389 \times 10^6/(115.8 - 93.67) = 153.14 \times 10^3 \text{ mm}^3$$

The distance x from the support where the maximum compressive bending stress f_2 occurs is obtained from Equation (7.46)

$$x = [L - (Z_c/Z_2)(l_o/2)]/2$$

$$= [15 - (2.49 \times 10^6/153.14 \times 10^3)(0.1149/2)]/2 = 7.05 \text{ m}$$

The first castellation is at $x = L/2 - 1.08D_s = 7.5 - 0.4938 = 7.006 \text{ m}$.

The maximum bending stress f_2 is obtained from Equation (7.45)

$$f_2 = wx(L - x)/(2Z_c) + w(L/2 - x)(l_o/2)/(2Z_2)$$

$$= (122/15)7006(15000 - 7006)/(2 \times 2.49 \times 10^6) + \ldots\ldots$$

$$\ldots\ldots(122/15)(15000/2 - 7006)(114.9/2)/(2 \times 153.14 \times 10^3)$$

$$= 91.47 + 0.73 = 92.20 < 95 \text{ N/mm}^2$$

therefore satisfactory. The stress of 92.2 N/mm^2 is only slightly greater than the value of 91.87 N/mm^2 that occurs at mid-span for a udl, which would be of sufficient accuracy for most design purposes. The bending tensile stresses in the bottom flange are less than $p_{bt} = 165$ N/mm^2 in this example and are therefore not critical.

The maximum compressive stress at the bottom of the tee section may occur at a different cross section of the beam. The distance x from the end of the beam is obtained from Equation (7.48). The value of Z_c' to be used in this equation is

$$Z_c' = I_c/(D_c/2 - h) = 857.6 \times 10^6/(344.4 - 115.8)$$

$$= 3.752 \times 10^6 \text{ mm}^3$$

From Equation (7.48) the distance of the critical section from the support

$$x = [L - (Z_c'/Z_1)(l_0/2)]/2$$

$$= [15 - (3.752 \times 10^6/36.18 \times 10^3)(0.1149/2)]/2$$

$$= 4.521 \text{ m from the support.}$$

Check the bending stress at the bottom of the tee section at a section at the centre line of the sixth castellation from centre span, i.e. x $= 7.5 - 6 \times 0.4938 = 4.537$ m from the support.

$$f_1 = wx(L - x)/(2Z_c') + w(L/2 - x)(l_0/2)/(2Z_1)$$

$$= (122//15)4537(15000 - 4537)/(2 \times 3.752 \times 10^6) + \ldots\ldots$$

$$\ldots(122/15)(15000/2 - 4537)(114.9/2)/(2 \times 36.18 \times 10^3)$$

$$= 51.45 + 19.13 = 70.58 \text{ N/mm}^2$$

This value is less than $f_{bc} = 92.2$ N/mm^2 calculated previously for the first castellation and is therefore not critical.

The maximum shear stress at the centroidal axis of the castellated section at the end of the beam is obtained from Equation (7.49).

The first moment of area of the tee section about the x-x axis of the CUB in Fig. 7.10 is

$$A\bar{y} = BT(D_c - T)/2 + t(D_c/2 - T)^2/2$$

$$= 191.3 \times 16(688.8 - 16)/2 + 9.9(344.4 - 16)^2/2$$

$$= 1.563 \times 10^6 \ mm^3$$

The gross second moment of area for the CUB section about the x-x axis is

$$I_x = I_{net} + tD_s^3/12 = 857.6 \times 10^6 + 9.9 \times 457.2^3/12$$

$$= 936.4 \times 10^6 \ mm^4$$

From Equation (7.49) the maximum shear stress

$$f_q = (QA\bar{y}/I_x t) \times (1.08D_s/l_o)$$

$$= [(122/2) \times 10^3 \times 1.563 \times 10^6/(936.4 \times 10^6 \times 9.9)] \times ..$$

$$...(1.08 \times 457.2/114.9) = 44.19 \ N/mm^2.$$

This is less than the maximum allowable value of $p_q = 115 \ N/mm^2$ given in BS 449 Table 10 (Appendix).

The maximum deflection for the applied load at mid-span from bending of the net cross section of the CUB about the x-x axis is

$$= 5WL^3/(384EI_x)$$

$$= 5 \times 110 \times 10^3 \times (15 \times 10^3)^3/(384 \times 210 \times 10^3 \times 857.6 \times 10^6)$$

$$\Delta = 26.84 \ mm.$$

This value will be increased by between 10% and 30% by the shear deformation of the castellation. If 30% is assumed $\Delta = 1.3 \times 26.84 = 34.89$ mm. The allowable value is $L/360 = 15 \times 10^3/360 = 41.67 > 34.89$ mm, therefore satisfactory.

The application of concentrated point loads at or near the centre line of holes is to be avoided and the support reaction should be distributed where possible, or the holes filled in.

The web bearing strength at the support from the method described in Section 2.6 is

$$P_b = [b_1 + k(T + r_b)\cot 30^\circ]tp_b$$

$$= [200 + 2(16 + 10.2)\sqrt{3}]9.9 \times 190 \times 10^{-3} = 546.9 \text{ kN.}$$ This is greater than the end shear force of 61 kN.

The web buckling strength at the support is related to the slenderness ratio of the web of the CUB as described in Section 2.6

$$l/r = \sqrt{3}d_c/t = \sqrt{3} \times 636.5/9.9 = 111.4$$

From BS 449 Table 17 (Appendix), $p_c = 68 \text{ N/mm}^2$

The buckling strength using the length of web between holes is

$$P_c = l_o tp_c = 114.9 \times 9.9 \times 68 \times 10^{-3} = 77.35 \text{ kN} > 61 \text{ kN.}$$

7.5   TAPERED BEAMS

A tapered beam is generally one where the depth varies linearly along the span and the areas of the flanges are constant. Tapered beams can be built-up from plates, or a UB can be flame cut at an angle to the centre line of the web and rewelded as shown in Fig. 7.1(e).

A tapered beam is useful for, reducing the depth of a beam at points of low bending moment, economising in material, providing useful shallow gradients to construction, e.g. roof slopes, and improving the aesthetic appearance. The reduction in self-weight of the beam however may be offset by the extra cost involved in manufacture.

The design of a tapered beam for a simple static structure is given in Section 7.5.1. The method of analysis of more complicated frameworks which include tapered sections is given by Just.

7.5.1   Minimum weight of tapered beams   The weight of steel in a tapered beam is related to the cross sectional area of the beam. The flange area is often constant and it is only the depth of the web that varies linearly as shown in Fig. 7.11. The depth of the beam at any cross section for design is controlled by the allowable bending stress, allowable shear stress and the deflection of the beam.

For a simply supported tapered beam carrying a central point load the design process is relatively simple. As can be seen from Fig. 7.11(a) the variation of the bending moment along the span is linear. The moment of resistance of the beam is approximately proportional to the depth of the beam which also varies linearly. The bending moment diagram and the depth are only coincident at centre span. The area of the flanges and the depth of the cross section at centre span are therefore determined from the bending moment at mid-span. The depth of the beam section at the support is determined from the shear stresses, bearing stresses, and buckling stresses in the web. The deflection of the beam must be checked and because the depth varies over the length a summation technique, such as the Mohr area-moment method, must be used if an accurate value is required.

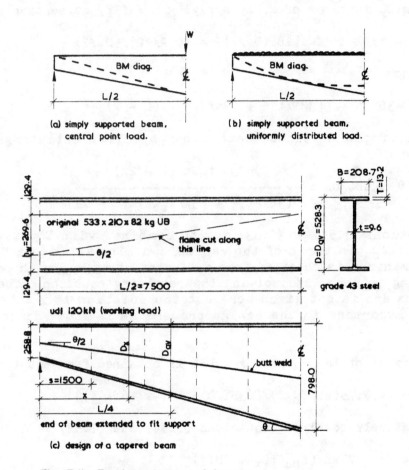

Fig. 7·11. Fabricated tapered beams.

The design process for a minimum weight beam of varying depth, simply supported and carrying a udl is more complicated. In this situation the maximum bending stress does not occur at mid-span as shown in Fig. 7.11(b), and it is necessary to develop a theory to determine the cross section at which this occurs for a minimum weight design. The self-weight of a tapered girder of average depth D_{av} is

$$W_{sw} = (2BT + D_{av}t)L\rho \tag{7.50}$$

It is convenient to express D_{av} in terms of D_x and $\tan\theta$. From Fig. 7.11(c)

$$\tan\theta = (D_{av} - D_x)/(L/4 - x)$$

rearranging $D_{av} = D_x + (L/4 - x)\tan\theta$ \hfill (7.51)

Expressing D_x in terms of x, by applying M = fI/y at section x.

$$wx(L - x)/2 \simeq f_{bc}[2BT(D_x/2)^2 + tD_x^3/12]/(D_x/2)$$

rearranging

$$D_x = -3BT/t + [(3BT/t)^2 + (3w/f_{bc}t)x(L - x)]^{1/2} \qquad (7.52)$$

The value of $\tan\theta$ can be determined by differentiating D_x with respect to x

$$\tan\theta = dD_x/dx = \frac{0.5(3w/f_{bc}t)(L - 2x)}{[(3BT/t)^2 + (3w/f_{bc}t)x(L - x)]^{1/2}} \qquad (7.53)$$

Equations (7.51) to (7.53) may be combined with Equation (7.50) to express W_{sw} in terms of the variable x. The value of x for which W_{sw} is a minimum is obtained from differentiating W_{sw} with respect to x, equating to zero, and solving the resulting equation. The algebra is tedious and is not given here but the solution is x = L/4. This value is important in the design process as is shown in the design example.

The value of $\tan\theta$ when x = L/4 is obtained from Equation (7.53)

$$\tan\theta = 0.75(wL/f_{bc}t)/[(3BT/t)^2 + (9w/16f_{bc}t)]^{1/2} \qquad (7.54)$$

or alternatively combining Equations (7.54) and (7.52)

$$\tan\theta = 0.75wL/[f_{bc}(tD_x + 3BT)] \qquad (7.55)$$

This value of $\tan\theta$ is also used in the design example.

Example 7.4 Tapered Beam Design a roof beam simply supported over a 15 m span and carrying an imposed udl of 120 kN as shown in Fig. 7.11. Use grade 43 steel, and assume the beam is restrained against lateral torsional buckling. Determine the size of a parallel flange beam and compare with a tapered beam.

If a UB section with parallel flanges is used the design is governed by deflection and not the bending stress. For a beam of uniform section the second moment of area required for an applied load deflection limit of L/360 = 15000/360 = 41.67 mm is

$$I = 5WL^3/(384E\Delta) = 5 \times 120 \times 15^3 \times 10^9/(384 \times 210 \times 41.67)$$

$$= 602.6 \times 10^6 \text{ mm}^4$$

From Steel Section Tables (Appendix) a 533 x 210 x 101 kg UB, the elastic section properties $I_{ex} = 616.6 \times 10^6$ mm^4 and $Z_{ex} = 2.298 \times 10^6$ mm^3. The self-weight of the beam = mLg = 101 x 15 x 9.81 x 10^{-3} = 14.86 kN.

Check the maximum bending stress at mid-span

$$f = M/Z_{ex} = (WL/8)/Z_{ex} = (120 + 14.86)15 \times 10^6/(8 \times 2.298 \times 10^6)$$

$$= 110 < 165 \text{ N/mm}^2 \text{ therefore satisfactory.}$$

Alternatively if a tapered section is used then the required second moment of area is approximately three quarters of the value for a uniform section

$$I \simeq 0.75 \times 602.6 \times 10^6 = 452 \times 10^6 \text{ mm}^4$$

Try a 533 x 210 x 82 kg UB, and from tables $I = 474.91 \times 10^6$ mm^4, $Z_{ex} = 1.798 \times 10^6$ mm^3. The self-weight of the beam = mLg = 82 x 15 x 9.81 x 10^{-3} = 12.07 kN. This is the section required at the quarter span as shown in the theory.

The maximum bending stress from applied load and self-weight at quarter span is

$$f_{bc} = M/Z_{ex} = wx(L - x)/(2Z_{ex}) = 0.75WL/(8Z_{ex})$$

$$= 0.75(120 + 12.07)15 \times 10^6/(8 \times 1.798 \times 10^6) = 103.3 \text{ N/mm}^2.$$

The slope of the tapered section from Equation (7.55) is

$$\tan \theta = 0.75W/[f_{bc}(tD_x + 3BT)]$$

Inserting $D_x = D_{av} = D$ from Structural Steel Tables for UB's in Appendix

$$\tan \theta = 0.75(120 + 12.07) \times 10^3/[103.3(9.6 \times 528.3 + 3 \text{ x}...$$

$$... 208.7 \times 13.2)]$$

$$= 0.07190$$

The slope of the cut in the web of the UB is $\theta/2$ and the rise h_w in Fig. 7.11(c) is

$$h_w = (L/2)\tan(\theta/2) = (15000/2)(0.07190/2) = 269.6 \text{ mm}$$

Check whether intermediate stiffeners are required at centre-span.

$d_1/t = (798 - 2 \times 13.2)/9.6 = 80.4 < 85$ therefore no stiffeners required according to Cl. 27f BS 449 (see Section 7.2.2). Check the average shear stress in the web at the support.

$$f_q' = Q/(Dt) = (W/2)/(Dt) = (120 + 12.07)10^3/(2 \times 258.8 \times 9.6)$$

$$= 26.58 < 100 \text{ N/mm}^2, \text{ for allowable values see Table 11 BS 449.}$$

The mid-span deflection is determined by applying the Mohr area-moment method to the elements of equal length s = 1500 mm shown in Fig. 7.11. At each value of x = 750, 2250, 3750, 5250, 6750 mm it is necessary to calculate the depth of the section.

$$D_x = 258.8 + x(798 - 258.8)/7500$$

The mean second moment of area of each section

$$I_x = tD_x^3/12 + 2(B - t)T(D_x - T)^2/4 + 2(B - t)T^3/12$$

$$= 9.6D_x^3/12 + 2(208.7 - 9.6)13.2(D_x - 13.2)^2/4 \ldots$$

$$\ldots + 2(208.7 - 9.6)13.2^3/12$$

The imposed loading bending moment at each section is

$$M_x = wx(L - x)/2 = 120 \times (15000 - x)10^3/(2 \times 15000)$$

$$= 4x(15000 - x) \text{ in Nmm units.}$$

These values are tabulated and the deflection is determined from

$$\Delta = (s/E) \quad \Sigma (M_x x/I_x)$$

x mm	D_x mm	I_x mm^4	M_x Nmm	$M_x x/I_x$
750	312.7	142.4 x 10^6	42.75 x 10^6	225.16
2250	420.6	277.7 x 10^6	114.75 x 10^6	929.73
3750	528.4	466.9 x 10^6	168.75 x 10^6	1355.35
5250	636.2	716.1 x 10^6	204.75 x 10^6	1501.10
6750	744.1	1031.7 x 10^6	222.75 x 10^6	1457.36

$$\Sigma(M_x x/I_x) = 5468.7$$

The theoretical central deflection due to the applied udl of 120 kN is

$$\Delta = (s/E)\sum (M_x x/I_x) = (1500/210 \times 10^3)5468.7 = 39.06 \text{ mm}$$

The allowable value is

$$\Delta = L/360 = 15000/360 = 41.67 > 39.06 \text{ mm therefore satisfactory.}$$

The bearing stresses and buckling at the support should be checked as shown in Section 2.6.

It should be noted that the size of beam determined in the previous calculation was controlled by the deflection at mid-span and not the maximum bending stress at quarter-span. If this beam size is controlled only by the maximum bending stress and not the deflection, then the design method is as follows.

The size of a uniform section is determined from the maximum bending moment at centre span.

$$Z = M/p = (WL/8)/p_{bc} = (120 + 12.07)15 \times 10^6/(8 \times 165)$$

$$= 1.5 \times 10^6 \text{ mm}^3$$

Use 533 x 210 x 82 kg UB, from Structural Steel Tables (Appendix) Z_{ex} = 1.798 x 10^6 mm^3

The size of a tapered section based on the bending moment at quarter span is

$$Z_{ex} = (0.75WL/8)/p_{bc} = 0.75(120 + 8.83)15 \times 10^6/(8 \times 165)$$

$$= 1.098 \times 10^6 \text{ mm}^3$$

Use 457 x 152 x 60 kg UB, from Structural Steel Tables (Appendix) Z_{ex} = 1.12 x 10^6 mm^3.

A further reduction in the self-weight of the beam can be obtained by an alternative design using a tapered girder built-up from plates with a minimum thickness of web of 6 mm but vertical stiffeners may be required at centre span.

REFERENCES

Bleich, F. (1952) - Buckling Strength of Metal Structures, McGraw-Hill.

Just, D.J. (January 1977) - Plane frameworks of tapering box and I section, ASCE (Struct), V103.

Kerensky, O.A., Flint, A.R. and Brown, W.C. (August 1956) - The basis for design of beams and plate girders in the revised BS 153, Proc. ICE, Vol 5, No 2, pp 396-461.

Longbotton, E. and Heyman, J. (1956) - Experimental verification of the strengths of plate girders designed in accordance with the revised BS 153, Proc. ICE., Pt III, V5.

Moore, R.L. (1942) - An investigation of the effectiveness of stiffeners on the shear-resistance of plate girder webs, Nat. Adv. Com. Aero, Tech Note 862.

Nethercot, D.A. and Kerdal, D. (September 1982) - Lateral torsional buckling of castellated beams, Str. Eng., V60B, No 3.

Rockey, K.C., Valtinat, G. and Tang, K.H. (December 1981) - The design of transverse stiffeners on webs loaded in shear - an ultimate load approach, Proc. ICE, Pt 2, V71.

Schilling, C.G. (December 1974) - Optimum proportions for I shaped beams, ASCE, 100, ST12.

Southwell, R.U. and Skan, S.W. (1924) - On the stability under shear from shear forces of a flat elastic strip, Proc. Roy. Soc, Ser. A, V105, 582.

Timoshenko, A. (1945) - Theory of bending, torsion and buckling of thin-welded members of open cross section, J. Franklin Inst. V 239.

Trahair, N.S. (1977) - The behaviour of design of steel structures Chapman and Hall.

Winter, G. (December 1941) - Lateral stability of unsymmetrical I beams and trusses in bending, Proc. ASCE, Vol 67, p 1851.

Appendix Tables

TABLE 1. ALLOWABLE EQUIVALENT STRESS p_e

NOTE. The increases permitted by Clauses 7 and 13 do not apply to these values.

Form	Grade	Thickness or diameter	p_e
		mm	N/mm^2
Rolled I-beams and channels	43	All	230
Plates, bars, Universal beams and columns, and sections other than above	43	Up to and including 40	230
		Over 40	210
Plates, sections and bars	50	Up to and including 65	320
		Over 65	$Y_s/1.1$
Plates, sections and bars	55	Up to and including 40	390
		Over 40	360
Hot rolled hollow sections	43	All	230
,, ,, ,, ,,	50	All	320
,, ,, ,, ,,	55	All	390

where Y_s = yield stress agreed with manufacturer, with a maximum value of 350 N/mm^2.

BS 449 : Part 2 : 1969

TABLE 2. ALLOWABLE STRESS p_{bc} OR p_{bt} IN BENDING
(See also Clauses 19 and 20, and Tables 3a, 3b and 3c)

Form	Grade	Thickness of material	p_{bc} or p_{bt}
		mm	N/mm^2
Rolled I-beams and channels	43	All	165
Compound girders composed of rolled I-beams or channels plated, with thickness of plate	43	Up to and including 40	165
		Over 40	150
Plates, flats, rounds, squares, angles, tees and any sections other than above	43	Up to and including 40	165
		Over 40	150
Plate girders with single or multiple webs	43	Up to and including 40	155
		Over 40	140
Universal beams and columns	43	Up to and including 40	165
		Over 40	150
Plates, flats, rounds, squares and other similar sections, rolled I-beams, double channels forming a symmetrical I-section which acts as an integral unit, compound beams composed of rolled I-beams or channels plated, single channels, angles and tees	50	Up to and including 65	230
		Over 65	$Y_s/1.52$
	55	Up to and including 40	280
		Over 40	260
Plate girders with single or multiple webs	50	Up to and including 65	215
		Over 65	$Y_s/1.63$
	55	Up to and including 40	265
		Over 40	245
Hot rolled hollow sections	43	All	165
,, ,, ,, ,,	50	All	230
,, ,, ,, ,,	55	All	280
Slab bases	All steels		185

where Y_s = yield stress agreed with manufacturer, with a maximum value of 350 N/mm^2.

28

TABLE 3a. ALLOWABLE STRESS p_{bc} IN BENDING (N/mm²) FOR BEAMS OF GRADE 43 STEEL

l/r_y	_____			D/T				
	10	15	20	25	30	35	40	50
90	165	165	165	165	165	165	165	165
95	165	165	165	163	163	163	163	163
100	165	165	165	157	157	157	157	157
105	165	165	160	152	152	152	152	152
110	165	165	156	147	147	147	147	147
115	165	165	152	141	141	141	141	141
120	165	162	148	136	136	136	136	136
130	165	155	139	126	126	126	126	126
140	165	149	130	115	115	115	115	115
150	165	143	122	104	104	104	104	104
160	163	136	113	95	94	94	91	94
170	159	130	104	91	85	82	82	82
180	155	124	96	87	80	76	72	71
190	151	118	93	83	77	72	68	63
200	147	111	89	80	73	68	64	59
210	143	105	87	77	70	65	61	55
220	139	100	84	74	67	63	58	52
230	134	95	81	71	64	59	55	49
240	130	92	78	69	61	56	52	47
250	126	90	76	66	59	54	50	44
260	122	88	74	64	57	52	48	42
270	118	86	72	63	55	50	46	40
280	114	84	70	60	53	48	44	39
290	110	82	68	58	51	46	42	37
300	106	80	66	56	49	44	41	36

Intermediate values may be obtained by linear interpolation.

NOTE. For materials over 40 mm thick the stress shall not exceed 150 N/mm².

TABLE 3b. ALLOWABLE STRESS p_{bc} IN BENDING (N/mm²) FOR BEAMS OF GRADE 50 STEEL

l/r_y	_____			D/T				
	10	15	20	25	30	35	40	50
80	230	230	230	230	230	230	230	230
85	230	230	230	227	227	227	227	227
90	230	230	228	220	220	220	220	220
95	230	230	222	212	212	212	212	212
100	230	230	215	204	204	204	204	204
105	230	226	209	196	196	196	196	196
110	230	221	203	188	188	188	188	188
115	230	216	196	181	181	181	181	181
120	230	211	190	173	173	173	173	173
130	230	202	177	157	157	157	157	157
140	225	193	165	142	142	142	142	142
150	219	183	152	126	126	126	126	126
160	213	174	139	112	110	110	110	110
170	207	165	126	106	97	94	94	94
180	201	155	114	101	91	85	80	77
190	195	146	109	96	86	80	75	68
200	189	136	104	91	82	75	70	64
210	183	127	100	87	77	71	66	60
220	177	118	96	83	74	67	63	56
230	171	112	92	79	70	64	59	53
240	165	108	89	76	67	61	56	50
250	159	105	86	73	64	58	53	47
260	153	102	83	70	62	56	51	45
270	147	99	80	68	59	53	49	43
280	141	96	77	65	57	51	47	41
290	135	93	75	63	55	49	45	39
300	129	90	72	61	53	47	43	37

Intermediate values may be obtained by linear interpolation.

NOTE. For materials over 65 mm thick the stress shall not exceed $\frac{y_s}{1.52}$ N/mm², where

y_s = yield stress agreed with manufacturer, with a maximum value of 350 N/mm².

TABLE 3c. ALLOWABLE STRESS p_{bc} IN BENDING (N/mm²) FOR BEAMS OF GRADE 55 STEEL

l/r_y	_____			D/T				
	10	15	20	25	30	35	40	50
75	280	280	280	280	280	280	280	280
80	280	280	280	277	277	277	277	277
85	280	280	277	267	267	267	267	267
90	280	280	269	257	257	257	257	257
95	280	278	261	247	247	247	247	247
100	280	272	253	238	238	238	238	238
105	280	266	245	228	228	228	228	228
110	280	260	237	218	218	218	218	218
115	280	254	228	208	208	208	208	208
120	280	248	220	198	198	198	198	198
130	273	236	204	178	178	178	178	178
140	266	224	188	158	158	158	158	158
150	258	212	171	138	138	138	138	138
160	250	200	155	121	119	119	119	119
170	243	188	139	114	103	99	99	99
180	235	176	123	107	97	89	84	89
190	227	164	117	101	91	83	78	71
200	219	152	111	96	86	78	73	66
210	212	140	106	91	81	74	69	62
220	204	128	102	87	77	70	65	58
230	196	120	97	83	73	66	61	54
240	189	116	94	79	70	63	58	51
250	181	112	90	76	67	60	55	49
260	173	108	87	73	64	57	53	46
270	165	105	84	70	61	55	50	44
280	158	101	81	68	59	53	48	42
290	150	98	78	65	57	51	46	40
300	142	95	76	63	55	49	44	38

Intermediate values may be obtained by linear interpolation.

NOTE. For materials over 40 mm thick the stress shall not exceed 260 N/mm².

TABLE 5. VALUES OF K_1

N	1·0	0·9	0·8	0·7	0·6	0·5	0·4	0·3	0·2	0·1	0·0
K_1	1·0	1·0	1·0	0·9	0·8	0·7	0·6	0·5	0·4	0·3	0·2

NOTE. Where the value of N calculated for the compression flange alone is smaller than that when both flanges are combined, this smaller value of N should be used.

TABLE 6. VALUES OF K_2

M	1·0	0·9	0·8	0·7	0·6	0·5	0·4	0·3	0·2	0·1	0·0
K_2	0·5	0·4	0·3	0·2	0·1	0·0	−0·2	−0·4	−0·6	−0·8	−1·0

TABLE 7. VALUES OF A AND B FOR CALCULATING C_s

where
$$A = \left(\frac{1675}{l/r_y}\right)^2 \Big/ \sqrt{\left\{1 + \frac{1}{20}\cdot\left(\frac{lT}{r_y D}\right)^2\right\}}$$
$$B = \left(\frac{1675}{l/r_y}\right)^2$$

l/r_y	D/T 8	10	12	14	16	18	20	25	30	35	40	50	60	80	100	B
40	2630	2353	2187	2081	2009	1958	1921	1862	1830	1810	1797	1781	1773	1764	1761	1754
45	2226	1965	1808	1706	1637	1587	1551	1494	1461	1442	1429	1413	1405	1396	1392	1385
50	1929	1683	1534	1436	1369	1321	1286	1229	1198	1178	1165	1150	1142	1133	1129	1122
55	1701	1470	1328	1235	1170	1123	1089	1034	1002	983	970	955	947	938	934	927
60	1522	1304	1169	1079	1017	972	938	884	854	835	822	807	799	790	786	779
65	1377	1172	1043	957	897	854	821	768	738	719	707	692	683	675	671	664
70	1258	1064	941	859	801	759	727	676	646	627	615	600	592	583	580	573
75	1158	974	857	778	723	682	651	601	571	553	541	526	518	510	506	499
80	1074	898	787	711	658	618	588	539	510	492	480	466	457	449	445	438
85	1001	834	727	655	603	565	536	488	460	442	430	415	407	399	395	388
90	938	778	676	607	557	520	491	445	417	400	388	373	365	357	353	346
95	882	730	632	565	517	481	454	408	381	364	352	338	330	322	318	311
100	833	687	593	529	482	447	421	376	350	333	321	307	299	291	287	281
110	750	616	529	469	425	393	368	325	300	283	272	258	251	243	239	231
120	682	558	477	421	380	350	326	286	261	246	235	221	213	206	202	195
130	626	510	435	383	344	315	293	255	231	216	205	192	184	177	173	166
140	578	470	400	351	315	287	266	229	207	192	182	169	161	154	150	143
150	537	436	370	324	290	264	243	209	187	173	163	150	143	135	132	125
160	502	407	345	301	268	244	225	191	171	157	147	135	128	120	116	110
170	471	382	322	281	250	227	208	177	157	143	134	122	115	107	104	97
180	444	359	303	264	234	212	195	164	145	132	123	111	104	97	93	87
190	420	339	286	248	221	199	182	153	135	122	113	102	95	88	84	78
200	398	321	271	235	208	188	172	144	126	114	105	94	87	80	77	70
210	379	305	257	223	197	178	162	135	118	106	98	87	81	74	70	64
220	361	291	245	212	187	169	154	128	111	100	92	81	75	68	65	58
230	345	278	233	202	179	161	146	121	105	94	86	76	70	63	60	53
240	330	266	223	193	170	153	139	115	100	89	82	71	65	59	55	49
250	317	255	214	185	163	146	133	110	95	85	77	67	61	55	51	45
260	304	245	205	177	156	140	128	105	91	80	73	64	58	51	48	42
270	293	236	197	170	150	135	122	101	86	77	70	60	55	48	45	38
280	282	227	190	164	145	130	118	97	83	73	66	57	52	45	42	34
290	272	219	183	158	139	125	113	93	79	70	64	55	49	43	40	33
300	263	211	177	153	134	120	109	89	76	67	61	52	47	41	38	31

TABLE 8. ALLOWABLE COMPRESSION STRESS p_{bc} IN BENDING FOR DIFFERENT VALUES OF C_s (but *see* Table 2)

C_s	p_{bc} for grade 43 steel		p_{bc} for grade 50 steel	p_{bc} for grade 55 steel	
	Up to and including 40 mm	Over 40 mm	Up to and including 65 mm	Up to and including 40 mm	Over 40 mm
N/mm²	N/mm²	N/mm²	N/mm²	N/mm²	N/mm²
20	11	11	11	11	11
30	16	16	16	17	17
40	21	21	22	22	22
50	26	25	27	27	27
60	30	30	32	32	32
70	35	34	37	37	37
80	39	38	41	42	42
90	43	42	46	47	47
100	47	46	51	52	52
110	51	49	55	57	56
120	54	53	59	62	61
130	58	56	64	66	65
140	61	59	68	71	70
150	64	62	72	75	74
160	67	65	76	79	78
170	70	67	79	84	82
180	73	70	83	88	86
190	75	72	86	92	90
200	78	74	90	96	94
210	80	76	93	100	97
220	82	78	96	103	101
230	84	80	99	107	105
240	86	82	102	111	108
250	88	83	105	114	111
260	89	85	108	117	114
270	91	86	110	121	117
280	93	87	113	124	120
290	95	89	117	129	124
300	97	91	120	133	129
310	99	93	123	137	133
320	101	95	126	141	136
330	103	97	129	145	140
340	105	99	132	149	143
350	107	100	135	153	146
360	109	102	138	156	150
370	110	103	140	159	153
380	112	105	142	162	155
390	113	106	145	165	158
400	115	107	147	168	161
420	117	110	151	174	166
440	120	112	155	179	170
460	122	114	159	184	175
480	125	116	162	188	179
500	127	118	165	192	183
520	129	120	169	196	186
540	131	122	171	200	190
560	132	123	174	204	193
580	134	125	177	207	196
600	136	126	179	210	199
620	137	128	181	213	202
640	139	129	184	216	204
660	140	130	186	219	207
680	141	131	188	222	209
700	142	132	190	224	212
720	144	133	192	227	214
740	145	135	193	229	216
760	146	136	195	231	218
780	147	136	197	233	220
800	148	137	198	235	222
850	150	140	202	240	226
900	153	142	205	245	230
950	155	143	208	249	234
1 000	156	145	211	252	237
1 050	158	147	214	256	240
1 100	160	148	216	259	243
1 150	161	149	219	262	246
1 200	163	150	221	265	249
1 300	165	150	225	270	253
1 400	165	150	228	275	258
1 500	165	150	230	279	260
1 600	165	150	230	280	260
1 700	165	150	230	280	260

TABLE 9. ALLOWABLE BEARING STRESS p_b

Form	Grade	p_b
		N/mm²
Plates, sections and bars	43	190
'' '' '' ''	50	260
'' '' '' ''	55	320
Hot rolled hollow sections*	43	190
'' '' '' ''	50	260
'' '' '' ''	55	320

* See Subclause 28c for provisions for stiffening tubes subject to bearing pressure.

TABLE 10. ALLOWABLE MAXIMUM SHEAR STRESS p_q

Form	Grade	Thickness or diameter	p_q
		mm	N/mm²
Plates, sections and bars	43	Up to and including 40	115
		Over 40	105
'' '' '' ''	50	Up to and including 65	160
		Over 65	$Y_s/2 \cdot 2$
'' '' '' ''	55	Up to and including 40	195
		Over 40	180
Hot rolled hollow sections	43	All	115
'' '' '' ''	50	All	160
'' '' '' ''	55	All	195

where Y_s = yield stress agreed with manufacturer, with a maximum value of 350 N/mm².

TABLE 11. ALLOWABLE AVERAGE SHEAR STRESS p_q' IN
UNSTIFFENED WEBS

(For stiffened webs see also b above and Tables 12a, 12b and 12c below)

Grade	Thickness	p_q'
	mm	N/mm²
43	Up to and including 40	100
	Over 40	90
50	Up to and including 65	140
55	Up to and including 40	170
	Over 40	160

TABLE 12a. ALLOWABLE AVERAGE SHEAR STRESS p_q' IN STIFFENED WEBS OF GRADE 43 STEEL

(see also Subclause 23b and Table 11)

d/t*	Stress p_q' (N/mm²) for different distances between stiffeners												
	0.3d	0.4d	0.5d	0.6d	0.7d	0.8d	0.9d	1.0d	1.1d	1.2d	1.3d	1.4d	1.5d
95	100	100	100	100	100	100	100	100	100	100	100	100	100
100	100	100	100	100	100	100	100	100	100	100	100	99	98
105	100	100	100	100	100	100	100	100	100	99	97	96	95
110	100	100	100	100	100	100	100	98	98	96	95	94	93
115	100	100	100	100	100	100	100	98	96	94	93	92	91
120	100	100	100	100	100	100	99	95	94	92	92	90	89
125	100	100	100	100	100	100	96	92	92	90	90	89	88
130	100	100	100	100	99	98	95	92	89	88	88	87	87
135	100	100	100	100	95	96	93	90	87	87	87	86	86
140	100	100	100	100	93	92	91	89	86	86	86	85	85
150	100	100	100	97	91	89	89	85	85	84	84	83	83
160	100	100	100	95	89	89	86	85	82	83	82	82	81
170	100	100	100	93	88	87	85	83	81	81	81	81	80
180	100	100	100	91	86	85	83	82	80	80	80	80	79
190	100	100	100	89	86	84	82	81					
200	100	100	95	87	85	83	81						
210	100	100	93	86	84	82							
220	100	100	91	86	82	81							
230	100	98	89	84	81								
240	100	95	87	83	81								
250	100	93	86	82	80								
260	100	86	85	81									
270	100	90	84	81									

The average stress shown in Table 11 shall not be exceeded.

NOTE. For materials over 40 mm thick, the stress shall not exceed 90 N/mm².
* Defined on page 46.

TABLE 12b. ALLOWABLE AVERAGE SHEAR STRESS p_q' IN STIFFENED WEBS OF GRADE 50 STEEL

(see also Subclause 23b and Table 11)

d/t*	Stress p_q' (N/mm²) for different distances between stiffeners												
	0.3d	0.4d	0.5d	0.6d	0.7d	0.8d	0.9d	1.0d	1.1d	1.2d	1.3d	1.4d	1.5d
80	140	140	140	140	140	140	140	140	140	140	140	140	140
85	140	140	140	140	140	140	140	140	140	140	140	138	136
90	140	140	140	140	140	140	140	140	140	137	135	134	132
95	140	140	140	140	140	140	140	138	135	133	131	130	129
100	140	140	140	140	140	140	140	135	132	130	128	127	126
105	140	140	140	140	140	140	136	131	129	127	126	124	124
110	140	140	140	140	140	138	132	128	126	125	123	122	121
115	140	140	140	140	140	134	129	126	124	122	121	120	119
120	140	140	140	140	140	132	127	124	122	120	119	118	118
125	140	140	140	140	140	129	125	122	120	119	118	117	116
130	140	140	140	140	133	127	123	120	118	117	116	116	115
135	140	140	140	138	130	123	121	118	117	116	114	114	114
140	140	140	140	136	128	123	120	117	116	115	114	113	113
150	140	140	140	131	124	120	117	115	114	113	112	111	111
160	140	140	137	127	121	117	115	113	112	111	110	110	110
170	140	140	132	124	119	115	113	111	110	110	109	109	109
180	140	138	129	121	116	113	111	110	109	109	108	108	107
190	140		126	119	115	112	110	109					
200	140	140	123	116	113	111	109						
210	140	140	121	114	112	110	109						
220	140	140	119	113	111	110							
230	140	138	117	112	110								
240	140	138	116	111	109								
250	140	122	114	110	108								
260	135	120	113	110									
270	132	118	112	109									

The average stress shown in Table 11 shall not be exceeded.

NOTE. For materials over 65 mm thick, the stress shall not exceed $0.4y_s$ N/mm², where y_s = yield stress agreed with manufacturer, with a maximum value of 350 N/mm².
* Defined on page 46.

TABLE 12c. ALLOWABLE AVERAGE SHEAR STRESS p_q' IN STIFFENED WEBS OF GRADE 55 STEEL

(see also Subclause 23b and Table 11)

d/t*	Stress p_q' (N/mm²) for different distances between stiffeners												
	0.3d	0.4d	0.5d	0.6d	0.7d	0.8d	0.9d	1.0d	1.1d	1.2d	1.3d	1.4d	1.5d
70	170	170	170	170	170	170	170	170	170	170	170	170	170
75	170	170	170	170	170	170	170	170	170	170	170	170	168
80	170	170	170	170	170	170	170	170	170	168	165	164	162
85	170	170	170	170	170	170	170	170	165	162	160	159	157
90	170	170	170	170	170	170	169	164	160	158	156	154	153
95	170	170	170	170	170	170	164	159	156	154	152	151	150
100	170	170	170	170	170	167	160	155	153	151	149	148	147
105	170	170	170	170	170	162	156	152	150	148	146	145	144
110	170	170	170	170	166	159	153	149	147	145	144	143	142
115	170	170	170	169	163	155	150	147	145	143	142	141	140
120	170	170	170	169	159	152	148	145	143	141	140	139	139
125	170	170	170	165	156	150	146	143	141	140	139	138	137
130	170	170	163	162	153	148	144	141	139	138	137	137	136
135	170	170	159	159	151	146	142	139	138	137	136	135	135
140	170	170	156	156	149	145	141	138	137	136	135	135	134
150	170	170	152	152	145	142	138	136	135	134	133	133	132
160	170	170	147	148	142	140							
170	170	170	153	145	137	136							
180	170	170	150	142	136	134							
190	170	165	147	140	133								
200	170	155	144	138	134								
210	170	152	142	136	133								
220	169	149	140	135	132								
230	165	147	138	134									
240	162	145	137	132									
250	158	143	135	132									
260	155	141	134										
270	153	139	133										

The average stress shown in Table 11 shall not be exceeded.

NOTE. For materials over 40 mm thick, the stress shall not exceed 160 N/mm².
* Defined on page 46.

TABLE 17a. ALLOWABLE STRESS p_c ON GROSS SECTION FOR AXIAL COMPRESSION

p_c (N/mm²) for grade 43 steel

l/r	0	1	2	3	4	5	6	7	8	9
0	155	155	154	154	153	153	153	152	152	151
10	151	151	150	150	149	149	148	148	148	147
20	147	146	146	146	145	145	144	144	144	143
30	143	143	142	142	141	141	141	140	140	139
40	139	138	138	137	137	136	136	136	135	134
50	133	133	132	132	131	130	129	128	127	126
60	126	125	124	123	122	121	120	119	118	117
70	115	114	113	112	111	110	108	107	106	105
80	104	102	101	100	99	97	96	95	94	92
90	91	90	89	87	86	85	84	83	81	80
100	79	78	77	76	75	74	73	72	71	70
110	69	68	67	66	65	64	63	62	61	61
120	60	59	58	57	56	56	55	54	53	53
130	52	51	50	49	49	48	48	47	47	46
140	46	45	44	43	43	43	42	42	41	41
150	40	40	39	38	38	38	38	37	37	36
160	36	35	35	34	34	34	33	33	33	32
170	32	32	31	31	31	30	30	30	29	29
180	29	28	28	28	27	27	27	27	26	26
190	26	26	25	25	25	25	25	24	24	24
200	24	23	23	23	23	22	22	22	22	22
210	21	21	21	21	21	21	20	20	20	20
220	20	19	19	19	19	19	19	18	18	18
230	18	18	18	18	17	17	17	17	17	17
240	17	17	16	16	16	16	16	16	16	15
250	15									
300	11									
350	8									

Intermediate values may be obtained by linear interpolation.
NOTE. For material over 40 mm thick, other than rolled I-beams or channels, and for Universal columns of thicknesses exceeding 40 mm, the limiting stress is 140 N/mm².

TABLE 17b. ALLOWABLE STRESS p_c ON GROSS SECTION FOR AXIAL COMPRESSION

p_c (N/mm²) for grade 50 steel

l/r	0	1	2	3	4	5	6	7	8	9
0	215	214	214	214	213	213	213	212	212	210
10	210	209	209	208	208	207	207	206	206	205
20	205	204	204	203	203	203	202	201	201	200
30	200	199	199	198	198	197	196	196	195	194
40	193	193	192	191	190	189	188	187	186	185
50	184	183	181	180	179	177	176	174	173	171
60	169	168	166	164	162	160	159	156	154	152
70	150	148	146	144	142	140	138	135	133	131
80	129	127	125	123	121	119	117	115	113	111
90	109	107	106	104	102	100	99	97	95	91
100	92	91	89	88	86	85	84	82	81	80
110	78	77	76	75	74	71	70	70	69	68
120	67	66	65	64	63	62	61	60	60	59
130	58	57	56	55	54	53	53	52	52	51
140	50	50	48	48	47	46	46	46	45	45
150	44	44	43	43	42	42	41	41	45	40
160	39	39	38	38	37	37	36	36	36	35
170	35	35	34	34	33	33	33	32	32	31
180	31	31	30	30	30	29	29	29	28	28
190	28	28	27	27	27	27	26	26	26	26
200	25	25	25	25	24	24	24	24	23	23
210	23	21	21	22	22	22	22	22	22	21
220	21	21	21	20	20	20	20	20	20	19
230	19	19	20	19	19	18	18	18	18	18
240	18	18	17	19	18	18	18	17	17	16
250	16									
300	11									
350	8									

Intermediate values may be obtained by linear interpolation.
NOTE. For material over 65 mm thick, the allowable stress p_c on gross section for axial compression shall be calculated in accordance with the procedure in Appendix B taking Y_s equal to the value of the yield stress agreed with the manufacturer, with a maximum value of 350 N/mm².

TABLE 17c. ALLOWABLE STRESS p_c ON GROSS SECTION FOR AXIAL COMPRESSION

p_c (N/mm²) for grade 55 steel

l/r	0	1	2	3	4	5	6	7	8	9
0	265	264	264	263	262	262	261	260	260	259
10	258	258	257	256	256	255	254	254	253	252
20	252	251	250	250	249	248	248	247	246	246
30	245	244	244	243	242	241	240	239	239	238
40	236	235	234	233	232	230	229	227	226	224
50	222	220	219	217	216	214	210	210	205	203
60	200	197	195	192	189	186	183	180	178	175
70	172	169	166	163	160	157	154	151	148	146
80	143	140	138	135	133	130	128	125	123	121
90	118	116	114	112	110	108	106	102	102	100
100	99	97	95	93	92	90	89	87	86	84
110	83	82	80	79	79	76	75	74	73	72
120	71	69	68	67	67	65	64	63	63	62
130	61	60	59	58	57	56	56	55	54	53
140	53	52	51	50	50	49	49	48	47	47
150	46	45	45	44	44	43	43	42	42	41
160	41	40	40	39	39	38	38	37	37	37
170	36	36	35	35	34	34	34	33	33	33
180	32	32	32	31	31	31	30	30	30	29
190	29	29	28	28	28	27	27	27	27	27
200	26	26	26	25	25	25	25	25	24	24
210	24	24	23	23	23	23	23	23	22	22
220	22	22	21	21	21	21	21	20	20	20
230	20	20	20	19	19	19	19	19	18	18
240	18	18	18	18	18	17	17	17	17	17
250	17									
300	12									
350	9									

Intermediate values may be obtained by linear interpolation.
NOTE. For material over 40 mm thick, other than rolled I-beams or channels, and for Universal columns of thicknesses exceeding 40 mm, the limiting stress is 245 N/mm².

TABLE 19. ALLOWABLE STRESS p_t IN AXIAL TENSION

Form	Grade	Thickness or diameter	p_t
		mm	N/mm²
Rolled I-beams and channels	43	All	155
Universal beams and columns	43	Up to and including 40	155
		Over 40	140
Plates, bars and sections other than above	43	Up to and including 40	155
		Over 40	140
Plates, bars and sections	50	Up to and including 65	215
		Over 65	* $Y_s/1.63$
,, ,, ,, ,,	55	Up to and including 40	265
		Over 40	245
Hot rolled hollow sections	43	All	155
,, ,, ,, ,,	50	All	215
,, ,, ,, ,,	55	All	265

* Where Y_s = yield stress agreed with the manufacturer with a maximum value of 350 N/mm².

TABLE 20. ALLOWABLE STRESSES IN RIVETS AND BOLTS (N/mm²)

Description of fasteners	Axial tension	Shear	Bearing
Power-driven rivets	100	100	300
Hand-driven rivets	80	80	250
Close tolerance and turned bolts	120	100	300
Bolts in clearance holes	120	80	250

TABLE 20A. ALLOWABLE BEARING STRESSES ON CONNECTED PARTS (N/mm²)

Descripton of fasteners	Material of connected part		
	Grade 43	Grade 50	Grade 55
Power-driven rivets Close tolerance and turned bolts	300	420	480
Hand-driven rivets Bolts in clearance holes	250	350	400

TABLE 21. EDGE DISTANCE OF HOLES

Diameter of hole	Distance to sheared or hand flame cut edge	Distance to rolled, machine flame cut, sawn or planed edge
mm	mm	mm
39	68	62
36	62	56
33	56	50
30	50	44
26	42	36
24	38	32
22	34	30
20	30	28
18	28	26
16	26	24
14	24	22
12 or less	22	20

* BS 4620, ' Rivets for general engineering purposes. Metric series

BENDING STRESSES (CASED BEAMS)

21. Beams and girders with equal flanges may be designed as cased beams when the following conditions are fulfilled:

1. The section is of single web and I-form or of double open channel form with the webs not less than 40 mm apart.

2. The beam is unpainted and is solidly encased in ordinary dense concrete, with 10 mm aggregate (unless solidity can be obtained with a larger aggregate), and of a works strength not less than 21 N/mm² at 28 days, when tested in accordance with BS 1881, ' Methods of testing concrete '.

3. The minimum width of solid casing is equal to $b + 100$ mm where b is the overall width of the steel flange or flanges in millimetres.

4. The surface and edges of the flanges of the beam have a concrete cover of not less than 50 mm.

5. The casing is effectively reinforced with wire to BS 4449, ' Hot-rolled steel bars for the reinforcement of concrete '. The wire shall be at least 5 mm diameter and the reinforcement shall be in the form of stirrups or binding at not more than 150 mm pitch, and so arranged as to pass through the centre of the covering to the edges and soffit of the lower flange.

The compressive stress in bending shall not exceed the value of p_{bc} obtained from Clauses 19 and 20; for this purpose the radius of gyration r_y may be taken as $0.2(b + 100$ mm), and D/T as for the uncased section.

The stress shall not, however, exceed 1½ times that permitted for the uncased section.

See Clauses 29 and 40 for solidly encased filler joists and grillage beams respectively.

NOTE. This clause does not apply to beams and girders having a depth greater than 1000 mm or a width greater than 500 mm or to box sections.

30 b. Cased struts. Struts of single I section or of two channels back to back in contact or spaced apart not less than 20 mm or more than half their depth and battened or laced in accordance with the requirements of Clauses 35 and 36 may be designed as cased struts when the following conditions are fulfilled:

1. The steel strut is unpainted and solidly encased in ordinary dense concrete, with 10 mm aggregate (unless solidity can be obtained with a larger aggregate) and of a works strength not less than 21 N/mm² at 28 days when tested in accordance with BS 1881, ' Methods of testing concrete '.

2. The minimum width of solid casing is equal to $b + 100$ mm, where b is the width overall of the steel flange or flanges in millimetres.

3. The surface and edges of the steel strut have a concrete cover of not less than 50 mm.

4. The casing is effectively reinforced with wire to BS 4449, ' Hot rolled steel bars for the reinforcement of concrete '. The wire shall be at least 5 mm in diameter and the reinforcement shall be in the form of stirrups or binding at not more than 150 mm pitch, so arranged as to pass through the centre of the covering of the edges and outer faces of the flanges and supported by and attached to longitudinal spacing bars not less than 4 in number.

The radius of gyration r of the strut section about the axis in the plane of its web or webs may be taken as $0.2 (b + 100)$ mm. The radius of gyration about its other axis shall be taken as that of the uncased section.

In no case shall the axial load on a cased strut exceed twice that which would be permitted on the uncased section, nor shall the slenderness ratio of the uncased section, measured over its full length centre-to-centre of connections, exceed 250.

In computing the allowable axial load on the cased strut the concrete shall be taken as assisting in carrying the load over its rectangular cross section, any cover in excess of 75 mm from the overall dimensions of the steel section of the cased strut being ignored. This cross section of concrete shall be taken as assisting in carrying the load on the basis of a stress equal to the allowable stress in the steel (as given in Table 17) divided by 0.19 times the numerical value of p_{bc} given in Table 2 for the grade of steel concerned.

NOTE. This clause does not apply to steel struts of overall sectional dimensions greater than 1000 mm × 500 mm, the dimension of 1000 mm being measured parallel to the web, or to box sections.

UNIVERSAL BEAMS
To: BS4 Part 1

Designation		Depth of section D	Width of section B	Thickness		Area of section	Moment of inertia		Radius of gyration		Elastic modulus		Plastic modulus	
Serial size	Mass per metre			Web t	Flange T		Axis x-x	Axis y-y	Axis x-x	Axis y-y	Axis x-x	Axis y-y	Axis x-x	Axis y-y
mm	kg	mm	mm	mm	mm	cm²	cm⁴	cm⁴	cm	cm	cm³	cm³	cm³	cm³
914 x 419	388	920.5	420.5	21.5	36.6	494.5	718742	45407	38.1	9.58	15616	2160	17657	3339
	343	911.4	418.5	19.4	32.0	437.5	625282	39150	37.8	9.46	13722	1871	15474	2890
914 x 305	289	926.6	307.8	19.6	32.0	368.8	504594	15610	37.0	6.51	10891	1014	12583	1603
	253	918.5	305.5	17.3	27.9	322.8	436610	13318	36.8	6.42	9507	871.9	10947	1372
	224	910.3	304.1	15.9	23.9	285.3	375924	11223	36.3	6.27	8259	738.1	9522	1162
	201	903.0	303.4	15.2	20.2	256.4	325529	9427	35.6	6.06	7210	621.4	8362	982.5
838 x 292	226	850.9	293.8	16.1	26.8	288.7	339747	11353	34.3	6.27	7986	772.9	9157	1211
	194	840.7	292.4	14.7	21.7	247.2	279450	9069	33.6	6.06	6648	620.4	7648	974.4
	176	834.9	291.6	14.0	18.8	224.1	246029	7792	33.1	5.90	5894	534.4	6809	841.5
762 x 267	197	769.6	268.0	15.6	25.4	250.8	239894	8174	30.9	5.71	6234	610.0	7167	958.7
	173	762.0	266.7	14.3	21.6	220.5	205177	6846	30.5	5.57	5385	513.4	6197	807.3
	147	753.9	265.3	12.9	17.5	188.1	168966	5468	30.0	5.39	4483	412.3	5174	649.0
686 x 254	170	692.9	255.8	14.5	23.7	216.6	170147	6621	28.0	5.53	4911	517.7	5624	810.3
	152	687.6	254.5	13.2	21.0	193.8	150319	5782	27.8	5.46	4372	454.5	4997	710.0
	140	683.5	253.7	12.4	19.0	178.6	136276	5179	27.6	5.38	3988	408.2	4560	637.8
	125	677.9	253.0	11.7	16.2	159.6	118003	4379	27.2	5.24	3481	346.1	3996	542.0
610 x 305	238	633.0	311.5	18.6	31.4	303.8	207571	15838	26.1	7.22	6559	1017	7456	1574
	179	617.5	307.0	14.1	23.6	227.9	151631	11412	25.8	7.08	4911	743.3	5521	1144
	149	609.6	304.8	11.9	19.7	190.1	124660	9300	25.6	6.99	4090	610.3	4572	936.8
610 x 229	140	617.0	230.1	13.1	22.1	178.4	111844	4512	25.0	5.03	3626	392.1	4146	612.5
	125	611.9	229.0	11.9	19.6	159.6	98579	3933	24.9	4.96	3222	343.5	3677	535.7
	113	607.3	228.2	11.2	17.3	144.5	87431	3439	24.6	4.88	2879	301.4	3288	470.2
	101	602.2	227.6	10.6	14.8	129.2	75720	2912	24.2	4.75	2515	255.9	2882	400.0
533 x 210	122	544.6	211.9	12.8	21.3	155.8	76207	3393	22.1	4.67	2799	320.2	3203	500.6
	109	539.5	210.7	11.6	18.8	138.6	66739	2937	21.9	4.60	2474	278.8	2824	435.1
	101	536.7	210.1	10.9	17.4	129.3	61659	2694	21.8	4.56	2298	256.5	2620	400.0
	92	533.1	209.3	10.2	15.6	117.8	55353	2392	21.7	4.51	2076	228.6	2366	356.2
	82	528.3	208.7	9.6	13.2	104.4	47491	2005	21.3	4.38	1798	192.2	2056	300.1
457 x 191	98	467.4	192.8	11.4	19.6	125.3	45717	2343	19.1	4.33	1956	243.0	2232	378.3
	89	463.6	192.0	10.6	17.7	113.9	41021	2086	19.0	4.28	1770	217.4	2014	337.9
	82	460.2	191.3	9.9	16.0	104.5	37103	1871	18.8	4.23	1612	195.6	1833	304.0
	74	457.2	190.5	9.1	14.5	95.0	33388	1671	18.7	4.19	1461	175.5	1657	272.2
	67	453.6	189.9	8.5	12.7	85.4	29401	1452	18.5	4.12	1296	152.9	1471	237.3
457 x 152	82	465.1	153.5	10.7	18.9	104.5	36215	1143	18.6	3.31	1557	149.0	1800	235.4
	74	461.3	152.7	9.9	17.0	95.0	32435	1012	18.5	3.26	1406	132.5	1622	209.1
	67	457.2	151.9	9.1	15.0	85.4	28577	878	18.3	3.21	1250	115.5	1441	182.2
	60	454.7	152.9	8.0	13.3	75.9	25464	794	18.3	3.23	1120	103.9	1284	162.9
	52	449.8	152.4	7.6	10.9	66.5	21345	645	17.9	3.11	949.0	84.6	1094	133.2
406 x 178	74	412.8	179.7	9.7	16.0	95.0	27329	1545	17.0	4.03	1324	172.0	1504	266.9
	67	409.4	178.8	8.8	14.3	85.5	24329	1365	16.9	4.00	1188	152.7	1346	236.5
	60	406.4	177.8	7.8	12.8	76.0	21508	1199	16.8	3.97	1058	134.8	1194	208.3
	54	402.6	177.6	7.6	10.9	68.4	18626	1017	16.5	3.85	925.3	114.5	1048	177.5
406 x 140	46	402.3	142.4	6.9	11.2	59.0	15647	539	16.3	3.02	777.8	75.7	888.4	118.3
	39	397.3	141.8	6.3	8.6	49.4	12452	411	15.9	2.89	626.9	58.0	720.8	91.08
356 x 171	67	364.0	173.2	9.1	15.7	85.4	19522	1362	15.1	3.99	1073	157.3	1212	243.0
	57	358.6	172.1	8.0	13.0	72.2	16077	1109	14.9	3.92	896.5	128.9	1009	198.8
	51	355.6	171.5	7.3	11.5	64.6	14156	968	14.8	3.87	796.2	112.9	894.9	174.1
	45	352.0	171.0	6.9	9.7	57.0	12091	812	14.6	3.78	686.9	95.0	773.7	146.7
356 x 127	39	352.8	126.0	6.5	10.7	49.4	10087	357	14.3	2.69	571.8	56.6	653.6	88.68
	33	348.5	125.4	5.9	8.5	41.8	8200	280	14.0	2.59	470.6	44.7	539.8	70.24
305 x 165	54	310.9	166.8	7.7	13.7	68.4	11710	1061	13.1	3.94	753.3	127.3	844.8	195.3
	46	307.1	165.7	6.7	11.8	58.9	9948	897	13.0	3.90	647.9	108.3	722.7	165.8
	40	303.8	165.1	6.1	10.2	51.5	8523	763	12.9	3.85	561.2	92.4	624.5	141.5
305 x 127	48	310.4	125.2	8.9	14.0	60.8	9504	460	12.5	2.75	612.4	73.5	706.1	115.7
	42	306.6	124.3	8.0	12.1	53.2	8143	388	12.4	2.70	531.2	62.5	610.5	98.24
	37	303.8	123.5	7.2	10.7	47.5	7162	337	12.3	2.67	471.5	54.6	540.5	85.66
305 x 102	33	312.7	102.4	6.6	10.8	41.8	6487	193	12.5	2.15	415.0	37.8	479.9	59.85
	28	308.9	101.9	6.1	8.9	36.3	5421	157	12.2	2.08	351.0	30.8	407.2	48.92
	25	304.8	101.6	5.8	6.8	31.4	4387	120	11.8	1.96	287.9	23.6	337.8	37.98
254 x 146	43	259.6	147.3	7.3	12.7	55.1	6558	677	10.9	3.51	505.3	92.0	568.2	141.2
	37	256.0	146.4	6.4	10.9	47.5	5556	571	10.8	3.47	434.0	78.1	485.3	119.6
	31	251.5	146.1	6.1	8.6	40.0	4439	449	10.5	3.35	353.1	61.5	395.6	94.52
254 x 102	28	260.4	102.1	6.4	10.0	36.2	4008	178	10.5	2.22	307.9	34.9	353.4	54.84
	25	257.0	101.9	6.1	8.4	32.2	3408	148	10.3	2.14	265.2	29.0	305.6	45.82
	22	254.0	101.6	5.8	6.8	28.4	2867	120	10.0	2.05	225.7	23.6	261.9	37.55
203 x 133	30	206.8	133.8	6.3	9.6	38.0	2887	384	8.72	3.18	279.3	57.4	313.3	88.05
	25	203.2	133.4	5.8	7.8	32.3	2356	310	8.54	3.10	231.9	46.4	259.8	71.39

3

UNIVERSAL COLUMNS

To BS4: Part 1

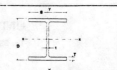

| Designation | | Depth of section D | Width of section B | Thickness | | Area of section | Moment of inertia | | Radius of gyration | | Elastic modulus | | Plastic modulus | |
Serial size	Mass per metre			Web t	Flange T		Axis x-x	Axis y-y	Axis x-x	Axis y-y	Axis x-x	Axis y-y	Axis x-x	Axis y-y
mm	kg	mm	mm	mm	mm	cm²	cm⁴	cm⁴	cm	cm	cm³	cm³	cm³	cm³
356 x 406	634	474.7	424.1	47.6	77.0	808.1	275140	98211	18.5	11.0	11592	4632	14247	7114
	551	455.7	418.5	42.0	67.5	701.8	227023	82665	18.0	10.9	9964	3951	12078	6058
	467	436.6	412.4	35.9	58.0	595.5	183118	67905	17.5	10.7	8388	3293	10009	5038
	393	419.1	407.0	30.6	49.2	500.9	146765	55410	17.1	10.5	7004	2723	8229	4157
	340	406.4	403.0	26.5	42.9	432.7	122474	46816	16.8	10.4	6027	2324	6994	3541
	287	393.7	399.0	22.6	36.5	366.0	99994	38714	16.5	10.3	5080	1940	5818	2952
Column core	235	381.0	395.0	18.5	30.2	299.8	79110	31008	16.2	10.2	4153	1570	4689	2384
	477	427.0	424.4	48.0	53.2	607.2	172391	68056	16.8	10.6	8075	3207	9700	4979
356 x 368	202	374.7	374.4	16.8	27.0	257.9	66307	23632	16.0	9.57	3540	1262	3977	1917
	177	368.3	372.1	14.5	23.8	225.7	57153	20470	15.9	9.52	3104	1100	3457	1668
	153	362.0	370.2	12.6	20.7	195.2	48525	17469	15.8	9.46	2681	943.8	2964	1430
	129	355.6	368.3	10.7	17.5	164.9	40246	14655	15.6	9.39	2264	790.4	2482	1196
305 x 305	283	365.3	321.8	26.9	44.1	360.4	78777	24545	14.8	8.25	4314	1525	5101	2337
	240	352.6	317.9	23.0	37.7	305.6	64177	20239	14.5	8.14	3641	1273	4245	1947
	198	339.9	314.1	19.2	31.4	252.3	50832	16230	14.2	8.02	2991	1034	3436	1576
	158	327.2	310.6	15.7	25.0	201.2	38740	12524	13.9	7.89	2368	806.3	2680	1228
	137	320.5	308.7	13.8	21.7	174.6	32838	10672	13.7	7.82	2049	691.4	2298	1052
	118	314.5	306.8	11.9	18.7	149.8	27601	9006	13.6	7.75	1755	587.0	1953	891.7
	97	307.8	304.8	9.9	15.4	123.3	22202	7268	13.4	7.68	1442	476.9	1589	723.5
254 x 254	167	289.1	264.5	19.2	31.7	212.4	29914	9796	11.9	6.79	2070	740.6	2417	1132
	132	276.4	261.0	15.6	25.3	167.7	22575	7519	11.6	6.68	1634	576.2	1875	878.6
	107	266.7	258.3	13.0	20.5	136.6	17510	5901	11.3	6.57	1313	456.9	1485	695.5
	89	260.4	255.9	10.5	17.3	114.0	14307	4849	11.2	6.52	1099	378.9	1223	575.4
	73	254.0	254.0	8.6	14.2	92.9	11360	3873	11.1	6.46	894.5	305.0	988.6	462.4
203 x 203	86	222.3	208.8	13.0	20.5	110.1	9462	3119	9.27	5.32	851.5	298.7	978.8	455.9
	71	215.9	206.2	10.3	17.3	91.1	7647	2536	9.16	5.28	708.4	246.0	802.4	374.2
	60	209.6	205.2	9.3	14.2	75.8	6088	2041	8.96	5.19	581.1	199.0	652.0	302.8
	52	206.2	203.9	8.0	12.5	66.4	5263	1770	8.90	5.16	510.4	173.6	568.1	263.7
	46	203.2	203.2	7.3	11.0	58.8	4564	1539	8.81	5.11	449.2	151.5	497.4	230.0
152 x 152	37	161.8	154.4	8.1	11.5	47.4	2218	709	6.84	3.87	274.2	91.78	310.1	140.1
	30	157.5	152.9	6.6	9.4	38.2	1742	558	6.75	3.82	221.2	73.06	247.1	111.2
	23	152.4	152.4	6.1	6.8	29.8	1263	403	6.51	3.68	165.7	52.95	184.3	80.87

UNIVERSAL BEARING PILES

To BS4: Part 1

| Designation | | Depth of section D | Width of section B | Thickness | | Area of section | Moment of inertia | | Radius of gyration | | Elastic modulus | | Plastic modulus | |
Serial size	Mass per metre			Web t	Flange T		Axis x-x	Axis y-y	Axis x-x	Axis y-y	Axis x-x	Axis y-y	Axis x-x	Axis y-y
mm	kg	mm	mm	mm	mm	cm²	cm⁴	cm⁴	cm	cm	cm³	cm³	cm³	cm³
356 x 368	174	361.5	378.1	20.4	20.4	222.2	51134	18444	15.2	9.11	2829	975.7	3194	1498
	152	356.4	375.5	17.9	17.9	193.6	43916	15799	15.1	9.03	2464	841.5	2764	1289
	133	351.9	373.3	15.6	15.6	169.0	37840	13576	15.0	8.96	2150	727.4	2398	1112
	109	346.4	370.5	12.9	12.9	138.4	30515	10900	14.8	8.87	1762	588.4	1950	897.3
305 x 305	223	338.0	325.4	30.5	30.5	284.8	52829	17572	13.6	7.85	3125	1080	3662	1682
	186	328.4	320.5	25.7	25.7	237.3	42643	14118	13.4	7.71	2597	880.4	3005	1365
	149	318.2	315.5	20.6	20.6	190.0	33013	10869	13.2	7.56	2075	688.8	2370	1063
	126	312.4	312.5	17.7	17.7	161.3	27484	8999	13.1	7.47	1760	575.8	1992	886.2
	110	307.9	310.3	15.4	15.4	140.4	23580	7689	13.0	7.40	1532	495.6	1722	761.0
	95	303.8	308.3	13.4	13.4	121.4	20112	6536	12.9	7.33	1324	423.7	1479	649.1
	88	301.7	307.2	12.3	12.3	111.8	18404	5960	12.8	7.30	1220	388.0	1358	593.8
	79	299.2	306.0	11.1	11.1	100.4	16400	5292	12.8	7.26	1096	345.9	1216	528.7
254 x 254	85	254.3	259.7	14.3	14.3	108.1	12264	4188	10.7	6.22	964.5	322.6	1090	496.0
	71	249.9	257.5	12.1	12.1	91.1	10153	3451	10.6	6.15	812.7	268.1	911.2	411.1
	63	246.9	256.0	10.6	10.6	79.7	8775	2971	10.5	6.11	710.9	232.1	792.5	355.3
203 x 203	54	203.9	207.2	11.3	11.3	68.4	4987	1683	8.54	4.96	489.2	162.4	552.8	249.7
	45	200.2	205.4	9.5	9.5	57.0	4079	1369	8.46	4.90	407.6	133.4	456.6	204.5

Note: An extensive range of hollow circular bearing piles is also available. Details may be obtained from BSC Tubes Division. All steel bearing piles are also listed, with safe loads, in 'Steel bearing piles' available from Constrado.

JOISTS
To BS4: Part 1

| Designation | | Depth of section D | Width of section B | Thickness | | Area of section | Moment of inertia | | Radius of gyration | | Elastic modulus | | Plastic modulus | |
Nominal size	Mass per metre			Web t	Flange T		Axis x-x	Axis y-y	Axis x-x	Axis y-y	Axis x-x	Axis y-y	Axis x-x	Axis y-y
mm	kg	mm	mm	mm	mm	cm²	cm⁴	cm⁴	cm	cm	cm³	cm³	cm³	cm³
254 x 203	81.85	254.0	203.2	10.2	19.9	104.4	12016	2278	10.7	4.67	946.1	224.3	1076	370.4
254 x 114	37.20	254.0	114.3	7.6	12.8	47.4	5092	270.1	10.4	2.39	401.0	47.19	460.0	79.30
203 x 152	52.09	203.2	152.4	8.9	16.5	66.4	4789	813.3	8.48	3.51	471.4	106.7	539.8	175.5
203 x 102	25.33*	203.2	101.6	5.8	10.4	32.3	2294	162.6	8.43	2.25	225.8	32.02	256.3	51.79
178 x 102	21.54*	177.8	101.6	5.3	9.0	27.4	1519	139.2	7.44	2.25	170.9	27.41	193.0	44.48
152 x 127	37.20	152.4	127.0	10.4	13.2	47.5	1818	378.8	6.20	2.82	238.7	59.65	278.6	99.85
152 x 89	17.09*	152.4	88.9	4.9	8.3	21.8	881.1	85.98	6.36	1.99	115.6	19.34	131.0	31.29
152 x 76	17.86	152.4	76.2	5.8	9.6	22.8	873.7	60.77	6.20	1.63	114.7	15.90	132.5	26.67
127 x 114	29.76	127.0	114.3	10.2	11.5	37.3	979.0	241.9	5.12	2.55	154.2	42.32	180.9	70.85
127 x 114	26.79	127.0	114.3	7.4	11.4	34.1	944.8	235.4	5.26	2.63	148.8	41.19	171.9	68.07
127 x 76	16.37	127.0	76.2	5.6	9.6	21.0	569.4	60.35	5.21	1.70	89.66	15.90	103.6	26.28
127 x 76	13.36*	127.0	76.2	4.5	7.6	17.0	475.9	50.18	5.29	1.72	74.94	13.17	85.23	21.29
114 x 114	26.79	114.3	114.3	9.5	10.7	34.4	735.4	223.1	4.62	2.54	128.6	39.00	151.2	65.63
102 x 102	23.07	101.6	101.6	9.5	10.3	29.4	486.1	154.4	4.06	2.29	95.72	30.32	113.4	50.70
‡102 x 64	9.65*	101.6	63.5	4.1	6.6	12.3	217.6	25.30	4.21	1.43	42.84	7.97	48.98	12.91
‡102 x 44	7.44	101.6	44.4	4.3	6.1	9.5	152.3	7.91	4.01	0.91	30.02	3.44	35.30	5.99
89 x 89	19.35	88.9	88.9	9.5	9.9	24.9	306.7	101.1	3.51	2.01	69.04	22.78	82.77	38.03
‡76 x 76	14.67	76.2	80.0	8.9	8.4	19.1	171.9	60.77	3.00	1.78	45.06	15.24	54.16	25.73
76 x 76	12.65	76.2	76.2	5.1	8.4	16.3	158.6	52.03	3.12	1 78	41.62	13.60	48.84	22.51

Note: Joists marked * have a 5° taper; all others taper 8°
‡These sections are only rolled to specific order.

CHANNELS
To BS4: Part 1

| Designation | | Depth of section D | Width of section B | Thickness | | Distance of c_y | Area of section | Moment of inertia | | Radius of gyration | | Elastic modulus | | Plastic modulus | |
Nominal size	Mass per metre			Web t	Flange T			Axis x-x	Axis y-y	Axis x-x	Axis y-y	Axis x-x	Axis y-y	Axis x-x	Axis y-y
mm	kg	mm	mm	mm	mm	cm	cm²	cm⁴	cm⁴	cm	cm	cm³	cm³	cm³	cm³
432 x 102	65.54	431.8	101.6	12.2	16.8	2.32	83.49	21399	628.6	16.0	2.74	991.1	80.14	1207	153.1
381 x 102	55.10	381.0	101.6	10.4	16.3	2.52	70.19	14894	579.7	14.6	2.87	781.8	75.86	932.7	144.4
305 x 102	46.18	304.8	101.6	10.2	14.8	2.66	58.83	8214	499.5	11.8	2.91	539.0	66.59	638.3	128.1
305 x 89	41.69	304.8	88.9	10.2	13.7	2.18	53.11	7061	325.4	11.5	2.48	463.3	48.49	557.1	92.60
254 x 89	35.74	254.0	88.9	9.1	13.6	2.42	45.52	4448	302.4	9.88	2.58	350.2	46.70	414.4	89.56
254 x 76	28.29	254.0	76.2	8.1	10.9	1.86	36.03	3367	162.6	9.67	2.12	265.1	28.21	317.4	54.14
229 x 89	32.76	228.6	88.9	8.6	13.3	2.53	41.73	3387	285.0	9.01	2.61	296.4	44.82	348.4	86.38
229 x 76	26.06	228.6	76.2	7.6	11.2	2.00	33.20	2610	158.7	8.87	2.19	228.3	28.22	270.3	54.24
203 x 89	29.78	203.2	88.9	8.1	12.9	2.65	37.94	2491	264.4	8.10	2.64	245.2	42.34	286.6	81.62
203 x 76	23.82	203.2	76.2	7.1	11.2	2.13	30.34	1950	151.3	8.02	2.23	192.0	27.59	225.2	53.32
178 x 89	26.81	177.8	88.9	7.6	12.3	2.76	34.15	1753	241.0	7.16	2.66	197.2	39.29	229.6	75.44
178 x 76	20.84	177.8	76.2	6.6	10.3	2.20	26.54	1337	134.0	7.10	2.25	150.4	24.72	175.4	48.07
152 x 89	23.84	152.4	88.9	7.1	11.6	2.86	30.36	1166	215.1	6.20	2.66	153.0	35.70	177.7	68.12
152 x 76	17.88	152.4	76.2	6.4	9.0	2.21	22.77	851.5	113.8	6.12	2.24	111.8	21.05	130.0	41.26
127 x 64	14.90	127.0	63.5	6.4	9.2	1.94	18.98	482.5	67.23	5.04	1.88	75.99	15.25	89.4	29.31
102 x 51	10.42	101.6	50.8	6.1	7.6	1.51	13.28	207.7	29.10	3.95	1.48	40.89	8.16	48.8	15.71
76 x 38	6.70	76.2	38.1	5.1	6.8	1.19	8.53	74.14	10.66	2.95	1.12	19.46	4.07	23.4	7.76

EQUAL ANGLES

To BS4848: Part 4

Designation		Mass per metre	Area of section	Distance of centre of gravity	Moment of inertia	Radius of gyration	Elastic modulus	Plastic modulus
Size A	Thickness t			c	Axis x-x, y-y	Axis x-x, y-y	Axis x-x, y-y	Axis x-x, y-y
mm	mm	kg	cm^2	cm	cm^4	.cm	cm^3	cm^3
250 × 250	35	128	163	7.49	9250	7.53	529	955.56
	32	118	150	7.38	8600	7.57	488	882.34
	28	104	133	7.23	7690	7.61	433	782.36
	25	93.6	119	7.12	6970	7.65	390	705.55
200 × 200	24	71.1	90.6	5.84	3330	6.06	235	426.20
	20	59.9	76.3	5.68	2850	6.11	199	361.01
	18	54.2	69.1	5.60	2600	6.13	181	327.55
	16	48.5	61.8	5.52	2340	6.16	162	293.49
150 × 150	18	40.1	51.0	4.37	1050	4.54	98.7	179.37
	15	33.8	43.0	4.25	898	4.57	83.5	151.85
	12	27.3	34.8	4.12	737	4.60	67.7	123.35
	10	23.0	29.3	4.03	624	4.62	56.9	103.77
120 × 120	15	26.6	33.9	3.51	445	3.62	52.4	95.26
	12	21.6	27.5	3.40	368	3.65	42.7	77.73
	10	18.2	23.2	3.31	313	3.67	36.0	65.60
	8	14.7	18.7	3.23	255	3.69	29.1	53.10
100 × 100	15	21.9	27.9	3.02	249	2.98	35.6	64.77
	12	17.8	22.7	2.90	207	3.02	29.1	53.03
	8	12.2	15.5	2.74	145	3.06	19.9	36.43
90 × 90	12	15.9	20.3	2.66	148	2.70	23.3	42.50
	10	13.4	17.1	2.58	127	2.72	19.8	36.03
	8	10.9	13.9	2.50	104	2.74	16.1	29.30
	7	9.61	12.2	2.45	92.6	2.75	14.1	25.77
	6	8.30	10.6	2.41	80.3	2.76	12.2	22.31
80 × 80	10	11.9	15.1	2.34	87.5	2.41	15.4	28.15
	8	9.63	12.3	2.26	72.2	2.43	12.6	22.95
	6	7.34	9.35	2.17	55.8	2.44	9.57	17.52
70 × 70	10	10.3	13.1	2.09	57.2	2.09	11.7	21.25
	8	8.36	10.6	2.01	47.5	2.11	9.52	17.37
	6	6.38	8.13	1.93	36.9	2.13	7.27	13.30
60 × 60	10	8.69	11.1	1.85	34.9	1.78	8.41	15.32
	8	7.09	9.03	1.77	29.2	1.80	6.89	12.57
	6	5.42	6.91	1.69	22.8	1.82	5.29	9.67
	5	4.57	5.82	1.64	19.4	1.82	4.45	8.15
50 × 50	8	5.82	7.41	1.52	16.3	1.48	4.68	8.55
	6	4.47	5.69	1.45	12.8	1.50	3.61	6.61
	5	3.77	4.80	1.40	11.0	1.51	3.05	5.58
	4	3.06	3.89	1.36	8.97	1.52	2.46	4.51
	3	2.33	2.96	1.31	6.86	1.52	1.86	3.42
45 × 45	6	4.00	5.09	1.32	9.16	1.34	2.88	5.28
	5	3.38	4.30	1.28	7.84	1.35	2.43	4.47
	4	2.74	3.49	1.23	6.43	1.36	1.97	3.63
	3	2.09	2.66	1.18	4.93	1.36	1.49	2.74
40 × 40	6	3.52	4.48	1.20	6.31	1.19	2.26	4.13
	5	2.97	3.79	1.16	5.43	1.20	1.91	3.50
	4	2.42	3.08	1.12	4.47	1.21	1.55	2.85
	3	1.84	2.35	1.07	3.45	1.21	1.48	2.16
25 × 25	5	1.77	2.26	0.80	1.20	0.73	0.71	1.30
	4	1.45	1.85	0.76	1.01	0.74	0.58	1.07
	3	1.11	1.42	0.72	0.80	0.75	0.45	0.83

Note: 100 x 100 x 10mm angle is also frequently rolled; as an ISO size its properties are given in Appendix A (Table A1) to BS4848:Part 4. Other non-standard sections, particularly other thicknesses of the standard range, may also be available. Enquiries should be made to BSC Sections.

6

UNEQUAL ANGLES

To BS4848: Part 4

Designation		Mass per metre	Area of section	Distance of centre of gravity		Moment of inertia		Radius of gyration		Elastic modulus		Plastic modulus	
Size A x B	Thickness t			c_x	c_y	Axis x-x	Axis y-y	Axis x-x	Axis y-y	Axis x-x	Axis y-y	Axis x-x	Axis y-y
mm	mm	kg	cm²	cm	cm	cm⁴	cm⁴	cm	cm	cm³	cm³	cm³	cm³
200 x 150	18	47.1	60.0	6.33	3.85	2376	1146	6.29	4.37	174.0	103.0	316.19	186.02
	15	39.6	50.5	6.21	3.73	2022	979	6.33	4.40	147.0	86.9	267.38	156.62
	12	32.0	40.8	6.08	3.61	1652	803	6.36	4.44	119.0	70.5	216.97	126.54
200 x 100	15	33.7	43.0	7.16	2.22	1758	299	6.40	2.64	137.0	38.4	240.46	72.25
	12	27.3	34.8	7.03	2.10	1440	247	6.43	2.67	111.0	31.3	195.68	57.82
	10	23.0	29.2	6.93	2.01	1220	210	6.46	2.68	93.2	26.3	164.91	48.16
150 x 90	15	26.6	33.9	5.21	2.23	761	205	4.74	2.46	77.7	30.4	138.77	56.53
	12	21.6	27.5	5.08	2.12	627	171	4.77	2.49	63.3	24.8	113.40	45.60
	10	18.2	23.2	5.00	2.04	533	146	4.80	2.51	53.3	21.0	95.83	38.20
150 x 75	15	24.8	31.6	5.53	1.81	713	120	4.75	1.94	75.3	21.0	131.45	40.59
	12	20.2	25.7	5.41	1.69	589	99.9	4.79	1.97	61.4	17.2	107.60	32.51
	10	17.0	21.6	5.32	1.61	501	85.8	4.81	1.99	51.8	14.6	91.04	27.11
125 x 75	12	17.8	22.7	4.31	1.84	354	95.5	3.95	2.05	43.2	16.9	77.36	31.42
	10	15.0	19.1	4.23	1.76	302	82.1	3.97	2.07	36.5	14.3	65.57	26.34
	8	12.2	15.5	4.14	1.68	247	67.6	4.00	2.09	29.6	11.6	53.33	21.19
100 x 75	12	15.4	19.7	3.27	2.03	189	90.2	3.10	2.14	28.0	16.5	50.97	30.27
	10	13.0	16.6	3.19	1.95	162	77.6	3.12	2.16	23.8	14.0	43.31	25.55
	8	10.6	13.5	3.10	1.87	133	64.1	3.14	2.18	19.3	11.4	35.31	20.69
100 x 65	10	12.3	15.6	3.36	1.63	154	51.0	3.14	1.81	23.2	10.5	41.91	19.41
	8	9.94	12.7	3.27	1.55	127	42.2	3.16	1.83	18.9	8.54	34.21	15.67
	7	8.77	11.2	3.23	1.51	113	37.6	3.17	1.83	16.6	7.53	30.23	13.77
80 x 60	8	8.34	10.6	2.55	1.56	66.3	31.8	2.50	1.73	12.2	7.16	22.17	13.08
	7	7.36	9.38	2.51	1.52	59.0	28.4	2.51	1.74	10.7	6.34	19.63	11.54
	6	6.37	8.11	2.47	1.48	51.4	24.8	2.52	1.75	9.29	5.49	17.02	9.96
75 x 50	8	7.39	9.41	2.52	1.29	52.0	18.4	2.35	1.40	10.4	4.95	18.87	9.14
	6	5.65	7.19	2.44	1.21	40.5	14.4	2.37	1.42	8.01	3.81	14.54	6.95
65 x 50	8	6.75	8.60	2.11	1.37	34.8	17.7	2.01	1.44	7.93	4.89	14.40	8.94
	6	5.16	6.58	2.04	1.29	27.2	14.0	2.03	1.46	6.10	3.77	11.13	6.85
	5	4.35	5.54	1.99	1.25	23.2	11.9	2.05	1.47	5.14	3.19	9.41	5.76

Note: Additional non-standard sizes may be available, especially other thicknesses of the standard range and certain sizes in the old Imperial range, namely 125 x 75 x 6.5 and 137 x 102 x 9.5, 7.9 and 6.4 (purlin angles) and 100 x 75 x 6.5. Enquiries should be made to BSC Sections.

T SECTIONS

Ts cut from Universal Beams and Columns listed in this publication are available in all sizes except where web thicknesses exceed 19mm (Grade 43 steel) or 15.9mm (Grade 50 steel) and 914 x 305 UB 201 kg and 224 kg (Grade 50 steel). See BS4: Part 1 for properties.

Rolled Ts are available in two serial sizes only, 51 x 51mm and 44 x 44mm, with two weights in each serial size. See BS4: Part 1 for properties. Enquiries should be made to BSC Sections & Commercial Steels.

BULB FLATS

These are now produced in metric sizes and are listed in BS4848: Part 5. Enquiries should be made to BSC Sections & Commercial Steels.

CIRCULAR HOLLOW SECTIONS

To BS4848: Part 2 unless marked †

Designation		Mass per metre	Area of section	Moment of inertia	Radius of gyration	Elastic modulus	Plastic modulus	Torsional constants	
Outside diameter D	Thickness t							J	C
mm	mm	kg	cm²	cm⁴	cm	cm³	cm³	cm⁴	cm³
21.3	3.2	1.43	1.82	0.77	0.650	0.72	1.06	1.54	1.44
26.9	3.2	1.87	2.38	1.70	0.846	1.27	1.81	3.41	2.53
33.7	2.6	1.99	2.54	3.09	1.10	1.84	2.52	6.19	3.67
	3.2	2.41	3.07	3.60	1.08	2.14	2.99	7.21	4.28
	4.0	2.93	3.73	4.19	1.06	2.49	3.55	8.38	4.97
42.4	2.6	2.55	3.25	6.46	1.41	3.05	4.12	12.9	6.10
	3.2	3.09	3.94	7.62	1.39	3.59	4.93	15.2	7.19
	4.0	3.79	4.83	8.99	1.36	4.24	5.92	18.0	8.48
48.3	3.2	3.56	4.53	11.6	1.60	4.80	6.52	23.2	9.59
	4.0	4.37	5.57	13.8	1.57	5.70	7.87	27.5	11.4
	5.0	5.34	6.80	16.2	1.54	6.69	9.42	32.3	13.4
60.3	3.2	4.51	5.74	23.5	2.02	7.78	10.4	46.9	15.6
	4.0	5.55	7.07	28.2	2.00	9.34	12.7	56.3	18.7
	5.0	6.82	8.69	33.5	1.96	11.1	15.3	67.0	22.2
76.1	3.2	5.75	7.33·	48.8	2.58	12.8	17.0	97.6	25.6
	4.0	7.11	9.06	59.1	2.55	15.5	20.8	118	31.0
	5.0	8.77	11.2	70.9	2.52	18.6	25.3	142	37.3
88.9	3.2	6.76	8.62	79.2	3.03	17.3	23.5	158	35.6
	4.0	8.38	10.7	96.3	3.00	21.7	28.9	193	43.3
	5.0	10.3	13.2	116	2.97	26.2	35.2	233	52.4
114.3	3.6	9.83	12.5	192	3.92	33.6	44.1	384	67.2
	5.0	13.5	17.2	257	3.87	45.0	59.8	514	89.9
	6.3	16.8	21.4	313	3.82	54.7	73.6	625	109
139.7	5.0	16.6	21.2	481	4.77	68.8	90.8	961	138
	6.3	20.7	26.4	589	4.72	84.3	112	1177	169
	8.0	26.0	33.1	720	4.66	103	139	1441	206
	10.0	32.0	40.7	862	4.60	123	169	1724	247
168.3	5.0	20.1	25.7	856	5.78	102	133	1712	203
	6.3	25.2	32.1	1053	5.73	125	165	2107	250
	8.0	31.6	40.3	1297	5.67	154	206	2595	308
	10.0	39.0	49.7	1564	5.61	186	251	3128	372
193.7	5.0†	23.3	29.6	1320	6.67	136	178	2640	273
	5.4	25.1	31.9	1417	6.66	146	192	2834	293
	6.3	29.1	37.1	1630	6.63	168	221	3260	337
	8.0	36.6	46.7	2016	6.57	208	276	4031	416
	10.0	45.3	57.7	2442	6.50	252	338	4883	504
	12.5	55.9	71.2	2934	6.42	303	411	5869	606
	16.0	70.1	89.3	3554	6.31	367	507	7109	734
219.1	5.0†	26.4	33.6	1928	7.57	176	229	3856	352
	6.3	33.1	42.1	2386	7.53	218	285	4772	436
	8.0	41.6	53.1	2960	7.47	270	357	5919	540
	10.0	51.6	65.7	3598	7.40	328	438	7197	657
	12.5	63.7	81.1	4345	7.32	397	534	8689	793
	16.0	80.1	102	5297	7.20	483	661	10590	967
	20.0	98.2	125	6261	7.07	572	795	12520	1143

CIRCULAR HOLLOW SECTIONS

To BS4848: Part 2 unless marked †

Designation		Mass per metre	Area of section	Moment of inertia	Radius of gyration	Elastic modulus	Plastic modulus	Torsional constants	
Outside diameter D	Thickness t							J	C
mm	mm	kg	cm²	cm⁴	cm	cm³	cm³	cm⁴	cm³
244.5	6.3	37.0	47.1	3346	8.42	274	358	6692	547
	8.0	46.7	59.4	4160	8.37	340	448	8321	681
	10.0	57.8	73.7	5073	8.30	415	550	10150	830
	12.5	71.5	91.1	6147	8.21	503	673	12290	1006
	16.0	90.2	115	7533	8.10	616	837	15070	1232
	20.0	111	141	8957	7.97	733	1011	17910	1465
273	6.3	41.4	52.8	4696	9.43	344	448	9392	688
	8.0	52.3	66.6	5852	9.37	429	562	11700	857
	10.0	64.9	82.6	7154	9.31	524	692	14310	1048
	12.5	80.3	102	8697	9.22	637	849	17390	1274
	16.0	101	129	10710	9.10	784	1058	21410	1569
	20.0	125	159	12800	8.97	938	1283	25600	1875
	25.0	153	195	15130	8.81	1108	1543	30250	2216
323.9	6.3†	49.3	62.9	7929	11.2	490	636	15858	980
	8.0	62.3	79.4	9910	11.2	612	799	19820	1224
	10.0	77.4	98.6	12160	11.1	751	986	24320	1501
	12.5	96.0	122	14850	11.0	917	1213	29690	1833
	16.0	121	155	18390	10.9	1136	1518	36780	2271
	20.0	150	191	22140	10.8	1367	1850	44280	2734
	25.0	184	235	26400	10.6	1630	2239	52800	3260
355.6	8.0	68.6	87.4	13200	12.3	742	967	26400	1485
	10.0	85.2	109	16220	12.2	912	1195	32450	1825
	12.5	106	135	19850	12.1	1117	1472	39700	2233
	16.0	134	171	24660	12.0	1387	1847	49330	2774
	20.0	166	211	29790	11.9	1676	2255	59580	3351
	25.0	204	260	35680	11.7	2007	2738	71350	4013
406.4	10.0	97.8	125	24480	14.0	1205	1572	48950	2409
	12.5	121	155	30030	13.9	1478	1940	60060	2956
	16.0	154	196	37450	13.8	1843	2440	74900	3686
	20.0	191	243	45430	13.7	2236	2989	90860	4472
	25.0	235	300	54700	13.5	2692	3642	109400	5384
	32.0	295	376	66430	13.3	3269	4497	132900	6539
457	10.0	110	140	35090	15.8	1536	1998	70180	3071
	12.5	137	175	43140	15.7	1888	2470	86290	3776
	16.0	174	222	53960	15.6	2361	3113	107900	4723
	20.0	216	275	65680	15.5	2874	3822	131400	5749
	25.0	266	339	79420	15.3	3476	4671	158800	6951
	32.0	335	427	97010	15.1	4246	5791	194000	8491
	40.0	411	524	114900	14.8	5031	6977	229900	10060
508	10.0†	123	156	48520	17.6	1910	2480	97040	3821
	12.5†	153	195	59760	17.5	2353	3070	119500	4705

RECTANGULAR HOLLOW SECTIONS (SQUARE)

To BS4848: Part 2 unless marked †

Designation		Mass per metre	Area of section	Moment of inertia	Radius of gyration	Elastic modulus	Plastic modulus	Torsional constants	
Size DxD	Thickness t							J	C
mm	mm	kg	cm²	cm⁴	cm	cm³	cm³	cm⁴	cm³
20 × 20	2.0	1.12	1.42	0.76	0.73	0.76	0.95	1.22	1.07
	2.5†	1.35	1.72	0.87	0.71	0.88	1.12	1.41	1.21
25 × 25	2.0†	1.43	1.82	1.59	0.94	1.27	1.56	2.52	1.81
	2.5†	1.74	2.22	1.85	0.91	1.48	1.86	2.97	2.09
	3.2†	2.15	2.74	2.14	0.88	1.71	2.21	3.49	2.38
30 × 30	2.5†	2.14	2.72	3.40	1.12	2.27	2.79	5.40	3.22
	3.0†	2.51	3.20	3.84	1.10	2.56	3.21	6.17	3.61
	3.2	2.65	3.38	4.00	1.09	2.67	3.37	6.45	3.75
40 × 40	2.5†	2.92	3.72	8.67	1.53	4.33	5.21	13.6	6.23
	3.0†	3.45	4.40	9.96	1.51	4.98	6.07	15.7	7.11
	3.2	3.66	4.66	10.4	1.50	5.22	6.40	16.5	7.43
	4.0	4.46	5.68	12.1	1.46	6.07	7.61	19.5	8.56
50 × 50	2.5†	3.71	4.72	17.7	1.94	7.07	8.38	27.4	10.2
	3.0†	4.39	5.60	20.5	1.91	8.20	9.83	32.0	11.8
	3.2	4.66	5.94	21.6	1.91	8.62	10.4	33.8	12.4
	4.0	5.72	7.28	25.5	1.87	10.2	12.5	40.4	14.5
	5.0	6.97	8.88	29.6	1.83	11.9	14.9	47.6	16.7
60 × 60	3.0†	5.34	6.80	36.6	2.32	12.2	14.5	56.9	17.7
	3.2	5.67	7.22	38.7	2.31	12.9	15.3	60.1	18.6
	4.0	6.97	8.88	46.1	2.28	15.4	18.6	72.4	22.1
	5.0	8.54	10.9	54.4	2.24	18.1	22.3	86.3	25.8
70 × 70	3.0†	6.28	8.00	59.6	2.73	17.0	20.0	92.1	24.8
	3.6	7.46	9.50	69.5	2.70	19.9	23.6	108	28.7
	5.0	10.1	12.9	90.1	2.64	25.7	31.2	142	36.8
80 × 80	3.0†	7.22	9.20	90.6	3.14	22.7	26.5	139	33.1
	3.6	8.59	10.9	106	3.11	26.5	31.3	164	38.5
	5.0	11.7	14.9	139	3.05	34.7	41.7	217	49.8
	6.3	14.4	18.4	165	3.00	41.3	50.5	261	58.8
90 × 90	3.6	9.72	12.4	154	3.52	34.1	40.0	237	49.7
	5.0	13.3	16.9	202	3.46	45.0	53.6	315	64.9
	6.3	16.4	20.9	242	3.41	53.9	65.3	381	77.1
100 × 100	4.0	12.0	15.3	234	3.91	46.8	54.9	361	68.2
	5.0	14.8	18.9	283	3.87	56.6	67.1	439	81.9
	6.3	18.4	23.4	341	3.81	68.2	82.0	533	97.9
	8.0	22.9	29.1	408	3.74	81.5	99.9	646	116
	10.0	27.9	35.5	474	3.65	94.9	119	761	134
120 × 120	5.0	18.0	22.9	503	4.69	83.8	98.4	775	122
	6.3	22.3	28.5	610	4.63	102	121	949	147
	8.0	27.9	35.5	738	4.56	123	149	1159	176
	10.0	34.2	43.5	870	4.47	145	178	1381	206
140 × 140	5.0†	21.1	26.9	814	5.50	116	136	1251	170
	6.3†	26.3	33.5	994	5.45	142	168	1538	206
	8.0†	32.9	41.9	1212	5.38	173	207	1889	249
	10.0†	40.4	51.5	1441	5.29	206	250	2269	294
150 × 150	5.0	22.7	28.9	1009	5.91	135	157	1548	197
	6.3	28.3	36.0	1236	5.86	165	194	1907	240
	8.0	35.4	45.1	1510	5.78	201	240	2348	291
	10.0	43.6	55.5	1803	5.70	240	290	2829	345
	12.5	53.4	68.0	2125	5.59	283	348	3372	403
	16.0	66.4	84.5	2500	5.44	333	421	4029	468
180 × 180	6.3	34.2	43.6	2186	7.08	243	283	3357	355
	8.0	43.0	54.7	2689	7.01	299	352	4156	434
	10.0	53.0	67.5	3237	6.92	360	429	5041	519
	12.5	65.2	83.0	3856	6.82	428	519	6062	613
	16.0	81.4	104	4607	6.66	512	634	7339	725
200 × 200	6.3	38.2	48.6	3033	7.90	303	353	4647	444
	8.0	48.0	61.1	3744	7.83	374	439	5770	545
	10.0	59.3	75.5	4525	7.74	452	536	7020	655
	12.5	73.0	93.0	5419	7.63	542	651	8479	779
	16.0	91.5	117	6524	7.48	652	799	10330	929
250 × 250	6.3	48.1	61.2	6049	9.94	484	559	9228	712
	8.0	60.5	77.1	7510	9.87	601	699	11511	880
	10.0	75.0	95.5	9141	9.78	731	858	14086	1065
	12.5	92.6	118	11050	9.68	884	1048	17139	1279
	16.0	117	149	13480	9.53	1078	1298	21109	1548
300 × 300	10.0	90.7	116	16150	11.8	1077	1254	24776	1575
	12.5	112	143	19630	11.7	1309	1538	30290	1905
	16.0	142	181	24160	11.6	1610	1916	37566	2327
350 × 350	10.0	106	136	26050	13.9	1489	1725	39840	2186
	12.5	132	168	31810	13.8	1817	2122	48869	2655
	16.0	167	213	39370	13.6	2250	2655	60901	3265
400 × 400	10.0	122	156	39350	15.9	1968	2272	60028	2896
	12.5	152	193	48190	15.8	2409	2800	73815	3530

RECTANGULAR HOLLOW SECTIONS

To BS4848: Part 2 unless marked †

Designation		Mass per metre	Area of section	Moment of inertia		Radius of gyration		Elastic modulus		Plastic modulus		Torsional constants	
Size D×B	Thickness t			Axis x-x	Axis y-y	Axis x-x	Axis y-y	Axis x-x	Axis y-y	Axis x-x	Axis y-y	J	C
mm	mm	kg	cm²	cm⁴	cm⁴	cm	cm	cm³	cm³	cm³	cm³	cm⁴	cm³
50 × 25	2.5†	2.72	3.47	10.5	3.44	1.75	1.00	4.25	2.75	5.41	3.25	8.41	4.62
	3.0†	3.22	4.1	12.2	3.89	1.73	0.98	4.88	3.11	6.30	3.77	9.64	5.21
	3.2†	3.41	4.34	12.8	4.05	1.72	0.97	5.11	3.24	6.64	3.96	10.1	5.42
50 × 30	2.5	2.92	3.72	12.0	5.30	1.80	1.19	4.81	3.53	6.01	4.16	11.7	5.74
	3.0†	3.45	4.40	13.9	6.04	1.78	1.17	5.54	4.03	7.01	4.83	13.5	6.52
	3.2	3.66	4.66	14.5	6.31	1.77	1.16	5.82	4.21	7.39	5.08	14.2	6.81
60 × 40	2.5†	3.71	4.72	23.1	12.2	2.21	1.61	7.71	6.10	9.43	7.09	25.0	9.74
	3.0†	4.39	5.60	26.9	14.1	2.19	1.59	8.96	7.04	11.1	8.29	29.2	11.2
	3.2	4.66	5.94	28.3	14.8	2.18	1.58	9.44	7.39	11.7	8.75	30.8	11.8
	4.0	5.72	7 28	33.6	17.3	2.15	1.54	11.2	8.67	14.1	10.5	36.6	13.7
80 × 40	3.0†	5.34	6.80	55.0	18.2	2.85	1.64	13.8	9.1	17.3	10.5	43.7	15.3
	3.2	5.67	7.22	58.1	19.1	2.84	1.63	14.5	9.56	18.3	11.1	46.1	16.1
	4.0	6.97	8.88	69.6	22.6	2.80	1.59	17.4	11.3	22.2	13.4	55.1	18.9
90 × 50	3.0†	6.28	8.00	85.4	33.8	3.27	2.05	19.0	13.5	23.4	15.5	76.4	22.4
	3.6	7.46	9.50	99.8	39.1	3.24	2.03	22.2	15.6	27.6	18.1	89.3	25.9
	5.0	10.1	12.9	130	50.0	3.18	1.97	28.9	20.0	36.6	23.9	116	32.9
100 × 50	3.0†	6.75	8.60	111	37.1	3.59	2.08	22.2	14.8	27.6	16.9	88.3	25.0
	3.2	7.18	9.14	117	39.1	3.58	2.07	23.5	15.6	29.2	17.9	93.3	26.4
	4.0	8.86	11.3	142	46.7	3.55	2.03	28.4	18.7	35.7	21.7	113	31.4
	5.0	10.9	13.9	170	55.1	3.50	1.99	34.0	22.0	43.3	26.1	135	37.0
	6.3†	13.4	17.1	202	64.2	3.44	1.94	40.5	25.7	52.5	31.3	160	43.0
100 × 60	3.0†	7.22	9.20	125	56.2	3.69	2.47	25.0	18.7	30.5	21.3	121	30.7
	3.6	8.59	10.9	147	65.4	3.66	2.45	29.3	21.8	36.0	25.1	142	35.6
	5.0	11.7	14.9	192	84.7	3.60	2.39	38.5	28.2	48.1	33.3	187	45.9
	6.3	14.4	18.4	230	99.9	3.54	2.33	46.0	33.3	58.4	40.2	224	53.9
120 × 60	3.6	9.72	12.4	230	76.9	4.31	2.49	38.3	25.6	47.6	29.2	183	43.3
	5.0	13.3	16.9	304	99.9	4.24	2.43	50.7	33.3	63.9	38.8	242	56.0
	6.3	16.4	20.9	366	118	4.18	2.38	61.0	39.4	78.0	46.9	290	66.0
120 × 80	5.0	14.8	18.9	370	195	4.43	3.21	61.7	48.8	75.4	56.7	401	77.9
	6.3	18.4	23.4	447	234	4.37	3.16	74.6	58.4	92.3	69.1	486	93.0
	8.0	22.9	29.1	537	278	4.29	3.09	89.5	69.4	113	83.9	586	110
	10.0	27.9	35.5	628	320	4.20	3.00	105	80.0	134	99.4	688	126
150 × 100	5.0	18.7	23.9	747	396	5.59	4.07	99.5	79.1	121	90.8	806	127
	6.3	23.3	29.7	910	479	5.53	4.02	121	95.9	148	111	985	153
	8.0	29.1	37.1	1106	577	5.46	3.94	147	115	183	137	1202	184
	10.0	35.7	45.5	1312	678	5.37	3.86	175	136	220	164	1431	215
160 × 80	5.0	18.0	22.9	753	251	5.74	3.31	94.1	62.8	117	71.7	599	106
	6.3	22.3	28.5	917	302	5.68	3.26	115	75.6	144	87.7	729	127
	8.0	27.9	35.5	1113	361	5.60	3.19	139	90.2	177	107	882	151
	10.0	34.2	43.5	1318	419	5.50	3.10	165	105	213	127	1041	175
200 × 100	5.0	22.7	28.9	1509	509	7.23	4.20	151	102	186	115	1202	172
	6.3	28.3	36.0	1851	618	7.17	4.14	185	124	231	141	1473	208
	8.0	35.4	45.1	2269	747	7.09	4.07	227	149	286	174	1802	251
	10.0	43.6	55.5	2718	881	7.00	3.98	272	176	346	209	2154	296
	12.5	53.4	68.0	3218	1022	6.88	3.88	322	204	417	249	2541	342
	16.0	66.4	84.5	3808	1175	6.71	3.73	381	235	505	297	2988	393
250 × 150	6.3	38.2	48.6	4178	1886	9.27	6.23	334	252	405	284	4049	413
	8.0	48.0	61.1	5167	2317	9.19	6.16	413	309	505	353	5014	506
	10.0	59.3	75.5	6259	2784	9.10	6.07	501	371	618	430	6082	606
	12.5	73.0	93.0	7518	3310	8.99	5.97	601	441	751	520	7317	717
	16.0	91.5	117	9089	3943	8.83	5.82	727	526	924	635	8863	851
300 × 200	6.3	48.1	61.2	7880	4216	11.3	8.30	525	422	627	475	8468	681
	8.0	60.5	77.1	9798	5219	11.3	8.23	653	522	785	593	10549	840
	10.0	75.0	95.5	11940	6331	11.2	8.14	796	633	964	726	12850	1016
	12.5	92.6	118	14460	7619	11.1	8.04	964	762	1179	886	15654	1217
	16.0	117	149	17700	9239	10.9	7.89	1180	924	1462	1094	19227	1469
400 × 200	10.0	90.7	116	24140	8138	14.5	8.39	1207	814	1492	916	19236	1377
	12.5	112	143	29410	9820	14.3	8.29	1471	982	1831	1120	23408	1657
	16.0	142	181	36300	11950	14.2	8.14	1815	1195	2285	1388	28835	2011
450 × 250	10.0	106	136	37180	14900	16.6	10.5	1653	1192	2013	1338	33247	1986
	12.5	132	168	45470	18100	16.5	10.4	2021	1448	2478	1642	40668	2407
	16.0	167	213	56420	22250	16.3	10.2	2508	1780	3103	2047	50478	2948

STRUCTURAL TEES
CUT FROM UNIVERSAL BEAMS
DIMENSIONS AND PROPERTIES

Serial size	Mass per metre	Cut from universal beam Serial size	Mass	Depth of section A	Width of section B	Thickness Web t	Thickness Flange T	Root radius r	Area of section
mm	kg	mm	kg/m	mm	mm	mm	mm	mm	cm²
305 × 457	127	914 × 305	253	459.2	305.5	17.3	27.9	19.1	161.2
	112		224	455.2	304.1	15.9	23.9	19.1	142.5
	101		201	451.5	303.4	15.2	20.2	19.1	128.0
292 × 419	113	838 × 292	226	425.5	293.8	16.1	26.8	17.8	144.2
	97		194	420.4	292.4	14.7	21.7	17.8	123.4
	88		176	417.4	291.6	14.0	18.8	17.8	111.9
267 × 381	99	762 × 267	197	384.8	268.0	15.6	25.4	16.5	125.3
	87		173	381.0	266.7	14.3	21.6	16.5	110.1
	74		147	376.9	265.3	12.9	17.5	16.5	93.9
254 × 343	85	686 × 254	170	346.5	255.8	14.5	23.7	15.2	108.2
	76		152	343.8	254.5	13.2	21.0	15.2	96.8
	70		140	341.8	253.7	12.4	19.0	15.2	89.2
	63		125	339.0	253.0	11.7	16.2	15.2	79.7
305 × 457	127	914 × 305	253	459.2	305.5	17.3	27.9	19.1	161.4
	112		224	455.2	304.1	15.9	23.9	19.1	142.6
	101		201	451.5	303.4	15.2	20.2	19.1	128.2
292 × 419	113	838 × 292	226	425.5	293.8	16.1	26.8	17.8	144.4
	97		194	420.4	292.4	14.7	21.7	17.8	123.6
	88		176	417.4	291.6	14.0	18.8	17.8	112.1
267 × 381	99	762 × 267	197	384.8	268.0	15.6	25.4	16.5	125.4
	87		173	381.0	266.7	14.3	21.6	16.5	110.2
	74		147	376.9	265.3	12.9	17.5	16.5	94.0
254 × 343	85	686 × 254	170	346.5	255.8	14.5	23.7	15.2	108.3
	76		152	343.8	254.5	13.2	21.0	15.2	96.9
	70		140	341.8	253.7	12.4	19.0	15.2	89.3
	63		125	339.0	253.0	11.7	16.2	15.2	79.8
305 × 305	119	610 × 305	238	316.5	311.5	18.6	31.4	16.5	151.9
	90		179	308.7	307.0	14.1	23.6	16.5	114.0
	75		149	304.8	304.8	11.9	19.7	16.5	95.1
229 × 305	70	610 × 229	140	308.5	230.1	13.1	22.1	12.7	89.2
	63		125	305.9	229.0	11.9	19.6	12.7	79.8
	57		113	303.7	228.2	11.2	17.3	12.7	72.2
	51		101	301.1	227.6	10.6	14.8	12.7	64.6
210 × 267	61	533 × 210	122	272.3	211.9	12.8	21.3	12.7	77.9
	55		109	269.7	210.7	11.6	18.8	12.7	69.3
	51		101	268.4	210.1	10.9	17.4	12.7	64.6
	46		92	266.6	209.3	10.2	15.6	12.7	58.9
	41		82	264.2	208.7	9.6	13.2	12.7	52.2

Values in the shaded area relate to Universal Beams with tapered flanges. See page 9.
Structural Tees used in compression shall conform to the requirements of clause 32a of BS 449

STRUCTURAL TEES
CUT FROM UNIVERSAL BEAMS
DIMENSIONS AND PROPERTIES

Serial size	Dimension Cx	Moment of inertia Axis x-x	Moment of inertia Axis y-y	Radius of gyration Axis x-x	Radius of gyration Axis y-y	Elastic modulus Axis x-x Flange	Elastic modulus Axis x-x Toe	Elastic modulus Axis y-y
mm	cm	cm⁴	cm⁴	cm	cm	cm³	cm³	cm³
305 × 457	12.0	32663	6255	14.2	6.23	2716	963.7	409.5
	12.2	29000	5212	14.3	6.05	2386	869.3	342.8
	12.6	26399	4315	14.4	5.81	2101	810.1	284.5
292 × 419	10.8	24636	5330	13.1	6.08	2272	777.2	362.9
	11.1	21353	4192	13.2	5.83	1922	690.4	286.7
	11.4	19559	3555	13.2	5.64	1718	644.3	243.8
267 × 381	9.91	17511	3849	11.8	5.54	1766	613.0	287.2
	10.0	15477	3187	11.9	5.38	1546	550.9	239.0
	10.2	13307	2601	11.9	5.16	1304	484.1	188.5
254 × 343	8.69	12025	3112	10.5	5.36	1384	463.2	243.3
	8.61	10928	2795	10.5	5.28	1246	416.2	218.8
	8.66	9926	2394	10.5	5.18	1146	389.1	188.7
	8.88	8983	1995	10.6	5.00	1011	359.1	157.7
305 × 457	12.0	32696	6659	14.2	6.42	2722	964.2	435.9
	12.1	29036	5611	14.3	6.27	2392	869.8	369.1
	12.5	26433	4713	14.4	6.06	2108	810.8	310.7
292 × 419	10.8	24660	5676	13.1	6.27	2277	777.6	386.5
	11.1	21381	4535	13.2	6.06	1928	690.9	310.2
	11.4	19589	3896	13.2	5.90	1723	644.8	267.2
267 × 381	9.90	17528	4087	11.8	5.71	1770	613.3	305.0
	9.99	15495	3423	11.9	5.57	1551	551.3	256.7
	10.20	13328	2734	11.9	5.39	1308	484.5	206.1
254 × 343	8.67	12036	3310	10.5	5.53	1388	463.4	258.8
	8.6	10738	2891	10.5	5.46	1249	416.5	227.2
	8.65	9939	2589	10.5	5.38	1149	389.3	204.1
	8.87	8997	2190	10.6	5.24	1015	359.5	173.1
305 × 305	7.11	12292	7919	9.00	7.22	1730	500.9	508.4
	6.65	8949	5706	8.86	7.08	1346	369.5	371.7
	6.43	7365	4650	8.80	6.99	1145	306.3	305.1
229 × 305	7.61	7745	2256	9.32	5.03	1018	333.3	196.1
	7.55	6911	1966	9.31	4.96	915.8	299.8	171.8
	7.60	6295	1720	9.34	4.88	827.8	276.6	150.7
	7.80	5710	1456	9.40	4.75	731.8	255.5	127.9
210 × 267	6.67	5182	1696	8.16	4.67	776.8	252.1	160.1
	6.60	4593	1469	8.14	4.60	696.2	225.4	139.4
	6.57	4282	1347	8.14	4.56	647.1	210.1	128.3
	6.66	3805	1156	8.14	4.56	594.9	194.4	113.3
	6.73	3517	1003	8.21	4.38	522.2	178.7	96.1

Values in the shaded area relate to Universal Beams with tapered flanges. See page 9.
Structural Tees used in compression shall conform to the requirements of clause 32a of BS 449

STRUCTURAL TEES
CUT FROM UNIVERSAL BEAMS
DIMENSIONS AND PROPERTIES

Serial size (mm)	Dimension Cx (cm)	Moment of inertia Axis x-x (cm⁴)	Moment of inertia Axis y-y (cm⁴)	Radius of gyration Axis x-x (cm)	Radius of gyration Axis y-y (cm)	Elastic modulus Axis x-x Flange (cm³)	Elastic modulus Axis x-x Toe (cm³)	Elastic modulus Axis y-y (cm³)
191 × 229	5.55	2978	1172	6.90	4.33	536.5	167.2	121.5
	5.49	2701	1043	6.89	4.28	491.5	152.7	108.7
	5.48	2452	935.7	6.89	4.23	453.0	141.5	97.8
	5.42	2246	835.6	6.88	4.19	414.5	128.8	87.7
	5.47	2037	725.9	6.90	4.12	372.7	118.3	76.5
152 × 229	6.03	2608	571.6	7.07	3.31	432.6	151.4	74.5
	5.98	2364	505.9	7.06	3.26	395.0	138.4	66.3
	5.99	2128	438.9	7.06	3.21	355.5	126.1	57.8
	5.82	1870	397.2	7.02	3.23	321.4	110.6	52.0
	6.03	1667	322.3	7.08	3.11	276.3	101.3	42.3
178 × 203	4.80	1758	772.4	6.08	4.03	366.1	111.0	86.0
	4.73	1574	682.3	6.07	4.00	332.6	100.0	76.3
	4.62	1383	599.4	6.03	3.97	295.5	88.1	67.4
	4.81	1282	508.3	6.12	3.85	266.6	83.7	57.2
140 × 203	5.05	1130	269.4	6.19	3.02	223.9	75.0	37.8
	5.27	968.3	205.6	6.26	2.89	183.6	66.4	29.0
171 × 178	4.01	1158	681.2	5.21	3.99	289.1	81.6	78.6
	3.95	979.1	554.5	5.21	3.92	248.1	70.0	64.4
	3.92	877.9	483.9	5.21	3.87	223.7	63.4	56.5
	4.02	791.8	406.1	5.27	3.78	196.9	58.3	47.5
127 × 178	4.41	719.6	178.3	5.40	2.69	163.2	54.4	28.3
	4.52	618.4	140.1	5.44	2.59	136.7	47.9	22.4
165 × 152	3.19	630.4	530.6	4.31	3.94	199.5	51.5	63.6
	3.09	541.0	448.5	4.29	3.90	175.3	44.1	54.1
	3.07	476.2	381.4	4.30	3.85	155.3	39.3	46.2
127 × 152	3.91	653.6	230.0	4.64	2.75	167.1	56.3	36.7
	3.85	567.8	194.1	4.62	2.70	147.5	49.5	31.2
	3.80	504.4	168.6	4.61	2.67	132.6	44.3	27.3
102 × 152	4.14	486.7	96.67	4.83	2.15	117.4	42.4	18.9
	4.23	426.7	78.39	4.85	2.08	101.0	38.0	15.4
	4.48	375.7	60.01	4.89	1.96	83.9	34.9	11.8
146 × 127	2.67	349.1	338.7	3.56	3.51	130.9	33.9	46.0
	2.58	296.7	285.6	3.54	3.47	115.2	29.0	39.0
	2.68	263.2	224.5	3.63	3.35	98.1	26.6	30.7
102 × 127	3.25	278.8	89.10	3.93	2.22	85.7	28.6	17.5
	3.35	252.6	73.87	3.96	2.14	75.3	26.6	15.5
	3.49	227.2	59.97	4.00	2.05	65.2	24.7	11.8
133 × 102	2.10	152.6	191.9	2.83	3.18	72.8	18.5	28.7
	2.12	133.7	154.8	2.88	3.10	63.0	16.6	23.2

Structural Tees used in compression shall conform to the requirements of clause 32a of BS 449

STRUCTURAL TEES
CUT FROM UNIVERSAL BEAMS
DIMENSIONS AND PROPERTIES

Serial size (mm)	Mass per metre (kg)	Cut from universal beam Serial size (mm)	Cut from universal beam Mass (kg/m)	Depth of section A (mm)	Width of section B (mm)	Thickness Web t (mm)	Thickness Flange T (mm)	Root radius r (mm)	Area of section (cm²)
191 × 229	49	457 × 191	98	233.7	192.8	11.4	19.6	10.2	62.6
	45		89	232.0	192.0	10.6	17.7	10.2	57.0
	41		82	230.1	191.3	9.9	16.0	10.2	52.3
	37		74	228.6	190.5	9.1	14.5	10.2	47.5
	34		67	226.8	189.9	8.5	12.7	10.2	42.7
152 × 229	41	457 × 152	82	232.5	153.5	10.7	18.9	10.2	52.2
	37		74	230.6	152.7	9.9	17.0	10.2	47.5
	34		67	228.6	151.9	9.1	15.0	10.2	42.7
	30		60	227.3	152.9	8.0	13.3	10.2	38.0
	26		52	224.9	152.4	7.6	10.9	10.2	33.2
178 × 203	37	406 × 178	74	206.4	179.7	9.7	16.0	10.2	47.5
	34		67	204.7	178.8	8.8	14.3	10.2	43.0
	30		60	203.2	177.8	7.8	12.8	10.2	38.0
	27		54	201.3	177.6	7.6	10.9	10.2	34.2
140 × 203	23	406 × 140	46	201.2	142.4	6.9	11.2	10.2	29.5
	20		39	198.6	141.8	6.3	8.6	10.2	24.7
171 × 178	34	356 × 171	67	182.0	173.2	9.1	15.7	10.2	42.7
	29		57	179.3	172.1	8.0	13.0	10.2	36.1
	26		51	177.8	171.5	7.3	11.5	10.2	32.3
	23		45	176.0	171.0	6.9	9.7	10.2	28.5
127 × 178	20	356 × 127	39	176.4	126.0	6.5	10.7	10.2	24.7
	17		33	174.2	125.4	5.9	8.5	10.2	20.9
165 × 152	27	305 × 165	54	155.4	166.8	7.7	13.7	8.9	34.2
	23		46	153.5	166.7	6.7	11.8	8.9	29.5
	20		40	151.9	165.1	6.1	10.2	8.9	25.8
127 × 152	24	305 × 127	48	155.2	125.2	8.9	14.0	8.9	30.4
	21		42	153.3	124.3	8.0	12.1	8.9	26.6
	19		37	151.9	123.5	7.2	10.7	8.9	23.7
102 × 152	17	305 × 102	33	156.3	102.4	6.6	10.8	7.6	20.9
	14		28	154.4	101.9	6.1	8.9	7.6	18.2
	13		25	152.4	101.6	5.8	6.8	7.6	15.7
146 × 127	22	254 × 146	43	129.8	147.3	7.3	12.7	7.6	27.6
	19		37	128.0	146.4	6.4	10.9	7.6	23.7
	16		31	125.7	146.1	6.1	8.6	7.6	20.0
102 × 127	14	254 × 102	28	130.2	102.1	6.4	10.0	7.6	18.1
	13		25	128.5	101.9	6.1	8.4	7.6	16.1
	11		22	127.0	101.6	5.8	6.8	7.6	14.2
133 × 102	15	203 × 133	30	103.4	133.8	6.3	9.6	7.6	19.0
	13		25	101.6	133.4	5.8	7.8	7.6	16.1

Structural Tees used in compression shall conform to the requirements of clause 32a of BS 449

STRUCTURAL TEES
CUT FROM UNIVERSAL COLUMNS
DIMENSIONS AND PROPERTIES

Serial size (mm)	Dimension Cx (cm)	Moment of inertia Axis x-x (cm⁴)	Moment of inertia Axis y-y (cm⁴)	Radius of gyration Axis x-x (cm)	Radius of gyration Axis y-y (cm)	Elastic modulus Axis x-x Flange (cm³)	Elastic modulus Axis x-x Toe (cm³)	Elastic modulus Axis y-y (cm³)
406 × 178	3.41	2886	15504	4.39	10.2	846.3	184.5	785.1
368 × 178	3.32	2500	11816	4.40	9.57	754.1	162.1	631.2
	3.10	2099	10235	4.31	9.52	677.4	137.0	550.1
	2.92	1765	8735	4.25	9.46	605.5	116.3	471.9
	2.73	1451	7278	4.19	9.39	531.1	96.4	395.2
305 × 152	3.04	1529	6262	3.90	7.89	502.8	114.9	403.2
	2.86	1291	5336	3.85	7.82	450.9	99.1	345.7
	2.69	1075	4503	3.79	7.75	399.9	82.5	293.5
	2.50	858.0	3634	3.73	7.68	343.1	66.6	238.5
254 × 127	2.72	886.7	3722	3.25	6.66	325.6	79.9	285.2
	2.47	683.0	2951	3.16	6.57	277.0	62.8	228.5
	2.24	534.7	2424	3.06	6.52	238.7	49.6	189.5
	2.06	418.7	1936	3.00	6.46	203.7	39.3	152.5
203 × 102	2.22	379.8	1560	2.63	5.32	171.0	42.7	149.4
	1.98	288.1	1268	2.52	5.28	145.3	32.7	123.0
	1.88	241.2	1071	2.52	5.19	128.3	28.1	99.5
	1.76	203.1	885.0	2.47	5.16	115.2	23.8	86.8
	1.71	179.6	769.4	2.47	5.11	105.3	21.2	75.7
152 × 76	1.55	94.66	354.3	2.00	3.87	61.2	14.5	45.9
	1.41	72.79	279.2	1.95	3.82	51.5	11.3	36.5
	1.43	61.10	201.7	2.03	3.68	42.7	9.88	26.5

Structural Tees used in compression shall conform to the requirements of clause 32a of BS 449.

STRUCTURAL TEES
CUT FROM UNIVERSAL COLUMNS
DIMENSIONS AND PROPERTIES

Serial size (mm)	Mass per metre (kg)	Cut from universal beam Serial size (mm)	Mass (kg/m)	Width of section B (mm)	Depth of section A (mm)	Thickness Web t (mm)	Thickness Flange T (mm)	Root radius r (mm)	Area of section (cm²)
406 × 178	118	356 × 406	235	395.0	190.5	18.5	30.2	15.2	149.9
368 × 178	101	356 × 368	202	374.4	187.3	16.8	27.0	15.2	129.0
	89		177	372.1	184.2	14.5	23.8	15.2	112.9
	77		153	372.1	181.0	12.6	20.7	15.2	97.6
	65		129	368.3	177.8	10.7	17.5	15.2	82.5
305 × 152	79	305 × 305	158	310.6	163.6	15.7	25.0	15.2	100.6
	69		137	308.7	160.3	13.8	21.7	15.2	87.3
	59		118	306.8	157.2	11.9	18.4	15.2	74.9
	49		97	304.8	153.9	9.9	15.4	15.2	61.6
254 × 127	66	254 × 254	132	261.0	138.2	15.6	25.1	12.7	83.9
	54		107	258.3	133.4	13.0	20.5	12.7	68.3
	45		89	255.9	130.2	10.5	17.3	12.7	57.0
	37		73	254.0	127.0	8.6	14.2	12.7	46.4
203 × 102	43	203 × 203	86	208.8	111.1	13.0	20.5	10.2	55.0
	36		71	206.2	108.0	10.3	17.3	10.2	45.5
	30		60	205.2	104.8	9.3	14.2	10.2	37.9
	26		52	203.9	103.1	8.0	12.5	10.2	33.2
	23		46	203.2	101.6	7.3	11.0	10.2	29.4
152 × 76	19	152 × 152	37	154.4	80.9	8.1	11.5	7.6	23.7
	15		30	152.9	78.7	6.6	9.4	7.6	19.1
	12		23	152.4	76.2	6.1	6.8	7.6	14.9

Structural Tees used in compression shall conform to the requirements of clause 32a of BS 449.

84

CASTELLATED UNIVERSAL BEAMS

DIMENSIONS AND PROPERTIES

Serial size Original	Serial size Castellated	Mass per metre	Depth of section Dc	Width of section B	Web t	Flange T	Depth between fillets dc	Area Gross	Area Net
mm	mm	kg	mm	mm	mm	mm	mm	cm²	cm²
914 × 419	1371 × 419	388	1377.8	420.5	21.5	36.6	1248.6	592.0	395.8
		343	1368.6	418.5	19.4	32.0	1248.6	525.8	348.1
914 × 305	1371 × 305	289	1383.8	307.8	19.6	32.0	1276.3	457.9	279.1
		253	1375.8	305.5	17.3	27.9	1276.3	401.4	243.5
		224	1367.5	304.1	15.9	23.9	1276.3	357.6	212.3
		201	1360.2	303.4	15.2	20.2	1276.3	325.4	186.8
838 × 292	1257 × 292	226	1270.0	293.8	16.1	26.8	1175.6	356.0	220.8
		194	1259.8	292.4	14.7	21.7	1175.6	308.6	185.1
		176	1254.0	291.6	14.0	18.8	1175.6	282.6	165.1
762 × 267	1143 × 267	197	1150.6	268.0	15.6	25.4	1062.1	310.0	191.0
		173	1143.0	266.7	14.3	21.6	1062.1	274.8	165.6
		147	1134.9	265.3	12.9	17.5	1062.1	237.0	138.7
686 × 254	1029 × 254	170	1035.8	255.8	14.5	23.7	953.4	266.0	166.7
		152	1030.5	254.5	13.2	21.0	953.4	238.7	148.5
		140	1026.4	253.7	12.4	19.0	953.4	221.1	135.7
		125	1020.8	253.0	11.7	16.2	953.4	199.6	119.2
914 × 419	1371 × 419	388	1377.7	420.5	21.5	36.6	1256.3	592.2	396.3
		343	1368.6	418.5	19.4	32.0	1256.3	526.3	348.7
914 × 305	1371 × 305	289	1383.8	307.8	19.6	32.0	1281.7	458.2	279.4
		253	1375.7	305.5	17.3	27.9	1281.7	401.8	243.8
		224	1367.5	304.1	15.9	23.9	1281.7	357.8	212.7
		201	1360.2	303.4	15.2	20.2	1281.7	325.8	187.1
838 × 292	1257 × 292	226	1270.0	293.8	16.1	26.8	1180.8	356.3	221.1
		194	1259.8	292.4	14.7	21.7	1180.8	308.9	185.4
		176	1254.0	291.6	14.0	18.8	1180.8	282.9	165.4
762 × 267	1143 × 267	197	1150.6	268.0	15.6	25.4	1066.8	310.3	191.3
		173	1143.0	266.7	14.3	21.6	1066.8	275.1	165.9
		147	1134.9	265.3	12.9	17.5	1066.8	237.2	138.9
686 × 254	1029 × 254	170	1035.8	255.8	14.5	23.7	958.0	266.2	166.9
		152	1030.5	254.5	13.2	21.0	958.0	238.9	148.7
		140	1026.4	253.7	12.4	19.0	958.0	222.3	136.0
		125	1020.8	253.0	11.7	16.2	958.0	199.9	119.4
610 × 305	915 × 305	238	937.8	311.5	18.6	31.4	842.0	360.5	247.1
		179	922.3	307.0	14.1	23.6	842.0	271.0	184.9
		149	914.4	304.8	11.9	19.7	842.0	226.4	153.9
610 × 229	915 × 229	140	921.8	230.1	13.1	22.1	852.1	218.3	138.4
		125	916.7	229.0	11.9	19.6	852.1	195.9	123.2
		113	912.1	228.2	11.2	17.3	852.1	178.5	110.4
		101	907.0	227.6	10.6	14.8	852.1	161.4	97.0
533 × 210	800 × 210	122	811.3	211.9	12.8	21.3	743.2	189.8	121.8
		109	806.2	210.7	11.6	18.8	743.2	169.4	107.8
		101	803.4	210.1	10.9	17.4	743.2	156.4	100.2
		92	799.8	209.3	10.2	15.6	743.2	144.9	90.7
		82	795.0	208.7	9.6	13.2	743.2	129.9	79.0

Depth Dc of castellated section = D + Ds/2
where D = actual depth of original section.
and Ds = serial depth of original section.
Values in the shaded area relate to Universal Beams with tapered flanges. See page 9.

CASTELLATED UNIVERSAL BEAMS

DIMENSIONS AND PROPERTIES

Serial size	Moment of inertia (net) Axis x-x	Axis y-y	Design radius of gyration Axis x-x	Axis y-y	Elastic modulus (net) Axis x-x	Axis y-y	Pitch of standard castellation 1.08 Ds	Ratio Dc/T
mm	cm⁴	cm⁴	cm	cm	cm³	cm³	mm	
1371 × 419	1661760	42437	60.0	9.41	24124	2018	987.6	37.7
	1450060	36217	59.6	9.25	21191	1731	987.6	42.8
1371 × 305	1161830	14763	58.8	6.48	16792	959.3	987.6	43.2
	1000740	12491	58.6	6.37	14551	817.8	987.6	49.2
	870091	10407	58.1	6.20	12725	684.5	987.6	57.3
	755547	8617	57.4	5.97	11110	568.1	987.6	67.4
1257 × 292	779754	10645	54.3	6.21	12280	724.8	905.3	47.4
	643948	8372	53.5	5.97	10223	572.7	905.3	58.0
	568295	7100	53.1	5.79	9064	486.8	905.3	66.7
1143 × 267	552645	7686	49.1	5.67	9606	573.6	823.0	45.3
	474157	6365	48.6	5.51	8297	477.3	823.0	52.9
	391778	4994	48.1	5.30	6904	376.5	823.0	64.8
1029 × 254	391967	6215	44.4	5.47	7568	486.0	740.7	43.9
	347066	5383	44.2	5.39	6730	423.1	740.7	49.1
	315224	4783	44.0	5.30	6142	377.0	740.7	54.1
	273698	3986	43.5	5.13	5362	315.1	740.7	63.1
1371 × 419	1664700	45369	60.0	9.73	24168	2158	987.6	37.7
	1453070	39121	59.6	9.61	21235	1870	987.6	42.8
1371 × 305	1163600	15581	58.8	6.65	16818	1013	987.6	43.2
	1009510	13298	58.6	6.57	14677	870.0	987.6	49.2
	871861	11207	58.1	6.43	12751	737.1	987.6	57.3
	757316	9413	57.4	6.24	11136	620.5	987.6	67.4
1257 × 292	781081	11338	54.3	6.40	12300	771.9	905.3	47.4
	645275	9058	53.6	6.21	10244	619.6	905.3	58.0
	569622	7783	53.1	6.06	9085	533.7	905.3	66.7
1143 × 267	553577	8162	49.1	5.83	9622	609.1	823.0	45.3
	475089	6837	48.7	5.71	8313	512.7	823.0	52.9
	392710	5461	48.1	5.54	6921	411.7	823.0	64.8
1029 × 254	392620	6612	44.4	5.64	7581	517.0	740.7	43.8
	347719	5776	44.2	5.58	6749	454.0	740.7	49.1
	315898	5173	44.0	5.50	6155	407.7	740.7	54.1
	274342	4374	43.5	5.37	5375	345.8	740.7	63.1
915 × 305	474680	15021	40.7	7.32	10124	1016	658.4	29.9
	349528	11404	40.4	7.17	7580	742.9	658.4	39.0
	288534	9296	40.2	7.09	6311	610.0	658.4	46.5
915 × 229	257999	4506	39.6	5.13	5598	391.6	658.4	41.6
	228006	3929	39.4	5.07	4975	343.1	658.4	46.7
	202743	3436	39.1	4.99	4446	301.1	658.4	52.7
	176124	2909	38.7	4.86	3884	255.6	658.4	61.4
800 × 210	175224	3388	34.8	4.75	4320	319.8	576.1	38.0
	153921	2934	34.6	4.69	3818	278.5	576.1	46.2
	142447	2691	34.6	4.65	3546	256.2	576.1	46.2
	128162	2390	34.4	4.60	3205	228.4	576.1	51.2
	110323	2003	34.4	4.48	2775	192.0	576.1	60.2

Values in the shaded area relate to Universal Beams with tapered flanges. See page 9.
Design radius of gyration is average of values for gross and net sections.

CASTELLATED UNIVERSAL BEAMS

DIMENSIONS AND PROPERTIES

Serial size Original	Castellated	Mass per metre	Depth of section Dc	Width of section B	Web t	Flange T	Depth between fillets dc	Gross	Net
mm	mm	kg	mm	mm	mm	mm	mm	cm²	cm²
457 × 191	686 × 191	98	696.0	192.8	11.4	19.6	636.5	151.4	89.1
		89	692.2	192.0	10.6	17.7	636.5	138.1	89.8
		82	688.8	191.3	9.1	16.0	636.5	127.2	81.9
		74	685.8	190.5	9.1	14.5	636.5	115.8	74.2
		67	682.2	189.9	8.5	12.7	636.5	104.8	66.1
457 × 152	686 × 152	82	693.7	153.5	10.7	18.9	635.6	128.9	80.1
		74	689.9	152.7	9.9	17.0	635.6	117.6	72.4
		67	685.8	151.9	9.1	15.0	635.6	106.3	64.5
		60	683.3	152.9	8.0	13.3	635.6	94.3	57.6
		52	678.4	152.4	7.6	10.9	635.6	83.8	49.2
406 × 178	609 × 178	74	615.9	179.7	9.7	16.0	563.7	114.6	75.3
		67	611.6	178.8	8.8	14.3	563.7	103.3	67.6
		60	609.6	177.8	7.8	12.8	563.7	91.9	60.2
		54	605.8	177.6	7.6	10.9	563.7	83.9	53.0
406 × 140	609 × 140	46	605.5	142.4	6.9	11.2	562.9	73.0	44.9
		39	600.5	141.8	6.3	8.6	562.9	62.3	36.6
356 × 171	534 × 171	67	541.8	173.2	9.1	15.7	490.1	101.6	69.3
		57	536.4	172.1	8.0	13.0	490.1	86.3	58.0
		51	533.4	171.5	7.3	11.5	490.1	77.5	51.6
		45	529.8	171.0	6.9	9.7	490.1	69.2	44.8
356 × 127	534 × 127	39	530.6	126.0	6.5	10.7	488.9	61.0	37.8
		33	526.3	125.4	5.9	8.5	488.9	52.4	31.3
305 × 165	458 × 165	54	463.3	166.8	7.7	13.7	418.1	80.2	56.6
		46	459.5	165.7	6.7	11.8	418.1	69.2	48.7
		40	456.2	165.1	6.1	10.2	418.1	60.8	42.2
305 × 127	458 × 127	48	462.8	125.2	8.9	14.0	417.0	74.4	47.3
		42	459.0	124.3	8.0	12.1	417.0	65.3	41.1
		37	456.2	123.5	7.2	10.7	417.0	58.5	36.5
305 × 102	458 × 102	33	465.1	102.4	6.6	10.8	428.3	51.8	31.7
		28	461.3	101.9	6.1	8.9	428.3	45.6	27.0
		25	457.2	101.6	5.8	6.8	428.3	40.3	22.5
254 × 146	381 × 146	43	386.6	147.3	7.3	12.7	345.9	64.4	45.8
		37	383.0	146.4	6.4	10.9	345.9	55.6	39.3
		31	378.5	146.1	6.1	8.6	345.9	47.7	33.3
254 × 102	381 × 102	28	387.3	102.1	6.4	10.0	352.1	44.3	28.1
		25	384.0	101.9	6.1	8.4	352.1	39.9	24.4
		22	381.0	101.6	5.8	6.8	352.1	35.8	21.0
203 × 133	305 × 133	30	308.4	133.8	6.3	9.6	273.9	44.4	31.6
		25	304.8	133.4	5.8	7.8	273.9	38.3	26.4

Depth Dc of castellated section = D + Ds/2.
where D = actual depth of original section
and Ds = serial depth of original section.

87

CASTELLATED UNIVERSAL BEAMS

DIMENSIONS AND PROPERTIES

Serial size castellated	Moment of inertia (net) Axis x-x	Axis y-y	Design radius of gyration Axis x-x	Axis y-y	Elastic modulus (net) Axis x-x	Axis y-y	Pitch of standard castellation 1.08 Ds	Ratio Dc/T
mm	cm⁴	cm⁴	cm	cm	cm³	cm³	mm	
686 × 191	105135	2340	30.0	4.40	3021	242.7	493.8	35.6
	94589	2084	29.9	4.35	2733	217.1	493.8	39.2
	85760	1870	29.7	4.31	2490	195.4	493.8	43.0
	77336	1670	29.7	4.27	2255	175.3	493.8	47.4
	68284	1451	29.5	4.20	2002	152.8	493.8	53.7
686 × 152	83571	1141	29.5	3.38	2410	148.7	493.8	36.7
	75048	1010	29.4	3.34	2176	132.3	493.8	40.6
	66316	876.3	29.2	3.28	1934	115.3	493.8	45.8
	59152	793.4	29.2	3.31	1731	103.8	493.8	51.3
	49777	643.8	28.8	3.20	1467	84.49	493.8	62.3
609 × 178	62965	1543	26.7	4.10	2044	171.8	438.9	38.6
	56197	1364	26.6	4.06	1835	152.5	438.9	42.8
	49793	1198	26.5	4.04	1634	134.7	438.9	47.7
	43276	1016	26.2	3.93	1429	114.4	438.9	55.7
609 × 140	36405	538.2	26.0	3.09	1202	75.61	438.9	54.2
	29104	410.8	25.6	2.96	969.4	57.95	438.9	69.5
534 × 171	44857	1361	23.6	4.05	1656	157.2	384.0	34.5
	37124	1108	23.4	3.98	1384	128.8	384.0	41.2
	32784	967.3	23.3	3.93	1229	112.8	384.0	46.4
	28102	811.7	23.1	3.84	1061	94.92	384.0	54.5
534 × 127	23474	356.1	22.8	2.74	884.8	56.54	384.0	49.7
	19164	280.0	22.5	2.65	728.3	44.66	384.0	61.9
458 × 165	26956	1061	20.4	3.98	1164	127.2	329.2	33.8
	22993	896.7	20.3	3.95	1001	108.2	329.2	38.9
	19773	762.5	20.2	3.90	866.9	92.36	329.2	44.9
458 × 127	21962	459.0	19.8	2.80	949.1	73.31	329.2	38.1
	18893	387.6	19.6	2.76	823.3	62.37	329.2	38.0
	16667	336.7	19.6	2.72	730.7	54.50	329.2	42.7
458 × 102	14864	193.0	19.7	2.20	639.2	37.70	329.2	43.2
	12477	156.5	19.5	2.13	541.0	30.73	329.2	52.0
	10152	119.8	19.1	2.02	444.1	23.57	329.2	66.9
381 × 146	15120	676.9	17.0	3.54	782.2	91.91	274.3	30.4
	12869	571.0	16.5	3.51	672.0	78.03	274.3	35.1
	10348	448.7	16.6	3.40	546.8	61.45	274.3	43.8
381 × 102	9198	177.9	16.6	2.26	474.9	34.85	274.3	38.7
	7859	147.5	16.3	2.19	409.3	28.96	274.3	46.0
	6641	119.7	16.1	2.11	348.6	23.57	274.3	55.8
305 × 133	6666	383.7	13.6	3.21	432.4	57.34	219.5	32.1
	5472	309.4	13.4	3.14	359.1	46.41	219.5	39.0

Design radius of gyration is average of values for gross and net sections.

BLACK BOLTS AND NUTS TO BS 4190

DIMENSIONS IN MILLIMETRES

Nominal size and thread diameter	Pitch of thread (coarse pitch series)	Max. width of head and nut		Max. height of head		Max. thickness of nut		Tensile stress area mm²	Minimum distance between centres
		Across flats	Across corners	Black	Faced on underside	Black	Faced one side		
6	1	10.00	11.5	4.375	4.24	5.375	5	20.1	15
8	1.25	13.00	15.0	5.875	5.74	6.875	6.5	36.6	20
10	1.5	17.00	19.6	7.45	7.29	8.45	8	58.0	25
12	1.75	19.00	21.9	8.45	8.29	10.45	10	84.3	30
16	2	24.00	27.7	10.45	10.29	13.55	13	157	40
20	2.5	30.00	34.6	13.90	13.35	16.55	16	245	50
(22)	2.5	32.00	36.9	14.90	14.35	18.55	18	303	55
24	3	36.00	41.6	15.90	15.35	19.65	19	353	60
(27)	3	41.00	47.3	17.90	17.35	22.65	22	459	67.5
30	3.5	46.00	53.1	20.05	19.42	24.65	24	561	75
(33)	3.5	50.00	57.7	22.05	21.42	26.65	26	694	82.5
36	4	55.00	63.5	24.05	23.42	29.65	29	817	90
(39)	4	60.00	69.3	26.05	25.42	31.80	31	976	97.5
42	4.5	65.00	75.1	27.05	26.42	34.80	34	1120	105
(45)	4.5	70.00	80.8	29.05	28.42	36.80	36	1300	112
48	5.0	75.00	86.6	31.05	30.42	38.80	38	1470	120
(52)	5.0	80.00	92.4	34.25	33.50	42.80	42	1760	130
56	5.5	85.00	98.1	36.25	35.50	45.80	45	2030	140

Sizes shown in brackets are non-preferred.

BLACK WASHERS TO BS 4320

DIMENSIONS IN MILLIMETRES

Nominal size of bolt	Flat				Taper
	Form E		Form F		
	Outside dia	Thickness	Outside dia	Thickness	
6	12.5	1.6			For taper washers see page 135.
8	17.0	1.6	21.0	1.6	
10	21.0	2.0	24.0	2.0	
12	24.0	2.5	28.0	2.5	Sizes shown in brackets are non-preferred.
16	30.0	3.0	34.0	3.0	
20	37.0	3.0	39.0	3.0	
(22)	39.0	3.0	44.0	3.0	
24	44.0	3.0	50.0	3.0	
(27)	50.0	4.0	56.0	4.0	
30	56.0	4.0	60.0	4.0	
(33)	60.0	5.0	66.0	5.0	
36	66.0	5.0	72.0	6.0	

H.S.F.G. BOLTS AND NUTS TO BS 4395: PART 1

DIMENSIONS IN MILLIMETRES

Nominal size and thread diameter	Pitch of thread (coarse pitch series)	Max. width of head and nut		Max. height		Washer face of head and nut		Max. radius under head	Tensile stress area mm²	Proof load	
		Across flats	Across corners	Head	Nut	Max. dia.	Max. thickness			tonne-force (1000kgf)	kilo-newtons
12*	1.75	22.0	25.4	10.45	11.55	22.0	0.4	1.0	84.3	5.04	49.4
16	2.0	27.0	31.2	10.45	15.55	27.0	0.4	1.0	157	9.39	92.1
20	2.5	32.0	36.9	13.90	18.55	32.0	0.4	1.2	245	14.64	144
22	2.5	36.0	41.6	14.90	19.65	36.0	0.4	1.2	303	18.11	177
24	3.0	41.0	47.3	15.90	22.65	41.0	0.5	1.2	353	21.10	207
27	3.0	46.0	53.1	17.90	24.65	46.0	0.5	1.5	459	23.88	234
30	3.5	50.0	57.7	20.05	26.65	50.0	0.5	1.5	561	29.19	286
36	4.0	60.0	69.3	24.05	31.80	60.0	0.5	1.5	817	42.51	418

*Not recommended. Only to be used for the lighter type of construction where practical conditions, such as material thickness, do not warrant the usage of a larger size bolt than M12.
There are other types of friction grip bolts available and reference should be made to BS 4395: Parts 2 and 3.

H.S.F.G. WASHERS TO BS 4395: PART 1

DIMENSIONS IN MILLIMETRES

Nominal size of bolt.	Round flat		Square taper		
	Outside diameter	Thickness	Length of side	Mean thickness	
				3° and 5°	8°
12	30	2.8	31.75	4.76	6.35
16	37	3.4	38.10	4.76	6.35
20	44	3.7	38.10	4.76	6.35
22	50	4.2	44.45	4.76	6.35
24	56	4.2	57.15	4.76	6.35
27	60	4.2	57.15	4.76	6.35
30	66	4.2	57.15	4.76	6.35
36	85	4.6	57.15	4.76	6.35

SPACING OF HOLES IN COLUMNS, BEAMS, JOISTS AND TEES

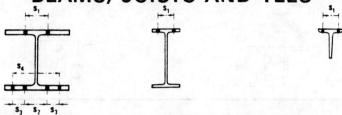

Nominal flange widths	Spacings				Recommended dia. of rivet or bolt	Actual b_{min}	Nominal flange widths	Spacing	Recommended dia. of rivet or bolt	Actual b_{min}
	S_1	S_2	S_3	S_4				S_1		
mm	mm	mm	mm	mm	mm	mm	mm	mm	mm	mm
419 to 368	140	140	75	290	24	362	146 to 114	70	20	130
330 to 305	140	120	60	240	24	312	102	54	12	98
do.	140	120	60	240	20	300	89	50		
292 to 203	140				24	212	76	40		
190 to 165	90				24	162	64	34		
152	90				20	150	51	30		

Note that the actual flange width for a universal section may be less than the nominal size and that the difference may be significant in determining the maximum diameter. The column headed b_{min} gives the actual minimum width of flange required to comply with Table 21 of BS 449.

The dimensions S_1 and S_2 have been selected for normal conditions but adjustments may be necessary for relatively large diameter fasteners or for particularly heavy weights in a given serial size.

SPACING OF HOLES IN CHANNELS

Nominal flange width	S_1	Recommended dia. of rivet or bolt
mm	mm	mm
102	55	24
89	55	20
76	45	20
64	35	16
51	30	10
38	22	

SPACING OF HOLES IN ANGLES

Nominal leg length	Spacing of holes						Maximum diameter of bolt or rivet		
	S_1	S_2	S_3	S_4	S_5	S_6	S_1	S_2 and S_3	S_4, S_5 and S_6
mm	mm	mm	mm	mm	mm	mm	mm	mm	mm
200		75	75	55	55	55		30	20
150		55	55					20	
125		45	50					20	
120		45	50					16	
100	55						24		
90	50						24		
80	45						20		
75	45						20		
70	40						20		
65	35						20		
60	35						16		
50	28						12		
45	25								
40	23								
30	20								
25	15								

Note that HSFG bolts may require adjustments to the backmarks shown due to the larger nut and washer dimensions.

Inner gauge lines are for normal conditions and may require adjustment for large diameters of fasteners or for thick members.

Outer gauge lines may require consideration in relation to a specific edge distance.

TABLE C. TORSION CONSTANTS, T

Universal beams

Serial size	Mass	T
mm	kg/m	N/mm²
914 × 419	388	285
	343	225
914 × 305	289	233
	253	180
	224	144
	201	120
838 × 292	226	187
	194	140
	176	118
762 × 267	197	210
	173	165
	147	124
686 × 254	170	223
	152	180
	140	154
	125	126
610 × 305	238	449
	179	261
	149	187
610 × 229	140	236
	125	191
	113	159
	101	131
533 × 210	122	287
	109	231
	101	202
	92	170
	82	138
457 × 191	98	319
	89	268
	82	225
	74	189
	67	155
457 × 152	82	309
	74	257
	67	209
	60	163
	52	129
406 × 178	74	279
	67	227
	60	182
	54	151
406 × 140	46	152
	39	110
356 × 171	67	333
	57	242
	51	197
	45	158
356 × 127	39	182
	33	135

Universal beams

Serial size	Mass	T
mm	kg/m	N/mm²
305 × 165	54	335
	46	253
	40	200
305 × 127	48	413
	42	323
	37	260
305 × 102	33	238
	28	184
	25	145
254 × 146	43	419
	37	317
	31	231
254 × 102	29	299
	25	244
	22	199
203 × 133	30	409
	25	306

Joists

Nominal size	Mass	T
mm	kg/m	N/mm²
254 × 203	81·85	1356
254 × 114	37·20	588
203 × 152	52·09	1494
203 × 102	25·33*	518
173 × 102	21·54*	533
152 × 127	37·20	2336
152 × 89	17·09*	606
152 × 76	17·86	968
127 × 114	29·76	2688
127 × 114	26·79	2030
127 × 76	16·37	1375
127 × 76	13·36*	768
114 × 114	26·79	3461
102 × 102	23·07	3966
102 × 64	9·65*	954
102 × 44	7·44	1101
89 × 89	19·35	5439
76 × 76	14·67	5893
76 × 76	12·65	3472

Note. Joist sections marked* are BS4 sizes and have 5° taper flanges. All other joists taper 8°.

Universal columns

Serial size	Mass	T
mm	kg/m	N/mm²
356 × 406	634	6517
	551	5142
	467	3869
	393	2839
	340	2182
	287	1597
	235	1102
Column core	477	4280
356 × 368	202	900
	177	696
	153	532
	129	384
305 × 305	283	3093
	240	2293
	198	1620
	158	1055
	137	807
	118	608
	97	421
254 × 254	167	2427
	132	1566
	107	1072
	89	755
	73	518
203 × 203	86	1640
	71	1145
	60	824
	52	634
	46	507
152 × 152	37	958
	30	651
	23	429

Glossary of Structural Terms

allowable stress - maximum stress permitted at working load

aspect ratio - ratio of the lengths of sides of a component

axial force - a tensile or compressive force applied at the centroid of the cross section of a member and parallel to the length of the member

back mark - distance to the centre line of a bolt hole from the outside corner of an angle

bar length - length of a standard steel section obtained from a steel stockist

batten - a short member linking two components of a compound structural member

bay - the space between two columns of a building framework

bearing - a support for a structure

bearing stress - a local compressive stress at a support or concentrated load

bending moment - a couple which produces rotation of members or structures

bending stress - a non-uniform stress produced by a bending moment

biaxial bending - bending moments applied simultaneously about the major x-x and minor y-y axis

bolt - a solid cylindrical fastener incorporating a fixed head at one end, washers, and a nut on a thread at the other end

boom - members situated at the top and bottom of a truss

bracing - a system of members designed to stiffen a structure, or prevent it from behaving as a mechanism, or becoming unstable

brittle fracture - a premature sudden tensile stress failure of a material associated with a defect and a temperature below the transition temperature

buckling - a form of failure associated with compressive stresses, which produces sudden movements at right angles to the length of the member

butt weld - a fusion of two components, one or both of which is attached by its cross section

cantilever beam - a beam built in to a rigid support at one end

castellated beam - an I section beam with hexagonal holes in the web

centroid (of a section) - the point about which the sum of first moments of area is zero

clearance - a small distance left intentionally between components of a structure to facilitate construction

cleat - a short length of angle used in a connection

close tolerance turned bolt - a special high strength bolt used in a bolt hole with 0.15 mm clearance to limit the slip in connections

coefficient of friction - the ratio of frictional force to normal force on sliding surfaces

collapse mechanism - the unique mechanism produced by the formation of plastic hinges in a structure at ultimate load, i.e. at collapse

column - a vertical strut

compatibility - the relationship between displacements or strains in a structure or component

compound beam - a standard beam section strengthened by the addition of flange plates, e.g. a UB with additional welded flange plates

compressive force - a force which reduces the dimension of the material in line with the force

connection - an assemblage of components which connect one member to another and which are generally capable of resisting forces

conservative - more safe than strictly necessary

construction depth - distance allowed, or available, for the structural component - generally applied to the overall depth of beams

continuous construction - a beam or column continued beyond a support without a break, or through a suitably designed splice

contraflexure - change from convex to concave curvature, or vice-versa

corrosion of steel - a reduction in cross sectional area of components due to atmospheric corrosion which results in pitting or flaking of the surface

critical stress - the nominal elastic compressive stress at buckling

curvature - the reciprocal of radius of curvature

dead load - a permanent load of constant magnitude and position, e.g. self-weight

deflection - movement of a member perpendicular to its length due to external forces

deformation - changes in dimension of a member relative to its original dimensions, due to external forces

distortion - a gross deformation where the cross section of a member changes shape

edge distance - distance from the centre line of a bolt hole to the edge of a component

effective length - the effective length of a strut is that length which, if the strut were pin-ended, would produce the same buckling load as the actual strut

effective width - width of a plate less than the real width over which, for simplicity, a uniform stress is assumed to act, to allow for shear lag, or local buckling

elastic material (or structure) - a material (or structure) which if initially loaded regains its original dimensions and geometry when unloaded

elastic limit - maximum load, or stress, to which a material (or structure) can be loaded and still retain the properties of an elastic material (or structure)

elongation - increase in length of a member due to a tensile force

end distance - distance from the centre line of a bolt hole to the end of a member

end plate - a steel plate usually welded to the end cross section of a member

equilibrium of forces - a situation where a structure, or part of a structure, subject to the action of external forces is at rest (or static), i.e. there is no translation or rotation

fabricated members - members, or components, strengthened by the addition of other components, usually plates

factor of safety - see safety factor

faying surface - friction surface in contact in a high strength friction grip connection

failure - degeneration of a material, or structure, so that it is unable to support loads

fatigue - premature loss of strength in a material, or structure, due to the action of oscillating loads

fillet (root fillet) - the curve at re-entrant corners in a rolled section, to assist in manufacture and reduce stress concentrations

fillet weld - a fusion of the surfaces of two components which are generally at right angles to each other

finishes - non-structural coating to floors and walls, e.g. plaster

flexural-torsional buckling - a mode of buckling where a member bends and twists

force - that which moves, or tends to move, a structure or component

fracture - disintegration of a material generally accompanied by cracking

friction grip bolt - a high strength bolt which is preloaded in tension by tightening the nut, and consequently the external loads are resisted by friction between surfaces

fulcrum - point about which a lever acts

gantry girder - a beam supporting the wheels of a travelling crane

gas cutting - a method of cutting steel members using oxy-acetylene

gauge - distance between consecutive lines of holes, measured perpendicular to the direction of stress

girder - horizontal member, or structure, acting as a beam

gross cross section - cross section of a member with no deduction for holes, threads, etc

gusset plate - a steel plate used to connect members of trusses at node points, or to support loads, e.g. in a bracket

hydrogen cracks - occur in welds due to the action of residual stresses when the cooling rate is too rapid and if sufficient hydrogen is present

hyperstatic structure - see statically indeterminate structure

imposed load - load that is not permanent, e.g. snow, wind, vehicles, etc.

initial tangent modulus - modulus of elasticity based on the tangent to the stress-strain curve at the origin

in-plane behaviour - the behaviour of a member or structure which deforms only in a selected plane of loading

instability - a condition of unstable equilibrium, usually manifested by sudden buckling or collapse

intumescent paint - a fire resistant paint that swells when heated, forming an insulating barrier

iso-static structure - see statically determinate structure

joint - see connection

kern - the area within a column section where the application of an eccentrically applied load produces compressive stresses only

lack of fit - unintentional gaps that occur between components or members due to errors, discrepancies and distortions

lamellar tearing - cracks in plates associated with welding of components which are at right angles to each other

lateral - in a plane perpendicular to the web of the section

lattice girder - a plane truss with parallel booms and square ends, used as a deep beam

limit state - a condition of the structure which is critical for design purposes, e.g. collapse, deflection, fire, fatigue, etc

linear elastic behaviour - behaviour of the material, or structure, in the linear elastic range, i.e. regains original geometry when unloaded

live load - any removable load except wind

lateral torsional buckling - a form of instability in beams involving lateral buckling combined with twisting

load - that which produces deformations in a component or structure

load factor - a factor (between 1 and 2) applied to the working load to produce the design load at collapse. Acts as a safety factor against collapse

local buckling - a mode of buckling that occurs locally in plate elements, e.g. flange and webs of beams

mechanism - a configuration of hinged members not capable of maintaining stable equilibrium under load

minimum guaranteed yield stress - the yield stress, determined from tensile tests of samples taken during the manufacture of steel, so that only 5% of the samples have yield strengths which are less

moment distribution - a method of determining the distribution of bending moments at the ends of members of a structure in the elastic stage of behaviour

moment of inertia - see second moment of area

net cross section - a reduced cross section of a member where there are holes, or a section across the threads for a bolt

neutral axis - an axis in the plane of the section, separating tension and compression zones

node - a point in a structure where several members intersect

out of plane behaviour - the behaviour of a member or structure perpendicular to the selected plane of loading

panel - repeated configuration of members in a truss

partition - an internal vertical wall used for subdividing a storey, but carrying no load other than its own weight; not part of the dead load on the structure

penetration - a term applied to a weld to indicate the extent of fusion with the parent material

permissible stress - see allowable stress

pinned joint - a connection with a low moment of resistance

pitch - spacing between bolt centres measured in the direction of stress

plastic - condition of a highly stressed material when large deformations occur

plastic design - a method of design which computes the collapse load based on the occurence of sufficient hinges for a collapse mechanism to form

plastic hinge - a point in a member where large rotations occur at collapse, due to local yielding of the material

plate girder - a beam fabricated from plates

polar second moment of area - second moment of area about the centroidal axis perpendicular to the plane of the section (polar axis)

post buckling strength - a reserve of strength after buckling which is possessed by some plate elements, e.g. webs of plate girders

preloaded - member or component which is prestressed before the external load is applied

principal axes - sectional axes about which the product moment of area is zero, and the second moments of area are a maximum and a minimum

profile - shape of the wall centre-line of a thin walled section

proof load - axial load induced in a HSFG bolt by a specified amount of prestressing

proof stress - stress at which a specified amount of plastic strain occurs in a tensile test

purlin - a horizontal beam supporting roof sheets

radius of gyration - the radius (r) at which the whole area (A) of a section can theoretically be concentrated, for the calculation of second moment of area (I), i.e. $I = Ar^2$

rafter - a beam spanning from ridge to eaves

redundant member or reaction - a member or reaction that is in excess of those required for a statically determinate structure, or part of a structure

redundant structure - see statically indeterminate structure

residual stress - a stress in unloaded material or an unloaded member which remains after a heating and/or rolling process

restraint - anything that effectively prevents movement at a point in a structure

rigid joint - a connection with a high moment of resistance

rivet - a solid cylindrical fastener incorporating two heads, one of which is formed in the connection when in a heated malleable condition

root of a thread - the bottom of the thread of a bolt, which is related to the minimum cross sectional area of a bolt

root radius - radius of a curve connecting two perpendicular faces of a member

safety factor - an ill-defined term originally relating the allowable design stress at working load to the failure stress

sag rod - a member introduced into a roof truss to prevent excessive deflection of a tie rod

saw cutting - a method of cutting steel members using a hardened steel or diamond saw

secant modulus - modulus of elasticity based on a line connecting the origin with a selected point on the stress-strain curve

second moment of area - the sum of elements of area multiplied by the square of their distance from a given axis

secondary stresses - stresses produced in a member or component due to forces considered to be secondary to the main forces, e.g. bending moments in members of a roof truss

semi-rigid joint - all connections are semi-rigid because there is no perfect pin joint, i.e. without friction, and there is no absolutely rigid joint

shape factor - the ratio between the plastic and elastic section moduli of a section

shear centre - a point in the cross sectional plane of a member through which a transverse force will not produce twisting of the member

shear force - a force which produces shear deformation

shear lag - a condition modifying bending and shear stresses in a beam, arising from the restraint of warping displacements

shear wall - a structural wall preventing sidesway in a building

sheeting rails – horizontal beams spanning between columns and used to support and fix sheeting

simply supported beam – a beam assumed to be supported on rollers near one end and by a pin at the other

single curvature – having no contraflexure

slender – having a high slenderness ratio

slenderness ratio – ratio of effective length to radius of gyration (l/r)

squash load – compressive load to produce uniform yielding across the section

stalk – the web of a structural tee

stanchion – a vertical strut of I cross section

statically determinate structure – one containing just sufficient bracing to preclude the formation of a mechanism, the internal and external forces being related by equations of equilibrium

statically indeterminate structure – one in which the relationship between internal and external forces depends on the elastic properties of the members as well as on equilibrium

steel skeleton – a structure composed of connected members

stiffness of a member – bending stiffness is the resistance of a member to bending, expressed as EI/L. The corresponding value for torsion is GJ/L

storey – the space between floors, or floor and roof, of a building

strain – change in length divided by the origin length

strain energy – the quantity of energy, or work, stored in a material when subjected to strain

strain hardening – the permanent increase in strength (and hardness) of a material, which occurs when it is stressed beyond the elastic limit

stress – force per unit area

stress concentration – a localised high stress produced in a material due to defects, holes, discontinuities, etc

strut - a member which supports a predominantly axial compressive load

table - the flange of a structural tee

tangent modulus - modulus of elasticity which is a tangent to a non-linear stress-strain curve

tensile force - a force which increases the dimension of a material in the direction of the force

tension field - a mode of shear transfer associated with the web of a plate girder

tie - a member which supports a predominantly axial tensile force

tolerance - a limit placed on unintentional inaccuracies that occur in dimensions when marking out for cutting or casting, etc

torque - moment applied about a polar axis

torque wrench - a spanner, calibrated to measure torque (or moment = force x distance), used to tighten high strength friction-grip bolts

toughness - the ability of material to absorb energy, measured approximately by the Charpy V-notch impact test

transition temperature - the temperature below which a brittle fracture is enhanced

transverse - in the plane of, or parallel to, the web of a section, and perpendicular to the longitudinal axis of the member

truss - an open triangulated framework of relatively slender members

ultimate load - the load at which the structure collapses, generally by forming a collapse mechanism

ultimate load design - a method of design for collapse of a member or structure when subjected to factored working loads

unsymmetrical bending - bending about an axis which is not a principal axis

warping - deformation perpendicular to the cross section i.e. where cross sections do not remain plane

work hardening - see strain hardening

working load - maximum load on a structure under normal working conditions

working load design - a method of design in which members are chosen to resist the working loads so that the maximum stress does not exceed the allowable stress

yield line - line across a plate or plate element of a structure, where the material is at the yield stress

yield stress - stress in a material at which large plastic strains occur, with little or no increase in stress

Index